T0310218

SYNTHETIC NATURAL GAS FROM COAL, DRY BIOMASS, AND POWER-TO-GAS APPLICATIONS

SYNTHETIC NATURAL GAS FROM COAL, DRY BIOMASS, AND POWER-TO-GAS APPLICATIONS

Edited by

TILMAN J. SCHILDHAUER
SERGE M.A. BIOLLAZ
Paul Scherrer Institut, Villigen/Switzerland

Published by John Wiley & Sons, Inc., Hoboken, New Jersey
Published simultaneously in Canada

For general information on our other products and services or for technical support, please contact our Customer Care Department within the United States at (800) 762-2974, outside the United States at (317) 572-3993 or fax (317) 572-4002.

Wiley also publishes its books in a variety of electronic formats. Some content that appears in print may not be available in electronic formats. For more information about Wiley products, visit our web site at www.wiley.com.

Library of Congress Cataloging-in-Publication Data:

Names: Schildhauer, Tilman J., editor. | Biollaz, Serge M.A., editor.
Title: Synthetic natural gas from coal, dry biomass, and power-to-gas applications /
    [edited by] Tilman J. Schildhauer, Serge M.A. Biollaz.
Description: Hoboken, New Jersey : John Wiley & Sons, 2016. |
    Includes bibliographical references and index.
Identifiers: LCCN 2016006837 (print) | LCCN 2016014453 (ebook) | ISBN 9781118541814 (cloth) |
    ISBN 9781119191254 (pdf) | ISBN 9781119191360 (epub)
Subjects: LCSH: Synthesis gas. | Coal gasification. | Biomass conversion. | Gas manufacture and works.
Classification: LCC TP360 .S96 2016 (print) | LCC TP360 (ebook) | DDC 660/.2844–dc23
LC record available at http://lccn.loc.gov/2016006837

Set in 10/12pt Times by SPi Global, Pondicherry, India

Printed in the United States of America

10   9   8   7   6   5   4   3   2   1

CONTENTS

LIST OF CONTRIBUTORS

Renato Baciocchi University of Rome Tor Vergata, Roma, Italy

Serge M.A. Biollaz Paul Scherrer Institut, Villigen, Switzerland

Jochen Brellochs Center for Solar Energy and Hydrogen Research (ZSW), Stuttgart, Germany

Giulia Costa University of Rome Tor Vergata, Roma, Italy

Volkmar Frick Center for Solar Energy and Hydrogen Research (ZSW), Stuttgart, Germany

Jörgen Held Renewable Energy Technology International AB, Lund, Sweden

Stefan Heyne Chalmers University of Technology, Göteborg, Sweden

Thomas Kienberger Montanuniversität Leoben, Leoben, Austria

Christian F.J. König Paul Scherrer Institut, Villigen, Switzerland

Lidia Lombardi Niccolò Cusano University, Roma, Italy

Maarten Nachtegaal Paul Scherrer Institut, Villigen, Switzerland

Luc P.L.M. Rabou Energieonderzoek Centrum Nederland, Petten, The Netherlands

Urs Rhyner AGRO Energie Schwyz, Schwyz, Switzerland

Tilman J. Schildhauer Paul Scherrer Institut, Villigen, Switzerland

Martin Seemann Chalmers University of Technology, Göteborg, Sweden

Michael Specht Center for Solar Energy and Hydrogen Research (ZSW), Stuttgart, Germany

Bernd Stürmer Center for Solar Energy and Hydrogen Research (ZSW), Stuttgart, Germany

Eric H.A.J. Van Dijk Energieonderzoek Centrum Nederland, Petten, The Netherlands

Bram Van der Drift Energieonderzoek Centrum Nederland, Petten, The Netherlands

Christiaan M. Van der Meijden Energieonderzoek Centrum Nederland, Petten, The Netherlands

Frédéric Vogel Paul Scherrer Institut, Villigen, Switzerland

Berend J. Vreugdenhil Energieonderzoek Centrum Nederland, Petten, The Netherlands

Christian Zuber Agnion Highterm Research GesmbH, Graz, Austria

Ulrich Zuberbühler Center for Solar Energy and Hydrogen Research (ZSW), Stuttgart, Germany

1

INTRODUCTORY REMARKS

Tilman J. Schildhauer

1.1 WHY PRODUCE SYNTHETIC NATURAL GAS?

The answer to this question [which may also explain why one should read a book on synthetic natural gas (SNG) production] changes with time.

During the years from 1950 to the early 1980s, SNG production was an important topic, mainly in the United States, in the United Kingdom, and in Germany. The interest was caused by a couple of reasons. In these countries, a relative abundance of coal and the expected shortage of natural gas triggered several industrial initiatives, partly funded by public authorities, to develop processes from coal to SNG. Due to the oil crisis during the 1970s, the use of domestic coal rather than the import of oil became a second motivation. A third motivation is the possibility to make domestic (low quality) energy reserves available de-centrally as an energy carrier for which the distribution infrastructure already exists and that allows for both clean and efficient use by consumers.

These boundary conditions lead in 1984 to the start-up of the 1.5 GW_{SNG} plant in Great Plains which is run by the Dakota Gas Company and converts lignite into SNG and many other products. This plant stayed the only commercial SNG production for nearly 30 years because, with the drop of the oil price in the mid1980s, the exploration of natural gas in the North Sea, and the gas pipelines between Russia and Europe, the interest in SNG from coal ceased.

Especially in the United States, the interest came back in the years after the turn of the millennium, now triggered by the again rising oil price and the meanwhile established use of CO_2 (which is an inherent by-product of coal to SNG plants) for

Synthetic Natural Gas from Coal, Dry Biomass, and Power-to-Gas Applications, First Edition.
Edited by Tilman J. Schildhauer and Serge M.A. Biollaz.
© 2016 John Wiley & Sons, Inc. Published 2016 by John Wiley & Sons, Inc.

1

enhanced oil recovery (EOR). Back then, a dozen coal to SNG projects were started, including EOR. Now, due to the rapidly increasing exploitation of the shale gas and the connected possibility for a significant reduction of CO_2 emission, all the projects in the United States have been stopped.

However, all the mentioned motivations for SNG production, that is, shortage of domestic natural gas, use of domestic coal reserves which are far away from the highly populated areas, and the possibility for clean and efficient combustion, still prevail in China. Therefore, China is now by far the most important market for the production of SNG from coal. Three large plants have started operation, and further plants are planned or under construction.

In Europe, several aspects triggered a reconsideration of SNG production about 15 years ago. Due to its cleaner combustion and inherently lower CO_2 emission, using natural gas in transportation (e.g., for CNG cars) is supported in many countries and has even been economically beneficial for the past few years due to the lower gas price. With the aim of the European Commission to replace up to 20% of European fuel consumption by biofuel, replacing natural gas partly with bio-methane becomes necessary. So far, bio-methane is mostly produced by up-grading biogas from anaerobic digestion. However, due to the limited amount of substrate, this pathway cannot be increased much more and other sources of bio-methane are sought.

Additionally, many European countries wish to use their domestic biomass resources for energy production in order to decrease CO_2 emissions and the import of energy. A major part of the biomass is ligno-cellulosic (mostly wood) and mainly used for heating, for example, in wood pellet heating. As the heat demand is generally decreasing due to better building insulation, the conversion of wood to high value forms of energy, that is, electricity and fuels, is of increasing interest. Like in the case of coal, conversion to fuels requires (so far) gasification as the first step. As shown by process simulations and the first demonstration plants, the conversion of wood to SNG can reach significantly higher efficiencies than conversion to liquid fuels.

Very recently, a third aspect began to gain greater importance, especially in Central Europe. Due to the increasing integration of stochastic renewable sources like photovoltaics and wind energy into electricity generation, the demand for balancing the electricity supply and the demand over spatial and temporal distances is increasing. For the future, even the seasonal storage of electricity may be necessary. Here, the production of SNG can play an important role. While the gasification of solid feedstocks is a more or less continuous process, the further conversion to electricity or SNG can be flexibly adjusted to the balancing needs of the electricity grid within so-called polygeneration schemes.

Moreover, in times where the electricity production from renewables exceeds the actual demand in the electricity grid (a situation that today occasionally is observed in Central Europe and is expected to be more common in future), producing SNG could utilize the excess electricity instead of curtailing photovoltaics or wind turbines. In so-called power to gas applications, hydrogen is produced from excess electricity by electrolysis of water and then converted to SNG by methanation of carbon oxides. As a source of carbon oxides, biogas, producer gas from (biomass) gasification, flue gas from industry, or even CO_2 from the atmosphere can be considered, opening a pathway to produce SNG without solid feedstock that can be stored or transported over long distances within the existing natural gas infrastructure.

1.2 OVERVIEW

This book aims at a suitable overview over the different pathways to produce SNG (Figure 1.1).

The first four chapters cover the main process steps during conversion of coal and dry biomass to SNG: gasification, gas cleaning, methanation, and gas upgrading. The main technology options will be highlighted and the impact of a technology choice for downstream processes and the complete process chain. In these chapters, especially in the chapter on methanation reactors, the state of the art coal to SNG processes are discussed in detail.

The following chapters describe a number of novel processes for the production of SNG with their specific combination of process steps as well as the boundary conditions for which the respective process was developed. These processes comprise those which are already in operation (e.g., the 20 MW_{SNG} bio-SNG production in Gothenburg, Sweden, or the 6 MW_{SNG} power to gas plant in Werlte, Germany) and processes which are still under development.

The *gasification* chapter covers the thermodynamics of gasification and presents both coal and biomass gasification technologies.

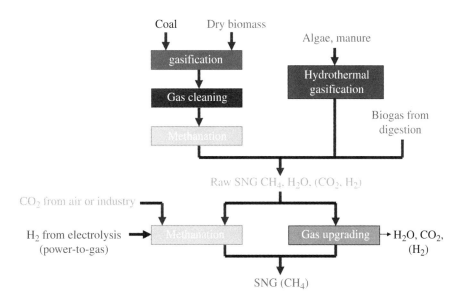

FIGURE 1.1 The different pathways to produce SNG.

The *gas cleaning* chapter discusses the impurities to be expected in gasification-derived producer gas, explains the state of the art gas cleaning technologies, and focuses on the innovative gas cleaning steps which are developed for hot gas cleaning.

The chapter on *methanation reactors* presents the chemical reactions proceeding inside the reactors, their thermodynamic limitation and their reaction mechanisms. Further, an overview of the different reactor types with their advantages and challenges is given covering coal to SNG, biomass to SNG and power to gas processes. The last section of this chapter focuses on the modeling and simulation of methanation reactors, including the necessary experiments to determine reaction kinetics and to generate data for model validation.

The chapter on *gas-upgrading* discusses technologies for gas drying, CO_2 and hydrogen removal based on adsorption, absorption, and membranes and includes a techno-economic comparison.

The chapter on the *GoBiGas project* ("Gothenburg Bio Gas") presents the boundary conditions and technologies applied in the 20 MW_{SNG} wood to SNG plant in Gothenburg, Sweden, which was commissioned in 2014.

The next chapter explains the development of the *power to gas* process at the Zentrum für solare Wasserstofferzeugung (ZSW), including the 6 MW_{SNG} plant in Werlte, Germany.

The chapter on *fluidised bed methanation* describes the process development at the Paul Scherrer Institut aiming at a flexible technology for efficiently converting wood to SNG and for hydrogen conversion within power to gas applications.

The following chapter presents the technologies developed at the *Energy Center of the Netherlands (ECN)* for efficient SNG production from wood, especially their allothermal gasification technology (MILENA) and their broad experience with gas cleaning.

The chapter on *hydrothermal gasification* discusses the unique technology allowing for the simultaneous catalytic gasification and methanation of wet biomass under super-critical conditions.

The chapter on *agnion's small scale SNG concept* focuses on two novel technologies that allow for significant process simplification, especially in small scale bio-SNG plants: the pressurized heatpipe reformer and the polytropic fixed bed methanation.

The last chapter offers a view on the research for even more simplified SNG processes, that is, for methanation steps that allow for *integrated desulfurization and methanation.*

The author of these lines wishes to express his gratitude, especially to the contributors of this book and to the persons at the publisher for their excellent work, but also to all colleagues, scientific collaborators, partners, friends and scientists in the community for many fruitful and interesting discussions. All of you bring the field forward and made this book possible.

2

COAL AND BIOMASS GASIFICATION FOR SNG PRODUCTION

Stefan Heyne, Martin Seemann, and Tilman J. Schildhauer

2.1 INTRODUCTION – BASIC REQUIREMENTS FOR GASIFICATION IN THE FRAMEWORK OF SNG PRODUCTION

Within the production of synthetic natural gas – basically methane – from solid feedstock such as coal or biomass the major conversion step is gasification, generating a product gas containing a mixture of permanent and condensable gases, as well as solid residues (e.g., char, ash). The gasification step can be conducted in different atmospheres and using different reaction agents. Figure 2.1 represents the basic pathway from solid fuel to methane, considering the main elements, carbon, hydrogen, and oxygen. It is obvious that an increase in hydrogen content is necessary for all feedstock illustrated. At the same time, the oxygen content needs to be reduced, in particular for biogenic feedstock that is oxygenated to a higher degree.

There exist different strategies or pathways for performing the conversion from feedstock towards methane within gasification, as illustrated in Figure 2.1. Adding steam as a gasification agent is common practice, not only due to the stoichiometric effect, but also for enhanced char gasification and temperature moderation within the reactor. H_2 addition is used in hydrogasification, leading to a higher initial methane

Synthetic Natural Gas from Coal, Dry Biomass, and Power-to-Gas Applications, First Edition.
Edited by Tilman J. Schildhauer and Serge M.A. Biollaz.
© 2016 John Wiley & Sons, Inc. Published 2016 by John Wiley & Sons, Inc.

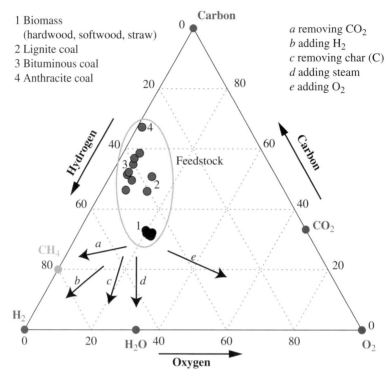

FIGURE 2.1 CHO diagram for coal and biomass. Feedstock composition based on [1, 2]. Data from Higman 2008 [1]; Phyllis2, database for biomass and waste, https://www.ecn.nl/ phyllis2 Energy research Centre of the Netherlands.

content in the product gas [3, 4]. CO_2 removal is an intrinsic part of the SNG production process; some gasification concepts using adsorptive bed material for direct CO_2 removal within the gasification reactor [5]. The addition of oxygen (common practice for all direct gasification technologies) actually leads to an increased need for CO_2 removal downstream of the reactor. For indirect gasification where ungasified char is combusted in a separate chamber for heat supply, the composition is changed towards methane via path (c) in Figure 2.1. Pretreatment technologies such as torrefaction decrease the oxygen content of the feedstock at the cost of increased energy demand.

2.2 THERMODYNAMICS OF GASIFICATION

For an increased understanding of the role of gasification for the overall SNG process, the basic thermodynamic aspects within gasification are discussed in the following. The gasification process is a series of different conversions involving both homogeneous and heterogeneous reactions. The basic steps from solid fuel to product gas are drying, pyrolysis or devolatilization, and gasification. Depending on the physical size of the fuel these different steps occur in a sequential order for small particles or overlap in bigger particles. The detailed description of the complete process is a complex task and the considerations made in the following therefore focus on specific parts of the process and their implications for the overall conversion, starting with some stoichiometric aspects for gasification.

2.2.1 Gasification Reactions

The major reactions occurring during the gasification step that commonly are considered relevant are:

$$C(s) + 0.5O_2(g) \leftrightarrow CO(g) - 111\,kJ/mol\,(partial\ oxidation) \tag{2.1}$$

$$CO(g) + 0.5O_2(g) \leftrightarrow CO_2(g) - 283\,kJ/mol\,(carbon\ monoxide\ combustion) \tag{2.2}$$

$$C(s) + O_2(g) \leftrightarrow CO_2(g) - 394\,kJ/mol\,(carbon\ combustion) \tag{2.3}$$

$$C(s) + CO_2(g) \leftrightarrow 2CO(g) + 172\,kJ/mol\,(reverse\ Boudouard\ reaction) \tag{2.4}$$

$$C(s) + H_2O(g) \leftrightarrow CO(g) + H_2(g) + 131\,kJ/mol\,(water\ gas\ reaction) \tag{2.5}$$

$$CO(g) + H_2O(g) \leftrightarrow H_2(g) + CO_2(g) - 41\,kJ/mol\,(water - gas\ shift\ reaction) \tag{2.6}$$

$$CH_4(g) + H_2O(g) \leftrightarrow 3H_2(g) + CO(g) + 206\,kJ/mol\,(steam\ reforming) \tag{2.7}$$

The char gasification reactions converting carbon into gaseous fuels [Equations (2.4) and (2.5)] are endothermic reactions requiring heat supply that is realized either by combusting part of the fuel in the same reactor (direct or autothermal gasification) or by indirect heat supply from an external combustion or heat source (indirect or allothermal gasification).

2.2.2 Overall Gasification Process – Equilibrium Based Considerations

Considering the overall process from coal or biomass to methane at the example of steam gasification, the reaction stoichiometry can be expressed as

$$C_x H_y O_z + a\,H_2O \rightarrow b\,CH_4 + c\,CO_2 \tag{2.8}$$

with $a = x - \dfrac{y}{4} - \dfrac{z}{2}$ $b = \dfrac{x}{2} + \dfrac{y}{8} - \dfrac{z}{4}$ $c = \dfrac{x}{2} - \dfrac{y}{8} + \dfrac{z}{4}$

Table 2.1 gives the coefficients for steam gasification to methane [Equation (2.8)] for different coal and biomass feedstock materials, allowing the calculation of the heat

TABLE 2.1 Composition and Overall Reaction Data for Steam Gasification for Different Feedstock Materials.

	Feedstock	Molar Composition	LHV [MJ/kg daf]	HHV [MJ/kg daf][c]	Reaction Coefficients for Equation (2.8) a	b	c	ΔH_r [MJ/kg daf Feedstock]	Methane Yield [kg CH$_4$/kg daf Feedstock]
Coal[a]	Brown coal –Rhein, Germany	$CH_{0.88}O_{0.29}$	26.2	27.3	0.632	0.537	0.463	−0.19	0.489
	Lignite – N. Dakota, USA	$CH_{0.72}O_{0.25}$	26.7	27.7	0.697	0.529	0.471	0.6	0.509
	Bituminous – typical, South Africa	$CH_{0.68}O_{0.08}$	34	35.1	0.792	0.567	0.433	1.21	0.654
	Anthracite –Ruhr, Germany	$CH_{0.47}O_{0.02}$	36.2	37.0	0.873	0.553	0.447	1.46	0.693
Biomass[b]	Willow wood – hardwood	$CH_{1.46}O_{0.65}$	18.5	19.9	0.310	0.520	0.480	−0.45	0.350
	Beech wood – hardwood	$CH_{1.47}O_{0.69}$	17.9	19.2	0.286	0.511	0.489	−0.71	0.333
	Fir – softwood	$CH_{1.45}O_{0.65}$	19.6	21.0	0.313	0.520	0.480	−1.58	0.350
	Spruce – softwood	$CH_{1.42}O_{0.68}$	18.4	19.7	0.304	0.508	0.492	−1.17	0.335
	Wheat straw	$CH_{1.46}O_{0.68}$	18.3	19.6	0.297	0.512	0.488	−0.84	0.338
	Rice straw	$CH_{1.43}O_{0.68}$	17.5	18.8	0.303	0.508	0.492	−0.23	0.335

[a] Taken from Higman and van der Burgt [1].
[b] Taken from Phyllis [2] – average data for material group.
[c] HHV [MJ/kg daf] = LHV [MJ/kg daf] + 2.44 · 8.94 · H [wt% daf]/100.

of reaction (based on the HHV of feedstock and methane, all reactants and products at 25 °C) as well as the stoichiometric methane yield per kilogram of feedstock. The heat of reaction for the overall reaction corresponds to well below 10% for all feedstock materials. For coal based feedstock with low oxygen content, the reaction is endothermic while for brown coal and biomass feedstock it is exothermic. The methane yield per kilogram feedstock also is considerably lower for biomass based feedstock due to the high oxygen content. One option to improve the methane yield would be the additional supply of hydrogen as for example considered in hydrogasification. The only technical conversion process capable of converting carbonaceous feedstock directly to methane and carbon dioxide according to the reaction in Equation (2.8) is gasification under supercritical conditions, so-called hydrothermal gasification [6–8]. This technology is capable of handling wet feedstock but not yet available at commercial scale and is not covered in this chapter. Common gasification technology converts the feedstock to a product gas, being a mixture of CO, CO_2, H_2, H_2O, CH_4, light and higher hydrocarbons, and trace components, followed by a downstream gas cleaning and methane synthesis step.

The major operating parameters for gasification are pressure and temperature. Equilibrium calculations for steam gasification (0.5 kg H_2O/kg daf feedstock) for a generic biomass are used to illustrate the basic trends for the product gas composition and heat demand with changing pressure and temperature. All feedstock is assumed to be converted to product gas. As can be observed from Figure 2.2, methane formation is favored by lower temperatures and higher pressures. Hydrogen and carbon monoxide formation increases with temperature and so does the endothermicity of the overall reaction. The theoretically exothermic reaction to CH_4 and CO_2 at 25 °C [similar to Equation (2.8)] turns into an endothermic reaction requiring heat supply at higher temperature.

Light hydrocarbons (represented by C_2H_4) and tars (represented by $C_{10}H_8$) are only formed to a very small extent according to equilibrium calculations. The amount of steam added for gasification will mainly influence the H_2/CO ratio via the water gas shift reaction, Equation (2.6). In reality, the equilibrium state is usually not reached in the shown temperature range, but actual product gas composition resulting from gasification is influenced by a number of other parameters as will be discussed in subsequent sections. An increase in gasification pressure favors methane formation predicted by equilibrium calculations; at 800 °C the methane molar fraction increases from 0 to about 15.5% from 1 to 30 bar. With more methane being formed, the endothermic heat of reaction is reduced by 66.6% from 1 to 30 bar. Again, no to very little formation of light hydrocarbons and tars is predicted by the equilibrium, even at higher pressures. At high pressures and moderate temperatures a mixture of basically CH_4, CO_2, and H_2O – representing Equation (2.8) – can be obtained. A process example is hydrothermal gasification which is operating at these conditions but still is at development state [6–8]. The above-mentioned trends are all under the assumption of complete conversion of feedstock to product gas. Many gasification concepts however have a considerable amount of char remaining unconverted, being removed with the ashes, or in indirect gasification, being converted in a separate combustion chamber for supplying the gasification heat. The carbon feedstock entering the gas phase

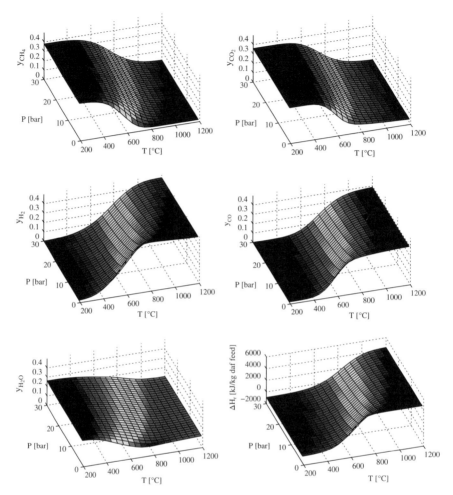

FIGURE 2.2 Pressure and temperature dependence of the molar concentration of major gas species and the heat of reaction for steam gasification (S/B = 0.5 kg H_2O/kg daf) of a generic biomass (C – 50 wt%, H – 6 wt%, O – 44 wt%: $CH_{1.43}O_{0.66}$) assuming complete carbon conversion, calculated by ASPEN PLUS.

during gasification will therefore be drastically changed when carbon conversion is incomplete. Even minor effects on the hydrogen and oxygen balance in the gas phase can be expected as, for example, biomass char still contains oxygen and hydrogen [9]. Figure 2.3 depicts the influence of pressure and temperature on carbon conversion; at temperatures below 800 °C, considerable amounts of solid carbon formation are predicted. This will in turn influence the equilibrium conditions in the gas phase as the carbon stock in gas phase is reduced. The major reactions affected are the Boudouard, water gas, and water–gas shift reactions, Equations (2.4) to (2.6). Incomplete carbon conversion leads to a decrease of CO concentration in the product gas, as well as a decrease in H_2 compared to equilibrium at complete conversion.

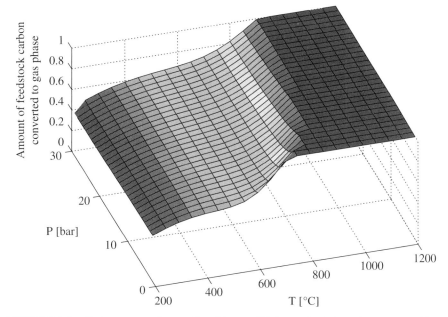

FIGURE 2.3 Carbon conversion predicted by equilibrium calculations for steam gasification (S/B = 0.5 kg H_2O/kg daf) of a generic biomass (C – 50 wt%, H – 6 wt%, O – 44 wt%: $CH_{1.43}O_{0.66}$).

2.2.3 Gasification – A Multi-step Process Deviating from Equilibrium

Equilibrium calculations, while useful for identifying trends with changing operating conditions, cannot however predict the performance of technical equipment to a full extent. They represent a boundary value that can be approached but never reached. The gasification process on a physical level is a multi-stage process starting with drying of the feedstock, followed by pyrolysis and gasification/combustion. The kinetics of the numerous homogeneous and heterogeneous reactions occurring – as well as the residence time and reactor setup – ultimately determine the product gas composition resulting from gasification. Figure 2.4 illustrates a simplified reaction network for the conversion from received fuel to product gas. The drying and primary pyrolysis (also referred to as devolatilization) steps are similar for all gasification technologies, whereas the extent of gasification reactions occurring as well as the approach to equilibrium are a strong function of the gasification medium and the reactor setup. During pyrolysis, a considerable amount of tars, a complex mixture of 1- to 5-ring aromatic hydrocarbons, is formed that will undergo various conversion pathways during gasification. In the final product gas tar can still represent a considerable amount of energy, for example, biomass steam gasification at 800 °C can result in more than 33 g tars/Nm³, corresponding to about 8% of the total product gas energy content on a lower heating value basis [10].

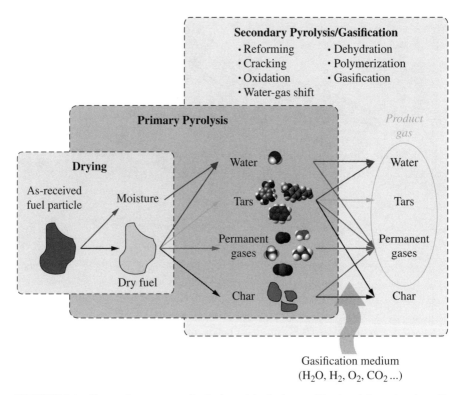

FIGURE 2.4 Conversion process of a fuel particle during gasification (adopted and modified from Neves et al. [9]).

The gas composition from primary pyrolysis represents the starting point for the gasification reactions. Neves et al. [9] conducted an extensive literature review of data on pyrolysis experiments and derived an empirical model for estimating gas species yields as well as tar and char yields and elemental composition as a function of peak pyrolysis temperature. Figure 2.5 illustrates the pyrolysis gas composition and the total gas and char yield based on Neves' model. In contrast to the equilibrium calculations for gasification represented in Figure 2.2, the model predicts considerable amounts of tars as well as a considerable amount of char produced from pyrolysis. Even light hydrocarbons are present in the pyrolysis gas. Generally speaking, the product distribution from pyrolysis as presented in Figure 2.5 undergoes conversion towards the equilibrium conditions during the final conversion step in Figure 2.4 – the gasification step.

The extent of conversion towards the equilibrium state is a function of a large number of parameters, such as pressure and temperature, reactor design, and associated residence time for gas and solids, as well as gas–solid mixing and the presence of catalytically active materials promoting specific reactions, among others.

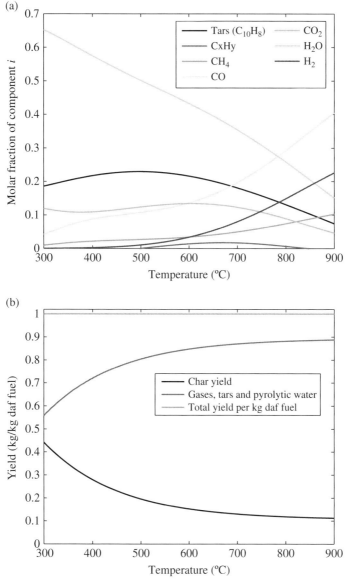

FIGURE 2.5   Pyrolysis gas molar composition (a) and overall mass yields (b) as a function of temperature (based on Neves [9]) for a generic biomass (C – 50 wt%, H – 6 wt%, O – 44 wt%: $CH_{1.43}O_{0.66}$).

2.2.4   Heat Management of the Gasification Process

As temperature is the major influencing parameter on the kinetics of the different gasification reactions, the thermal management of the gasification reactor is of particular importance. The conversion steps from solid fuel to product gas as illustrated in Figure 2.4 occur at different temperature levels. A qualitative representation of the temperature profile for a fuel particle over time is illustrated in Figure 2.6a. After particle heat-up the moisture is evaporated. The dry fuel particle is further heated, releasing pyrolysis gases, and finally the char particle is gasified. Heat for gasification needs to be supplied by the hot environment. Particle combustion indicated by the dashed line occurs at a particle temperature above environment due to the exothermic nature of the combustion reactions. For complete conversion of a biomass fuel with initial moisture content of 20 wt% the distribution of the heat demand for conversion on the different processes is illustrated in Figure 2.6b. The gasification heat demand is dominant but even pyrolysis and drying

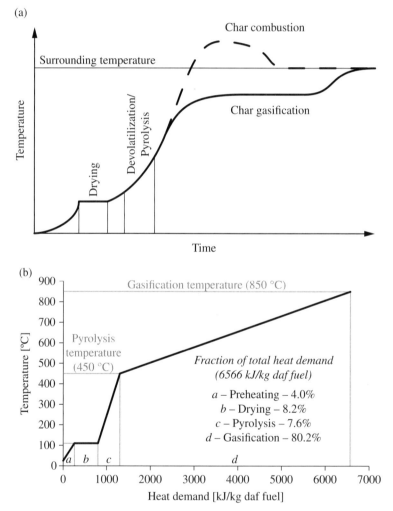

FIGURE 2.6   (a) Qualitative representation of the temperature evolution of a fuel particle during gasification. (b) Steam gasification heat demand profile for full conversion of a generic biomass (C – 50 wt%, H – 6 wt%, O – 44 wt%: $CH_{1.43}O_{0.66}$) at equilibrium (S/B ratio = 0.5, steam supply at 400 °C, 20 wt% initial biomass moisture).

represent a considerable share of the specific heat demand for conversion. Of course in reality, pyrolysis and gasification rather occur over a certain temperature range instead of the fixed temperatures used for the calculation (Neves' model for pyrolysis [9] and equilibrium calculations for the gasification). In addition, the processes are not strictly sequential but partly occur in parallel within a gasification reactor. Nevertheless the gasification heat demand will be largest and above all requires the highest temperature level.

All heat for the conversion process within the gasification unit needs to be supplied by combustion of part of the fuel or additional external fuel at high temperature. Changing the initial moisture content of the biomass will reduce the drying heat demand [b in Figure 2.6b] and therefore improve the conversion efficiency. External drying also allows for using a heat source at lower temperature, improving the conversion process from an exergy perspective. Even the pyrolysis process can be conducted in a separate reactor with a separate heat source. For these considerations in relation to the thermal management of the gasification process, temperature–heat load graphs can be a useful tool for identifying improvements for the energy efficiency of the gasification process. Figure 2.7 presents such a graph as an example of an indirect gasification process modelled using Neves' pyrolysis model and gasification at equilibrium. It is assumed that the amount of char combusted is set to ensure that the combustion heat covers the heat demand for fuel heat-up, drying, pyrolysis, and gasification. The supply of steam to gasification (400 °C) and hot air to combustion (400 °C) is an additional heat demand that needs to be covered. The thick curve in Figure 2.7 represents the aggregation of all the above-mentioned heat demands, whereas the dashed curve is a representation of all heat sources, namely the combustion

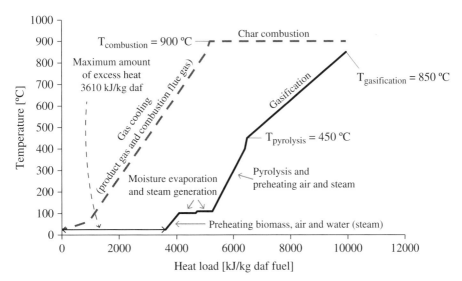

FIGURE 2.7 Temperature heat load curve for indirect steam gasification of a generic biomass (C – 50 wt%, H – 6 wt%, O – 44 wt%: $CH_{1.43}O_{0.66}$). S/B ratio = 0.5, air and steam supply at 25 °C and heated to 400 °C, 20 wt% initial biomass moisture.

of char and the cooling of hot product gas and combustion flue gases to ambient temperature. It is obvious that, for ideal heat transfer, the process has a considerable amount of excess heat, allowing for a further increase in conversion to product gas by more efficient use of the heat and reducing the char combustion. Converting more solid fuel to product gas will increase the heat demand while at the same time the heat supply will decrease as less char is burnt. Even considering the overall SNG process these curves can be used for a holistic analysis of the heat integration of sinks and sources, including the operations up- and downstream of the gasification step. Heat from the methanation reaction might, for example, be used for biomass drying or for regeneration of an amine solution used for downstream CO_2 removal.

The slope of the heat demand curves for pyrolysis and gasification also is dependent on the gas composition that actually is obtained during the different processes. A deviation from equilibrium conditions for the gasification will result in considerable changes of the conversion heat demand. Two parameters commonly used in thermal conversion processes are used to illustrate the influence of different parameters on the energy performance of the gasification process. The first parameter is the relative air to fuel ratio λ defining the amount of air (oxygen) actually supplied to the reaction in relation to the amount necessary for complete combustion according to stoichiometry:

$$\lambda = \frac{\dot{n}_{air(O_2),actual}}{\dot{n}_{air(O_2),stoichiometric}} \qquad (2.7)$$

The second parameter commonly used in gasification is the chemical efficiency η_{ch}, relating the chemical energy content of the product gas to the fuel chemical energy. η_{ch} can be defined on both a lower and a higher heating value basis, but in order to avoid confusion with respect to moisture content, the higher heating value is used here:

$$\eta_{ch,HHV} = \frac{\sum_i \dot{n}_{i,PG}.HHV_i}{\dot{n}_{fuel}.HHV_{fuel}} \qquad (2.8)$$

Figure 2.8 shows λ and $\eta_{ch,HHV}$ for the base case as illustrated in Figure 2.7, as well as the influence of changes in different operating parameters. Increasing the feed temperature for both steam to gasification (Point 1 in Figure 2.8) and air to combustion (Point 3 in Figure 2.8) increases the chemical efficiency and reduces λ as less char needs to be burnt. Reducing the incoming moisture content of the biomass from 20 to 10 wt% (Point 4 in Figure 2.8) results in a remarkable effect, reducing the necessary air to fuel ratio and gives a relative increase of the chemical efficiency of 3.2%. Assuming heat losses from the gasification unit (point 6 in Figure 2.8) corresponding to 2% of the thermal input on a higher heating value basis on the other hand, considerably increases the air to fuel ratio and leads to a relative decrease of the chemical efficiency by 3%. All changes except for points 5, 7, and 8 represent thermal

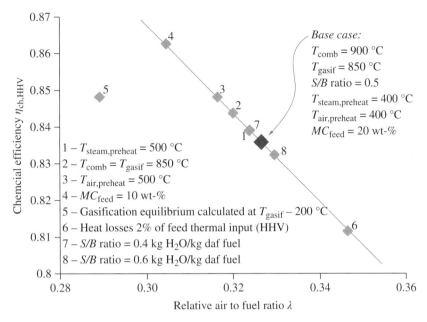

FIGURE 2.8 Influence of different parameters on chemical efficiency and relative air to fuel ratio for gasification.

improvements basically changing the balance between combustion and gasification based on the energy balance. This only marginally changes the gas composition and in consequence the specific heat demand for gasification. This is why the points follow a linear relationship between λ and $\eta_{ch,HHV}$ indicated by the gray line. Even changes in the steam to biomass ratio (Points 7 and 8) are along this line. Theoretically, the water-gas-shift reaction should be influenced by this parameter but the S/B ratio is considerably above the stoichiometric minimum (0.23 kg H_2O/kg daf fuel according to equation 2.8 for the given biomass feedstock) for all cases. The change in steam addition for the investigated cases only influences the thermal balance in consequence. When assuming a change in the resulting gas composition from gasification, as has been done for point 5 in Figure 2.8 however, the reaction enthalpy of the gasification process is changed and λ and η_{ch} influenced differently. The gas composition for the gasification step has been calculated for equilibrium at a temperature 200 °C below the actual gasification temperature, resulting in increased methane formation and a reduced endothermicity of the conversion process (Figure 2.2).

Changes in the gasification temperature in consequence have a twofold effect on the performance; one being basically of thermal nature related to temperature levels and heat demands, and the other being associated to chemical conversion, kinetics, and approach to equilibrium. Low temperature leads to decreased thermal losses, increased methane content, but also to higher production of tars from pyrolysis and slower kinetics for the gasification reactions, leading to high tar contents even in the product gas, generally. At higher temperatures, heat losses are more pronounced; the composition of the product gas is shifted towards CO and H_2, increasing the endo-thermal character of the reaction, and in consequence the heat demand, and to lower methane contents. On the other hand, tars are usually not present in high temperature gasification either, reducing the need (costs) for downstream upgrading equipment.

2.2.5   Implication of Thermodynamic Considerations for Technology Choice

The thermodynamic considerations mentioned above form the basis for the technical design of gasification reactors. Within the framework of SNG production, gasifica-tion aims at a high product gas yield with high methane concentration, as this favors overall process conversion efficiency. To achieve this, different conflicting objectives have to be optimized. High methane yield, as favored by low temperature, is accom-panied by a high yield of higher hydrocarbons and tars, increasing the demand for downstream gas cleaning. In addition, lower temperatures have a negative effect on the carbon conversion as gasification reaction rates are slower. This in turn penalizes the product gas yield. Pressurized operation also favors methane formation and sometimes is favorable for large scale production due to an increased volume-specific throughput (smaller equipment size for a given thermal input). Even downstream compression of the product gas prior to methanation can be avoided that way. On the other hand, feeding of solid fuels in pressurized vessels often is associated with oper-ational problems and may lead to an increased need of inertial gas in the feeding system that will influence the gasification kinetics and increase the downstream gas upgrade demand. An exergy based comparison between indirect atmospheric and direct pressurized biomass gasification indicates that there is no clear benefit of pres-surization of the gasification vessel as the gain in efficiency is compensated for by the increased gas upgrade energy demand within SNG production, rendering both indirect atmospheric and pressurized direct gasification equally suitable from an overall SNG process thermodynamic perspective [11]. The choice of gasification technology in conclusion needs to be evaluated and optimized within the overall SNG process framework. In the following paragraphs, different gasification technol-ogies will be presented and the particular aspects with respect to SNG production will be highlighted.

2.3   GASIFICATION TECHNOLOGIES

From a technological viewpoint, there basically exist three different gasification reactor types that are used at large scale: fixed (or moving) bed reactors, entrained flow reactors, and fluidized bed reactors. Figure 2.9 gives a representation of the major technologies and the feedstock and gasification agents used in connection with the respective technologies. Coal is mainly used in entrained flow gasification or fixed bed units, whereas biomass gasification is mostly done in fluidized bed reac-tors. The distinction is not sharp, though, as indicated by the thin gray arrows. An exemption, for example, is black liquor from the pulping process where entrained flow gasification is the preferred technology. Considering the gasification agents,

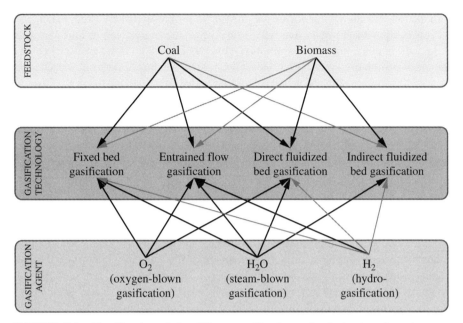

FIGURE 2.9 Classification of the different gasification technologies based on the most common feedstock and gasification agents.

steam is used in all gasification reactors. Indirect gasification is the only gasification technology not using oxygen as gasification agent as the heat is supplied from an external heat source. Hydrogen addition for hydrogasification is mainly used in coal gasification in entrained flow units but is not limited to this technology.

In the following paragraphs, the different gasification concepts for SNG production developed and constructed will be presented and discussed with respect to their feedstock requirements, heat management and optimization potential with respect to product gas yield, char conversion, methane content, and tar generation, where applicable. Specific issues particular to each technology within the framework of SNG production will be discussed, as well as pilot and industrial scale projects that have been proposed or realized.

2.3.1 Entrained Flow

In entrained flow gasifiers, the feedstock is fed concurrently with the gasification agent and is gasified while being transported (entrained) together with the produced gas through the reactor. This results in very short residence times (maximum 5 s, but usually even lower) and therefore necessitates high reaction temperatures above 1300 °C [1]. Such high temperatures avoid completely the production of tars and hydrocarbons; therefore, the producer gas consists mainly of carbon monoxide, hydrogen, and steam. In most cases, the gasification agent is pure oxygen supplied from an air separation unit (ASU) [12]. Avoiding nitrogen is necessary for SNG production and allows for lower volumetric flow rates and therefore smaller, less costly equipment; however, the ASU in the process is connected to an investment cost and internal energy consumption penalty. Entrained flow gasifiers can be operated at high pressures (several tens of bars) and can be fed with pulverized feedstock (<0.1 mm) or slurries, which offers a large flexibility. Due to the high operating temperatures, which are inherently above the ash melting temperature, this reactor type has to cope with the formation of slag. Some gasifier types (e.g., Shell, PRENFLO™, or Siemens) make use of it by forming a protective layer of slag flowing down the membrane wall. This however restricts the choice of feedstock to a certain minimum ash content and properties range. Other entrained flow gasifier types (e.g., the GE gasifier or the E-Gas™ technology by Conoco Philips) protect their walls against the heat by a refractory lining which allows for extreme flexibility with respect to ash content and properties, but may be connected to some ceramic corrosion problems. In any case, low ash-containing coals are preferred.

Entrained flow gasifiers are built on a large scale for several commercial applications by several suppliers; further development, especially in China (e.g., the East China University for Science and Technology [13] and for the Chinese market) is ongoing. Most entrained flow gasifiers are applied for conversion of coal in integrated gasification combined cycle (IGCC) plants and in the production of methanol. In the field of biomass gasification, entrained flow gasifiers have been chosen for bio-diesel and bio-kerosene production within the BioTFuel project in France (Uhde's PRENFLO™ gasification with Fischer–Tropsch synthesis [14]) and for the production of dimethyl ether within the BioLiq® process in Germany (decentralized straw pyrolysis for the production of bio-crude and subsequent central gasification in a high pressure entrained flow gasifier, followed by methanol/DME synthesis) [15].

Due to the reaction temperatures, the formation of methane in the producer gas is suppressed in most entrained flow gasifiers. This and the high exit temperatures leading to a low cold gas efficiency limit the overall process chain efficiency from coal to SNG and hamper the application of entrained flow gasifiers. Only within the POSCO project in Korea, was the E-Gas™ technology chosen to be combined with a Lurgi-Rectisol® scrubbing and a fixed bed methanation by Haldor Topsoe (TREMP®) for producing SNG from coal and petcoke. The reason for choosing this gasification process is the second gasification stage, where secondary coal is fed and reacts with the producer gas from the first high temperature stage, see Figure 2.10. This endothermic reaction causes a so-called chemical quench, leading to lower exit temperatures (900–1000 °C), and thus, to a higher cold gas efficiency and to the beneficial formation of methane [16].

2.3.2 Fixed Bed

While fixed bed biomass gasifiers (both counter-current and co-current operation) usually use air as the gasification agent and are thus rather applied for electricity production, all so far commissioned or operating coal to SNG plants (one in the USA, three in China) apply oxygen/steam blown countercurrent fixed bed coal gasification. The main reason is the relatively high amount of methane in the producer gas

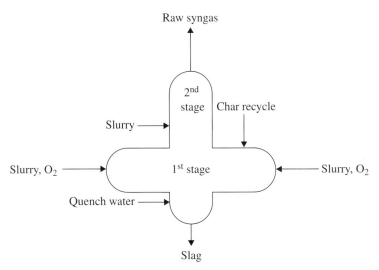

FIGURE 2.10 Two stage coal gasification process by Phips Conoco (E-Gas™), from [16]. Adapted from Cliff Keeler TL. POSCO Gwangyang Project for Substitute Natural Gas (SNG) 2010.

which is highly favorable for the overall process chain efficiency. In countercurrent fixed bed gasification, coal particles (from below 1 cm to a few centimeters in size) are fed via a lock hopper system into the pressurized gasifier (typical 20–24 bar), where they form a slowly downwards moving bed. Oxygen and steam are injected from the bottom, thus establishing a hot combustion zone converting char in the bottom of the reactor. The hot combustion flue gas moves upwards through the coal particle bed. The downward moving coal particles experience an increasing temperature level and undergo consecutively drying, pyrolysis, and gasification of the char formed in the pyrolysis layer. In consequence, the producer gas contains steam (both unreacted steam and from feedstock drying), tars from pyrolysis as well as gasification and combustion products, among them 5–10% methane. The original design was developed by Lurgi from the 1930s to the 1950s and for a long time was the only proven gasification technology for lignite and bituminous coals. In Lurgi's dry bottom gasifier (applied by the Dakota Gas Company in the Great Plains SNG plant, USA, but also in SASOL's syngas plants in South Africa) the coal bed is supported by a rotating grate and oxygen and a high amount of steam and oxygen are fed from below. The steam limits the combustion temperature and avoids melting the ash which therefore can be removed dry through the grate. The drawbacks of the high steam addition are loss of thermal efficiency, the formation of relatively high CO_2 amounts leading to larger downstream equipment, and a large amount of condensed water enriched with soluble organic impurities (such as phenols) which necessitates an appropriate wastewater treatment.

In consequence, the countercurrent fixed bed gasifier was further developed by Lurgi and British Gas Company (BGC) to decrease the steam/oxygen ratio significantly from about nine to a value between one and two [17]. As a result of the lower steam amount, temperatures in the combustion zone rise and the ash is converted to slag. The design of the bottom part has therefore been changed such that the molten slag can flow through a hole into a water quench section below the gasification chamber, while the steam and oxygen are fed through a number of tubes ("tuyeres") into the bottom part of the bed, see Figure 2.11 from [17]. Through these tubes, coal fines and/or condensed tars could also be fed, allowing enhancement of the carbon conversion of the system. Due to the significantly lower steam feed, higher throughput, lower CO_2 content, lower exit temperature, and thus higher thermal efficiency could be achieved (around 80 instead of 62% [17]). The so-called British Gas/Lurgi (BGL) slagging gasifier was chosen for SNG projects in the United Kingdom and the United States [17] which however were not realized.

2.3.3 Direct Fluidized Bed

In direct fluidized bed gasification the heat for gasification is supplied in the same unit by partial combustion of the feedstock. Bed material – catalytically active or

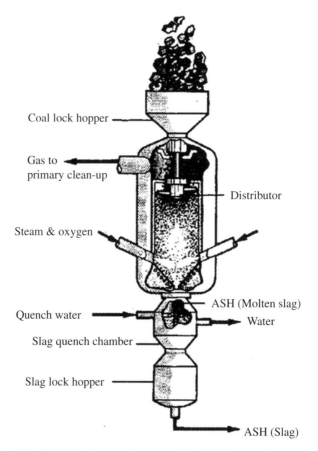

FIGURE 2.11 The British Gas/Lurgi slagging gasifier [17]. *Source*: Sharman RB, Lacey JA, Scott JE. The British Gas/lurgi Slagging Gasifier: A Springboard Into Synfuels: British Gas; 1981.

inert – can be used to ensure an even temperature distribution within the reactor and improved fuel mixing, but a number of coal gasification processes actually operate without bed material, the inert ash being sufficient for a particle inventory of the fluidized bed. The reactor design can be a bubbling fluidized bed or a circulating fluidized bed, the latter having a larger superficial gas velocity inside the reactor, leading to entrainment of particles that are separated via a cyclone and recycled to the fluidized bed via a return leg. Direct fluidized bed gasification can be operated under pressurized conditions, differentiating it from indirect fluidized bed gasification that is more complex to operate at pressures above ambient. The scale of the process is in the several hundred megawatt range based on the thermal input, being intermediate between entrained flow and indirect fluidized bed gasification. In the following, industrial scale technologies and demonstration plants for direct gasification will be presented. The coal based technologies are restricted to those that have been discussed for SNG production. For an extensive review of coal based direct fluidized bed gasification, the reader is referred to Higman [1], from where a large part of the information on coal based fluidized bed gasification presented in the following is taken.

2.3.3.1 Winkler Gasification

- Pressurized operation in the range of 30 bar demonstrated.
- Technology has been used for hydrogasification process.

The atmospheric fluidized bed process developed by Winkler and patented in 1922 was among the first to use oxygen as a fluidization and oxidation medium. It can be operated on a large number of fuels, with commercial plants having used a large spectrum of coal grades, from brown coal to bituminous coal. No additional bed material in addition to the coal particles itself and eventually limestone for sulfur control is used. The gasifier has to be operated below the ash melting point of coal in order to avoid sintering. Most commercial plants have been operated in the range of 950–1050 °C. The secondary oxygen/air supply above the bed improves the conversion of entrained particle fines. Unconverted char – that can be in the range of 20% of the carbon feed – is collected at the bottom to be burnt in an auxiliary boiler. Today the atmospheric Winkler process is no longer operated. The process was further developed in the 1970s for pressurized operation at up to 30 bar and is still today licensed by ThyssenKrupp Uhde as the high temperature Winkler (HTW) process, with a number of commercial plants in operation [18]. The operating temperature range is 700–950 °C with a carbon conversion of above 95%, according to ThyssenKrupp Uhde. Figure 2.12 is an illustration of the basic flowsheet for both the original Winkler gasifier and the HTW gasification process.

The HTW gasification process was also part of a hydrogasification process concept investigated by Rheinbraun AG in the late 1970s [19]. The process also has been applied for synthesis gas production suitable for methanol synthesis in the demonstration plant in Berrenrath, Germany [18], in the 1980s and is considered for a biomass to methanol project in Sweden [20].

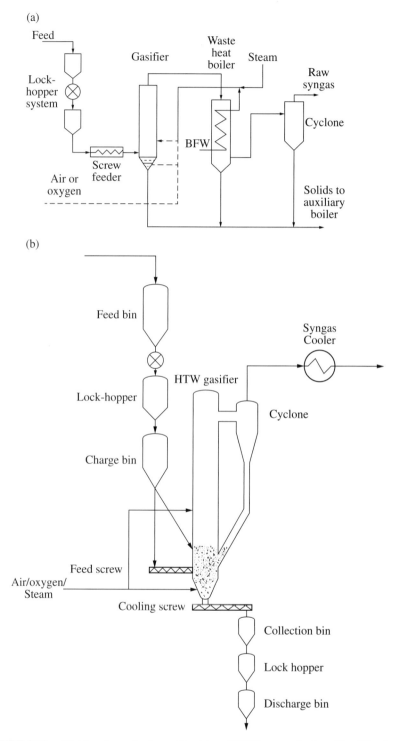

FIGURE 2.12   Winkler atmospheric gasifier (a) and HTW pressurised gasifier (b) schemes (reproduced from [1]). *Source*: Highman 2008 [1]. Reproduced with permission of Elsevier.

2.3.3.2   Kellogg–Rust–Westinghouse (KRW) Gasification

- Ash agglomeration process.
- Pressurized operation.

The Kellogg–Rust–Westinghouse (KRW) gasification process as illustrated in Figure 2.13 is a so-called agglomerating process where the ash in the coal feedstock

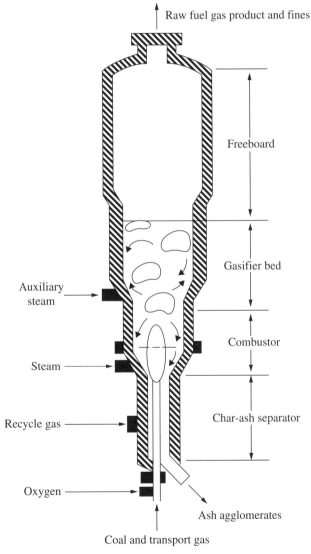

FIGURE 2.13   Schematic drawing of the KRW gasifier (reproduced from [3]). *Source*: Woodcock 1987 [3]. Reproduced with permission of Elsevier.

is heated to or above its melting point in the lower part of the fluidized bed. This causes the formation of larger ash agglomerates that can be withdrawn at the bottom of the gasifier. The agglomerate particles are mildly cooled by steam and/or recycle gas injection. The aim of this concept is to increase the carbon conversion. About 85% of the ash is collected at the bottom while the overall char conversion is typically 90–95% [3]. Another potential advantage of such a process is the reduction of problems with leachable ash in the downstream gas processes due to its removal at the bottom. A process development unit designed for pressures up to 16 bar and temperatures up to 1000 °C was operated by KRW [3].

2.3.3.3   Foster Wheeler

- Pressurized operation in the range of 20 bar demonstrated.
- More than 9000 operating hours of oxygen blown pressurized gasification for 12 MW gasifier.

At Värnamo, Sweden, a pressurized direct fluidized bed gasifier was operated at 18 bar on various biomass fuels as part of a demonstration programme [21]. Foster Wheeler together with Sydkraft (now E.ON) was the supplier of the gasification technology. The plant was successfully operated during approximately 3600 h in IGCC mode with air as the gasification agent and using the product gas in a gas turbine. From 18 MW_{th} of biomass input, about 6 MW of electricity and 9 MW of district heat could be generated. Both ceramic and metal hot gas filters were tested in the plant, both showing good filtration results, but the ceramic filters were subject to mechanical breakdown causing plant shutdown. Plans existed for conversion to oxygen blown operation with downstream fuel synthesis mainly focusing on hydrogen production, but these plans were never fulfilled. As of today the plant is not in operation [22].

NSE biofuels – a joint venture between Neste Oil and Stora Enso – together with the Finnish research institute VTT and Foster Wheeler conducted a biomass to liquid project at the Varkaus pulp mill in Finland. A pressurized 12 MW_{th} gasifier had accumulated more than 9000 h of operation since 2009 with an availability of 96% in 2011. Even hot gas cleaning was operated successfully with an availability of 93% and more than 5500 h of operation [23]. At VTT, a pilot plant of 0.5 MW_{th} was operated in parallel as part of VTTs ultra clean gas (UCG) project to support the technical development. After declined funding support from the European Union's NER300 program, NSE Biofuels decided to abandon the project in 2012.

2.3.3.4   Andritz Carbona/GTI

- Commercial units for >50 MW_{th} available.
- Test facility at GTI for gasification and catalytic tar reforming development.
- Catalytic tar reforming demonstrated for air blown operation.

Carbona as a part of the Andritz group provides a direct gasification technology based on the U-Gas and Renugas concepts developed by the Gas Technology Institute (GTI). A 20 MW$_{th}$ air blown pressurized gasifier operating at 0.5–2.0 barg is in operation at a CHP plant in Skive, Denmark, generating 6 MW of electricity and 12 MW of district heat by using the gas in stationary cogeneration engines. Hot gas cleaning using a catalytic tar reformer has been successfully demonstrated in Skive. The plant has an availability of 80–85%. At GTI in Chicago, United States, a pilot plant with oxygen blown pressurized gasification is in operation for testing and demonstrating biofuel synthesis process chains, in particular BtL and SNG processes [24]. Carbona's gasification technology is the core of the Bio2G project being pursued by the energy service company E.ON with the purpose of building a large scale Bio-SNG production plant with capacity of 200 MW$_{SNG}$ in Sweden. Targeted start of operation is 2018, but the company awaits a decision of the Swedish government on future subsidy plans for renewable transportation fuels [25].

2.3.4 Indirect Fluidized Bed Gasification

In indirect gasification the heat demand for the conversion from solid fuel to product gas is supplied by an external heat source – usually a combustion unit – with the help of some heat transfer medium. At large scale (>50 MW$_{th}$ input) the common technology for indirect gasification is a circulating fluidized bed with the bed material being the heat transfer medium between the combustion and gasification section, but at small to medium scale (1–10 MW$_{th}$ range) technologies exist using, for example, immersed pipes filled with liquid metal as heat transfer medium between two completely separated reaction chambers. An important difference for the latter technology is the independent control of the fuel input to the two reaction chambers. In circulating fluidized bed gasification systems, there is a large interdependency between the gasification and combustion chambers; the system is in theory inherently auto-stabilizing, as a decrease in gasification temperature leads to a lower char conversion and in consequence more char being transferred to the combustion chamber. This in turn leads to an increased heat release in the combustion chamber with the heat being transferred to the gasification chamber, leading to a temperature rise. In real practice the gasification temperature can also be controled by an additional fuel supply or product gas recycle to the combustion chamber. For gasification with completely separated gasification and combustion chambers, however, the fuel feed to the two vessels can be controled more independently.

The basic idea with indirect gasification is to provide a medium to high calorific product gas not diluted by nitrogen without the need for oxygen supply. The air separation unit associated to entrained flow or direct gasification stands for a considerable part of the investment and operational costs of a gasification system. Indirect gasification is generally considered to only be operated at atmospheric pressure, limiting to some extent the possibilities for scale-up. Biomass gasification, being intrinsically limited in scale due to supply chain logistic limitations, hardly exceeds the scale at which indirect gasification is technically feasible. The majority of the technical options for SNG production from indirect gasification presented in the following therefore are based on biomass.

2.3.4.1   Fast Internally Circulating Fluidized Bed Gasification

- Gasification temperature range of 800–850 °C for standard operation.
- Atmospheric gasification.
- Demonstrated 1 MW$_{SNG}$ production.
- Olivine as standard bed material reducing tar content to some extent while having good mechanical stability.
- Gas cleaning (tars) using rapeseed methyl ester (RME) as a standard solution.

A prominent technology for biomass gasification is the fast internally circulating fluidized bed (FICFB) gasifier developed at TU Vienna with its first 8 MW$_{th}$ pilot plant at Güssing, Austria, started up in 2001 [26]. The initial process concept of the FICFB gasifier is the efficient cogeneration of heat and power in stationary gas engines, with several biomass CHP plants in the 10 MW$_{th}$ range in operation in Austria and Germany [27]. But the Güssing plant even has been used for the demonstration of a 1 MW Bio-SNG production unit using a side stream of the product gas [28]. The estimated chemical efficiency from biomass to Bio-SNG η_{ch} for this concept is in the range of 66% [29], larger scale units reaching 68% [28]. A general flowsheet of the FICFB gasifier as well as the block flow sheet of the integrated Bio-SNG demonstration plant is illustrated in Figure 2.14. The tar cleaning technology developed in association to the FICFB gasification is oil-based washing with spent washing oil being burnt in the combustion zone of the gasifier. The consumption of rapeseed methyl ester for tar cleaning is a non-negligible contribution to the process' fuel and energy consumption. The FICFB gasification is commonly operated with calcined Olivine, a magnesium iron silicate, as bed material.

The FICFB gasification technology that is licensed by repotec [31] is even the core of the GoBiGas project in Sweden aiming at a production of 20 MW of Bio-SNG from 32 MW$_{th}$ forest residues with operation started by the end of 2013 [32, 33]. The first tar cleaning step for the GoBiGas plant is RME oil scrubbing as in the Güssing plant, but the remainder of the Bio-SNG process is different in technology to the Bio-SNG demonstration in Güssing and supplied by Haldor Topsøe.

Another project aiming at Bio-SNG production from woody feedstock using the FICFB gasification technology is the Gaya project in France with a thermal feedstock input of 0.6 MW$_{th}$. The project is designed as a development and testing platform that is to be run during a 5-year testing period with the aim of identifying favorable process options and conditions for scale-up to industrial size Bio-SNG processes [34, 35].

2.3.4.2   Adsorption Enhanced Reforming Gasification

- Demonstrated with FICFB technology.
- Lower gasification temperature range of 600–700 °C.

- CaO-containing bed material for CO_2 removal during gasification by in situ carbonation.
- Lower tar content compared to standard gasification.
- Higher dust content due to increased bed material attrition.

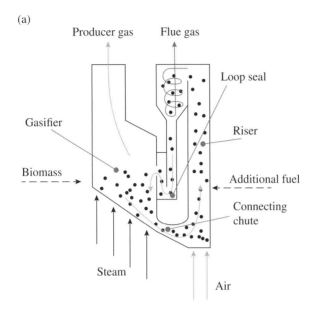

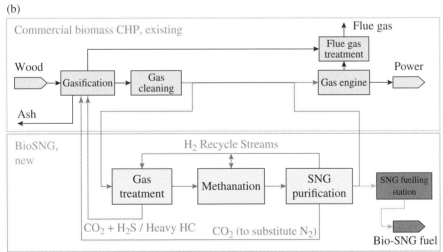

FIGURE 2.14 Dual bed fluidized steam gasifier concepts. (a) Fast internally circulating fluidized bed (FICFB) gasifier (8 MW_{th}; reproduced from [30]). *Source*: Pfeifer 2009 [30]. Reproduced with permission of Elsevier. (b) Bio-SNG process concept as integrated into the Güssing CHP plant (reproduced from [28]). *Source*: Bio-SNG. Bio-SNG – Demonstration of the production and utilization of synthetic natural gas (SNG) from solid biofuels. Bio-SNG project, 2009 TREN/05/FP6EN/S07.56632/019895.

A further gasification technology that is based on the FICFB concept is the adsorption enhanced reforming (AER) gasification technology that was successfully tested at the Güssing plant in 2007/2008 [5]. The AER concept uses a CaO-containing bed material and operates in the lower temperature range for fluidized bed gasification (600–700 °C). This allows in situ carbonation of the bed material forming $CaCO_3$, and CO_2 is thus removed from the product gas in the gasification reactor according to the following reaction:

$$CaO(s) + CO_2(g) \leftrightarrow CaCO_3(s) - 179 \, kJ/mol \qquad (2.9)$$

The reaction being highly exothermic, the heat supply to the gasification reactor is enhanced. For the carbonation to proceed to a satisfactory degree requires both lower temperature and longer residence times in the gasifier compared to standard gasification. The lower temperature in the gasifier makes the technology of particular interest for difficult fuels with high ash content and increased risk for bed agglomeration at higher temperatures. The lower temperature also favors methane formation according to the equilibrium relations. Removal of CO_2 in situ furthermore pushes the water gas shift reaction [Equation (2.6)] to the right hand side, leading to an increased hydrogen yield. On the combustion side, the carbonation reaction is reversed, forming CaO and the CO_2 is released with the flue gases. This increases the heat demand in the combustion unit as the inverse reaction [Equation (2.9)] is endothermic. A comparison of conventional and AER gasification in a pilot scale unit demonstrates a substantial increase of H_2 concentration from 37.7 to 73.9 mol% in the dry product gas and a change of the H_2/CO ratio from 1.3 to 12 [30]. The reduced mechanical stability of the bed material leads to increased dust formation, but the CaO-containing calcite used for the AER gasification has a higher activity to tar reforming, leading to a considerably lower tar content in the product gas compared with olivine as bed material (reduction from 3.5 to 1.4 g/Nm³) [30]. Considering the chemical energy output from the gasifier for the same thermal fuel input, the product gas chemical energy is reduced by 37% for AER gasification in relation to standard FICFB gasification. This is due to the increased energy demand for bed material regeneration in the combustion zone [reverse reaction; Equation (2.9)] requiring more char to be burnt. The CO_2 removal directly in the gasification vessel certainly has beneficial effects on gas composition with respect to downstream synthesis applications. But the energy demand for CO_2 removal is similar to or higher than the demand for conventional downstream CO_2 removal in an amine based absorption unit [179 kJ/mol CO_2 for Equation (2.9) compared to about 3 MJ/kg CO_2 (132 kJ/mol CO_2)]. In addition, the heat needs to be supplied at the high combustion temperature of about 900 °C in contrast to a temperature level for amine regeneration in the range of 140 °C. But beneficial effects such as reduced downstream equipment need (no shift necessary) as well as equipment size due to the lower product gas volume stream still make this concept in particular interesting for biofuel synthesis applications.

Based on successful tests in the Güssing demonstration plant [5], a 10 MW$_{th}$ plant based on AER gasification was planned to be built in Germany for combined heat and power generation. The plant was to serve as a research and development platform for the generation of biofuels, in particular H_2 and SNG from slip streams in a similar manner as the Güssing plant [36]. Due to drastic increases in wood fuel prices, the project however was abandoned for economic reasons [37].

2.3.4.3 MILENA Gasification

- Gasification and combustion integrated in single vessel.
- Oil-scrubbing above water dew-point as commonly associated tar removal technology.
- Bio-SNG process chain demonstrated at pilot scale.
- Pressurization considered for scale-up.

The Energy Research Centre of the Netherlands (ECN) has demonstrated Bio-SNG production based on its MILENA indirect gasification technology at pilot scale (0.8 MW_{th} input). The major difference compared to the FICFB gasifier is the fact that both combustion and gasification are integrated in a single vessel. Biomass is pyrolyzed and partly gasified with steam in an inner annular space in a fluidized bed. The unconverted char is burnt with air in the lower part of the vessel surrounding the gasification reactor for supplying the necessary heat [38]. ECN in collaboration with the HCV group is going to build a plant in Alkmaar for the gasification of 12 MW_{th} of waste wood for the production of Bio-SNG based on the MILENA technology. The plant will use the major part of the cleaned product gas for the production of green electricity using a boiler connected to a steam power cycle, while a minor product gas stream will be upgraded to Bio-SNG. The plant is scheduled for start-up in 2014 [39]. A schematic flowsheet of the Bio-SNG process design based on the MILENA technology is illustrated in Figure 2.15. The methanation technology employed is a multi-stage fixed bed technology. For scaling up the process, pressurization of the gasification reactor is considered feasible by ECN due to the single vessel design [39].

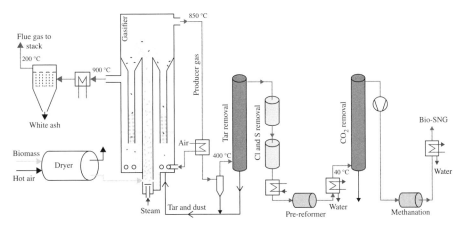

FIGURE 2.15 Simplified scheme of the MILENA Bio-SNG process scheme (reproduced from [38]). Adapted from Van Der Meijden CM et al. 2009.

2.3.4.4　Batelle/Silvagas Gasification

- Demonstrated at 40 MW_{th} input.
- Atmospheric gasification.

The Rentech–Silvagas (former FERCO Silvagas) indirect gasification process at atmospheric conditions developed in the United States is a DFBG gasification concept with two interconnected circulating fluidized bed reactors. It was initially developed at Batelle's Columbus Laboratory as part of the US Department of Energy's Biomass Power Program in the late 1970s. The process has been successfully demonstrated in a CHP plant in Burlington during 2000–2002 at a design scale of about 40 MW_{th} that was even operated with a thermal input of about 60 MW_{th} on a lower heating value basis, using the product gas for co-firing in the boiler. No activities on downstream fuel synthesis are reported however [40].

2.3.4.5　Chalmers Indirect Gasifier

- Retrofit concept of existing FB boilers.
- Gasifier purely dedicated to research.

The rationale of the concept behind the Chalmers indirect gasifier is to retrofit existing fluidized bed boilers by the addition of a gasification reactor to the circulation loop. As fluidized bed boilers are common all over the world it is a cost-effective alternative to new-build, stand-alone gasifiers. Depending on the size of the added gasifier, the result resembles either a pure indirect gasifier or a polygeneration plant with heat, power, and gas production. The concept was demonstrated in 2007 at the Chalmers 12 MW CFB boiler where a 2 MW gasifier was integrated via two loop seals into the circulation of hot bed material (Figure 2.16) [41, 42]. The gasifier is entirely dedicated to research and no production of fuel is currently planned; instead the raw gas is reintroduced to the boiler and combusted. The gasifier has been up and running for over 15 000 h and no negative effect on the functionality of the boiler has been reported.

2.3.4.6　Heatpipe Reformer

- Small to medium scale processes (<10 MW_{th}).
- Pressurized operation of up to 4 bar proven.
- Liquid metal in heat pipes for heat transfer.

An indirect gasification concept completely separating the gasification and combustion chambers is the heatpipe reformer (HPR) developed at the Technical University of Munich and licensed by Agnion Energy Inc. [49]. The heat for gasification is transferred to the gasification chamber (reformer) via heat pipes filled with liquid metals (sodium). The technology is designed for small scale and distributed applications. Pressurization of the gasification process at 4 bar was successfully

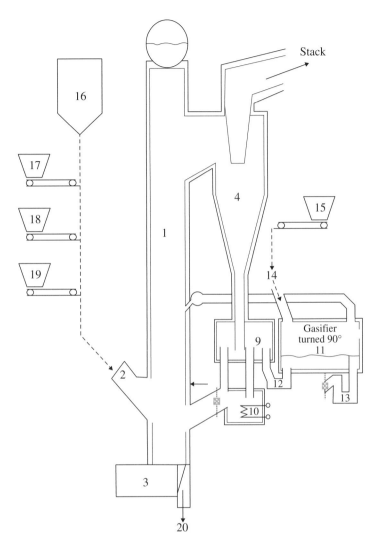

FIGURE 2.16 Scheme of the Chalmers CFB boiler after the retrofit with an add-on gasifier.

realized during a monitoring program. The technology has been demonstrated in a 500 kW$_{th}$ pilot unit and a commercial biomass CHP plant of 1.3 MW$_{th}$ scale. A declared target process is decentralized synthesis of biofuels in general and Bio-SNG in particular. In a conceptual study for Bio-SNG production a cold gas efficiency $\eta_{ch,LHV}$ of 68% has been determined for a 1.3 MW$_{th}$ input scale. The basic design of the HPR and the SNG process concept is illustrated in Figure 2.17.

A process concept with similarities to the HPR called Wood Roll is licensed by Cortus Energy. The technology employs separate vessel for all steps in gasification – drying, pyrolysis, and gasification – with heat for the drying being supplied by

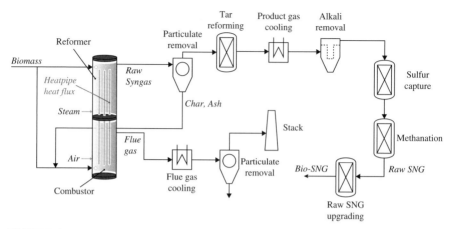

FIGURE 2.17   Bio-SNG process concept based on heatpipe reformer (based on [43, 44]). *Source*: Gallmetzer 2012 [43]. Reproduced with permission of Springer.

burning part of the pyrolysis gas. The remainder of the pyrolysis gas is burnt in single-ended recuperative burners immersed in the gasification chamber, providing the heat for gasification. The remaining char from pyrolysis is gasified with steam producing a nitrogen-free syngas suitable for fuel synthesis. Separating the different subprocesses allows heat provision at relevant temperature levels, in theory allowing a process with high exergy efficiency (refer to Figure 2.7). A 5 MW_{th} plant is commissioned for providing fuel gas to a lime kiln in Köping, Sweden, with estimated start of operation in June 2014 [45].

2.3.5   Hydrogasification and Catalytic Gasification

2.3.5.1   Hydrogasification

- No downstream methanation step necessary.
- Two-stage gasification concept with internal hydrogen generation.

In hydrogasification the stoichiometric conversion from coal to methane according to Figure 2.1 is aiming at a direct reaction of the carbon feedstock to methane with hydrogen according to the overall reaction

$$C + 2H_2 \rightarrow CH_4 \tag{2.10}$$

A technical design for the hydrogasification process for coal has been developed by Rheinbraun AG based on the HTW gasifier, as illustrated in Figure 2.18. The gasification process is conducted in two fluidized bed reactors. The upper bed uses hydrogen as a fluidizing agent for coal gasification, producing a methane-rich stream that is upgraded to SNG quality. A cryogenic separation is used for recycling hydrogen to the reactor. The generation of hydrogen is realized in the lower fluidized

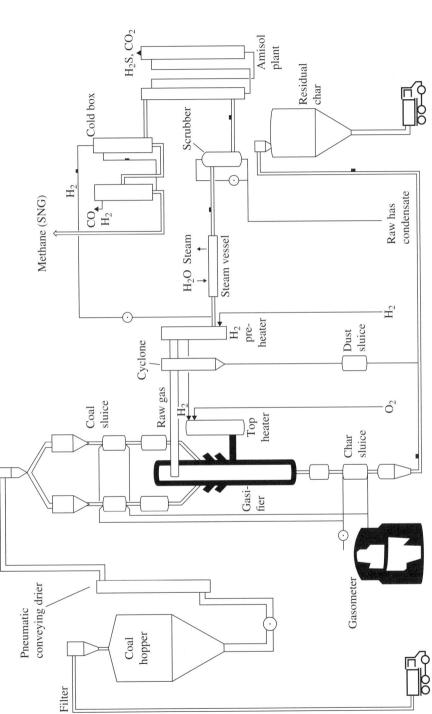

FIGURE 2.18 Rheinbraun AG hydrogasification process for SNG from coal based on HTW gasification (redrawn from [3]). *Source:* Woodcock 1987 [3]. Reproduced with permission of Elsevier.

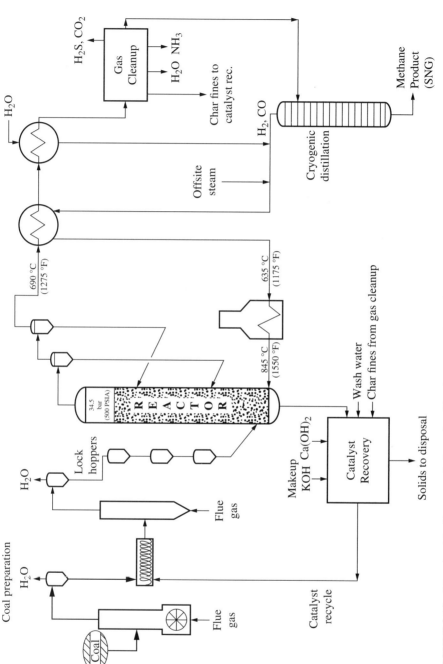

FIGURE 2.19 Flowsheet for Exxon catalytic coal gasification process (reproduced from [47]). *Source:* Gallagher 1980 [47]. Reproduced with permission of Elsevier.

bed reactor, using the residual char from hydrogasification and a mixture of steam and oxygen. Methane reforming is used to achieve the desired hydrogen concentration in the gas stream entering the hydrogasification reactor. Similar concepts have been developed and tested, for example, the Hydrane or Hygas gasifier, the latter developed by the Institute of Gas Technology (now Gas Technology Institute; GTI). For the Hydrane gasifier it is stated that about 50% of the carbon in coal feedstock is needed for hydrogen generation, while about 35% of the unconverted carbon leaving the first gasification reactor is converted in the hydrogasification step [46].

2.3.5.2   Catalytic Gasification

- Gasification at lower temperature (about $700\,°C$) results in higher methane yield.
- No downstream methanation necessary.
- Weak acid potassium salts proven catalyst for coal gasification.

The most prominent example for catalytic coal gasification is Exxon's process for SNG production from coal in a fluidized bed using alkali metal salts as catalyst. Figure 2.19 illustrates the basic process layout. Suitable catalysts identified during the development at pilot scale were potassium hydroxide (KOH), potassium carbonate (K_2CO_3), and potassium sulfide (K_2S). The major effect of the catalyst is to promote gas phase shift and methanation reactions [46] as well as an increased rate for steam gasification [47]. Comparable carbon conversion rates to $925\,°C$ operation in a standard gasifier can be obtained at an operating temperature of $700\,°C$ for a catalytic gasification unit. This clearly favors methane yield according to equilibrium (Figure 2.2). No methanation step is required downstream but gas upgrade for the removal of carbon monoxide and hydrogen (recycled to gasifier) and acid gas removal (carbon dioxide, hydrogen sulfide) is needed to obtain pipeline quality SNG.

REFERENCES

[1] Higman C, van der Burgt MJ. Gasification, 2nd edn. Gulf Professional Publishing/ Elsevier, Oxford, UK, 435 pp.; 2008.

[2] Phyllis – Database for biomass and waste: ECN; 2013 [2013-08-01]. Available from: http://www.ecn.nl/phyllis2/ (accessed 12 December 2015).

[3] Woodcock KE, Hill VL. Coal gasification for SNG production. *Energy.* **12**(8/9):663–687; 1987.

[4] Mozaffarian M, Zwart RWR. *Production of Substitute Natural Gas by Biomass Hydrogasification*. Proceedings of First World Conference and Exhibition on Biomass for Energy and Industry. pp. 1601–1604; 2000.

[5] Koppatz S, Pfeifer C, Rauch R, Hofbauer H, Marquard-Moellenstedt T, Specht M. H2 rich product gas by steam gasification of biomass with in situ CO2 absorption in a dual fluidized bed system of 8 MW fuel input. *Fuel Processing Technology* **90**(7/8):914–921; 2009.

[6] Gassner M, Vogel F, Heyen G, Maréchal F. Optimal process design for the polygenera-
     tion of SNG, power and heat by hydrothermal gasification of waste biomass: Process
     optimisation for selected substrates. *Energy and Environmental Science* **4**(5):1742–
     1758; 2011.

[7] Waldner MH, Vogel F. Renewable production of methane from woody biomass by
     catalytic hydrothermal gasification. *Industrial and Engineering Chemistry Research*
     44(13):4543–4551; 2005.

[8] Vogel F, Waldner MH, Rouff AA, Rabe S. Synthetic natural gas from biomass by catalytic
     conversion in supercritical water. *Green Chemistry* **9**(6):616–619; 2007.

[9] Neves D, Thunman H, Matos A, Tarelho L, Gómez-Barea A. Characterization and
     prediction of biomass pyrolysis products. *Progress in Energy and Combustion Science*
     37(5):611–630; 2011.

[10] Lind F, Israelsson M, Seemann M, Thunman H. Manganese oxide as catalyst for tar
     cleaning of biomass-derived gas. *Biomass Conversion and Biorefinery* **2**(2):133–140;
     2012.

[11] Heyne S, Thunman H, Harvey S. Exergy-based comparison of indirect and direct bio-
     mass gasification technologies within the framework of Bio-SNG production. *Biomass
     Conversion and Biorefinery* **3**:36–42; 2013.

[12] Harris DJ, Roberts DG. *Coal Gasification and Conversion.* In: Osborne D (ed.) The Coal
     Handbook: Towards Cleaner Production, 2. Woodhead Publishing, London, pp. 427–
     454; 2013.

[13] Li C. *Current Development Situation of Coal to SNG in China.* IEA workshop, Peikin,
     China; 2014.

[14] ThyssenKrupp. *Uhdes PRENFLO™–Verfahren wird für gemeinsames Forschungs- und
     Entwicklungsprojekt BioTfueL in Frankreich ausgewählt.* Available from: http://www.thys-
     senkrupp.com/de/presse/art_detail.html&eid=TKBase_1267695470819_934277290
     (accessed 12 December 2015).

[15] Bioliq. Home page. Available at: www.bioliq.de; 2012 (accessed 12 December 2015).

[16] Cliff Keeler TL. *POSCO Gwangyang Project for Substitute Natural Gas (SNG).* POSCO,
     Gwangyang; 2010.

[17] Sharman RB, Lacey JA, Scott JE. *The British Gas/lurgi Slagging Gasifier: A Springboard
     Into Synfuels.* British Gas, London; 1981.

[18] Uhde T. *Gasification Technologies.* ThyssenKrupp Uhde, p. 24; 2012.

[19] Lambertz J, Brungel N, Ruddeck W, Schrader L. *Recent Operational Results of the High-
     Temperature Winkler and Hydrogasification Process.* Conference on Coal Gasification
     Systems and synthetic Fuels for Power Generation, San Francisco, CA. Electric Power
     Research Institute, Palo Alto, CA; 1985.

[20] VarmlandsMetanol AB. *Uhde Gasification Selected for World's First Commercial
     Biomass-to-Methanol Plant for VarmlandsMetanol AB.* VarmlandsMetanol AB, Hagfors,
     Sweden.; 2012.

[21] Ståhl K. *The Värnamo Demonstration Plant – A Demonstration Plant for CHP
     Production, based on Pressurized Gasification of Biomass. Demonstration programme
     1996–2000.* European Commission, Swedish Energy Agency, Sydkraft AB, Stokholm;
     2001.

[22] Ståhl K, Waldheim L, Morris M, Johnsson U, Gårdmark L. *Biomass IGCC at Värnamo,
     Sweden – Past and Future.* GCEP Energy Workshop. Stanford, USA; 2004.

[23] Hannula I, Kurkela E. *Biomass Gasification – IEA Task 33 Country Report – Finland.* IEA Task 33 Meeting. Istanbul, Turkey; 2012.

[24] Patel J, Salo K, Horvath A, Jokela.*ANDRITZ Carbona Biomass Gasification Process for Power and Bio Fuels.* 21st European Biomass Conference and Exhibition, ETA–Florence Renewable Energies, Copenhagen; 2013.

[25] Fredriksson Möller B, Ståhl K, Molin A. *Bio2G – A Full-scale Reference Plant for Production of Bio-Sng (Biomethane) Based on Thermal Gasification of Biomass in Sweden.* 21st European Biomass Conference and Exhibition. ETA-Florence Renewable Energies, Copenhagen, Denmark; 2013.

[26] Hofbauer H, Rauch R, Loeffler G, Kaiser S, Fercher E, Tremmel H. *Six Years Experience with the FICFB-Gasification Process.* 12th European Conference and Technology Exhibition on Biomass for Energy, Industry and Climate Protection. ETA, Amsterdam, Netherlands, pp. 982–985; 2002.

[27] Rauch R. *Biomass Steam Gasification – A Platform for Synthesis Gas Applications.* In: Agency IE (ed.). IEA Bioenergy Conference, Vienna. International Energy Agency, Vienna; 2012.

[28] Bio-SNG. *Bio-SNG – Demonstration of the Production and Utilization of Synthetic Natural Gas (SNG) from Solid Biofuels.* Bio-SNG Project TREN/05/FP6EN/ S07.56632/019895, Bio-SNG, Malmo, Sweden; 2009.

[29] Rehling B, Hofbauer H, Rauch R, Aichernig C. BioSNG – process simulation and comparison with first results from a 1-MW demonstration plant. *Biomass Conversion and Biorefinery* **1**(2):111–119; 2011.

[30] Pfeifer C, Puchner B, Hofbauer H. Comparison of dual fluidized bed steam gasification of biomass with and without selective transport of CO_2. *Chemical Engineering Science* **64**(23):5073–5083; 2009.

[31] repotec – *Renewable Power Technologies 2013.* Available from: http://www.repotec.at/ inde2.php/technology.html (accessed 14 December 2015).

[32] GoBiGas. *GoBiGas – Gothenburg Biomass Gasification.* Göteborg Energi AB; 2012. Available from: http://gobigas.goteborgenergi.se/En/Start (accessed 14 December 2015).

[33] Gunnarsson I. *The GoBiGas Project.* International Seminar on Gasification – Gas Quality, CHP and New Concepts. Swedish Gas Center, Malmö, Sweden; 2011.

[34] Mambré V. *The GAYA Project.* International Seminar on Gasification – Feedstock, Pretreatment and Bed Material. Swedish Gas Centre, Göteborg; 2010.

[35] Perrin M. *Biomass Gasification Technology as an Opportunity to Produce Green Gases – the GDF SUEZ Vision.* International Seminar on Gasification – Process and System Integration. Swedish Gas Technology Centre Ltd, Stockholm; 2012.

[36] Marquard-Möllenstedt T, Specht M, Brellochs J, et al. *Lighthouse Project: 10 MWth Demonstration Plant for Biomass Conversion to SNG and Power via AER.* 17th European Biomass Conference and Exhibition. ETA-Florence Renewable Energies and WIP-Renewable Energies, Hamburg; 2009.

[37] Bomm M. *The "Lighthouse" Tipped Over.* Südwest Presse,Dusseldorf, Germany; 2011.

[38] Van Der Meijden CM, Veringa HJ, Vreugdenhil BJ, Van Der Drift A, Zwart RWR, Smit R. *Production of Bio-Methane from Woody Biomass.* Energy Research Centre of the Netherlands, Report ECN-M-09-086, Contract ECN-M-09-086, Energy Research Centre of the Netherlands, Petten, Netherlands; 2009.

[39] van der Meijden CM, Könemann JW, Sierhuis W, van der Drift A, Rietveld G. *Wood to Bio-Methane Demonstration Project in the Netherlands.* 21st European Biomass Conference and Exhibition. ETA-Florence Renewable Energies, Copenhagen, Denmark; 2013.

[40] Paisley MA, Overend RP, Welch MJ, Igoe BM. *FERCO's Silvagas Biomass Gasification Process Commercialization Opportunities for Power, Fuels, and Chemicals.* Second World Conference on Biomass for Energy, Industry and Climate Protection, ETA-Florence and WIP-Munich, Rome, Italy; 2004.

[41] Larsson A, Seemann M, Neves D, Thunman H. Evaluation of performance of industrial-scale dual fluidized bed gasifiers using the Chalmers 2–4-MWth gasifier. *Energy and Fuels* **27**(11):6665–6680; 2013.

[42] Thunman H, Seemann MC. *First Experiences with the New Chalmers Gasifier.* Proceedings of the 20th International Conference on Fluidized Bed Combustion. Tsinghua University Press, Beijing; 2009.

[43] Gallmetzer G, Ackermann P, Schweiger A, et al. The agnion heatpipe-reformer – operating experiences and evaluation of fuel conversion and syngas composition. *Biomass Conversion and Biorefinery* **2**(3):207–215; 2012.

[44] Gröbl T, Walter H, Haider M. Biomass steam gasification for production of SNG – Process design and sensitivity analysis. *Applied Energy.* **97**:451–461; 1012.

[45] Cortus Energy. *Home page*; 2013. Available from: http://www.cortus.se (accessed 16 August 2015).

[46] Probstein RF, Hicks RE. *Hydrogasification and Catalytic Gasification. Synthetic Fuels.* McGraw-Hill. New York, pp. 189–201; 1982.

[47] Gallagher Jr JE, Euker Jr CA. Catalytic Coal Gasification For SNG Manufacture. *International Journal of Energy Research* **4**(2):137–147; 1980.

[48] Neste Oil Corp. *Neste Oil and Stora Enso to end their biodiesel project and continue cooperation on other bio products.* Press Release, Neste Oil Corporation; 2012. Available from: www.nesteoil.com (accessed 16 August 2015).

[49] ENTRADE Group. *Agnion Energy Inc. is currently subject to insolvency proceedings and the ENTRADE Group acquired all shares in May 2013.* Press Release, ENTRADE Group; 2013. Available from: http://biomassmagazine.com/articles/9051/the-entrade-group-to-acquire-agnion-energy/ (accessed 22 August 2013).

3

GAS CLEANING

Urs Rhyner

3.1 INTRODUCTION

Thermochemical conversion of biomass or coal by gasification produces mainly gases such as H_2, H_2O, CO, CO_2, and C_xH_y but also carbon rich particulate matter. Impurities such as tars, sulfur compounds, alkali, halide, nitrogenous compounds, and trace elements are also present in biomass-derived producer gases. Particulate matter, tars, and contaminants reduce the performance of downstream equipment such as catalysts, internal combustion engines, or gas turbines. Efficient and effective gas cleaning is needed to reduce impurities to an acceptable level given by specific downstream applications.

Depending on the kind of feedstock, gasifier technology, and operation conditions, different amounts and compounds of contaminants and trace elements are present in the producer gas [1–3]. Seven groups of impurities can be discerned:

1. Particulate matter,
2. Tars,
3. Sulfur-containing compounds,
4. Halides,
5. Alkali,
6. Nitrogen-containing compounds,
7. Others.

Synthetic Natural Gas from Coal, Dry Biomass, and Power-to-Gas Applications, First Edition.
Edited by Tilman J. Schildhauer and Serge M.A. Biollaz.
© 2016 John Wiley & Sons, Inc. Published 2016 by John Wiley & Sons, Inc.

The amount of impurities entering the gasification system depends on the feed-stock. Coal contains on average more sulfur (S) than grass or wood [4]. Higher amounts of organic sulfur compounds were found for producer gas derived from grass than from wood pellets (bark free) [5]. Regarding wood, organic sulfur is more abundant in the bark than in the core of the trunk [6]. Alkalis, halides, and nitrogen are more abundant in grass than in wood or coal [4].

The gasifier technology and the operating conditions influence the signature of contaminants that can be found in the producer gas. The lowest tar content (close to $0\,g\,m^{-3}$) can be found in entrained flow gasifier systems where gasifier exit tempera-tures are the highest. Updraft fixed bed gasifiers with low exit temperatures produce more tars ($10–150\,g\,m^{-3}$) than downdraft fixed bed gasifiers ($0.01–6.0\,g\,m^{-3}$) where the producer gases pass the combustion zone before exiting the gasifier [7]. At a lower air to fuel ratio (lambda value) more tars are built than at a higher ratio [5, 8–12].

Downstream units require specific producer gas quality. Except filter units, any equipment downstream of the gasifier can tolerate only a certain degree of particulate matter without failure or loss of performance due to blockages. Especially porous layers with the same pore sizes as the particles (micron to submicron) but also gas turbines get damaged by particulate matter. Internal combustion engines and gas tur-bines are more tolerant to catalyst poisons such as sulfur but the combination with sodium (Na) or potassium (K) can lead to severe corrosion of the turbine [4]. Nickel (Ni), copper (Cu), cobalt (Co), or iron (Fe) as used in catalysts for methanation, liquid fuel synthesis, and fuel cells are prone to sulfur poisoning.

3.2 IMPURITIES

3.2.1 Particulate Matter

Solid particles are always present in gasifier-derived product gas. They can cause fouling, erosion, and corrosion of equipment. The particles include unconverted bio-mass (char, soot) and inorganic compounds (ash) but can also include bed material or catalytic active material (whole particles or attrition products) if applied in the gas-ifier bed. Efficient gasifier technologies reach carbon conversion efficiencies up to 99%, leaving small amounts of char in the product gas. Untreated wood contains up to $2\,wt\%$ of inorganic materials while agricultural residues can contain up to $20\,wt\%$. Ash from biomass mainly consists of K_2O, SiO_2, Cl, and P_2O_5 salts. Operating tem-perature influences the amount and composition of particulate matter. Depending on temperature, compounds can be in gaseous, liquid, or solid form (e.g., alkali). The particle sizes range from microns (μm) to submicrons. Emissions of particulate matter are regulated by a certain size and level (e.g., PM10 relates to particles with an aerodynamic diameter smaller than $10\,\mu m$).

The following technologies are used for particle removal, ordered by their maximum operation temperature: wet scrubbers, electrostatic precipitators, cyclones, barrier filters. Filtration efficiencies up to 99.999% [13] are reached by barrier filters (e.g., ceramic filter elements).

3.2.2 Tars

Tars can cause damage to equipment due to condensation, soot formation, or inhibition of catalytically active centers or sorption materials. The amount of tar in the producer gas strongly depends on the gasifier technology, the operating temperature, the steam to carbon ratio, and the application of catalysts in the gasifier. Tar content can vary up to one order of magnitude [7]. Depending on the energy conversion system, tars can be considered as fuel (e.g., solid oxide fuel cell; SOFC) [14].

Harsh sampling conditions directly downstream of gasifiers often prevent a proper characterization of the producer gas. Particle filters and gas cooling is often needed before tar species can be quenched or measured. Gas cleaning devices upstream of the sampling point have to be carefully considered when comparing literature data regarding tar content of producer gas. The so-called "tar protocol" explains a standardized method to sample and analyze tar compounds [15–17]. An alternative sampling method for continuous online measurements was developed by Paul Scherrer Institut [18,19].

Many definitions of tar exist and the variety of tar compounds found in biomass gasifiers is large. The Energy Research Center of the Netherlands (ECN) has developed an extensive database with information about more than 50 common tar compounds, as well as calculation procedures for estimating the tar dew point [20]. The "tar protocol" defines tars as "all hydrocarbons with molecular weights greater than that of benzene" [16]. Devi et al. [21] suggest five different tar classes: Class 1: GC undetectable, Class 2: Heterocyclic aromatics, Class 3: Light aromatic (1 ring), Class 4: Light PAHs (2 or 3 rings), Class 5: Heavy PAH compounds (4–7 rings). Based on the complexity of the molecules, Milne et al. classified tars as primary, secondary, alkyl tertiary, and condensed tertiary tars [8].

At the high operating temperatures of biomass gasifiers, tars hardly exist according to thermodynamic equilibrium calculations [22]. Limited residence time and slow conversion could be some of the reasons for the existence of tars.

Three methods are discerned to remove or reduce the amount of tars in producer gas: physical tar removal, noncatalytic tar decomposition, and catalytic tar decomposition.

3.2.3 Sulfur Compounds

Sulfur compounds corrode metal surfaces [23] and when burned, sulfur dioxide (SO_2), a regulated pollutant, is generated by oxidation. Even in small concentrations, sulfur is known as a poison for catalysts containing Ni, Cu, Co, or Fe as used in methanation, liquid fuel synthesis, and fuel cells. The most abundant compound is H_2S, followed by COS. Sulfur-containing hydrocarbons are often neglected though the sum of them can build a substantial fraction of the total sulfur concentration. Sulfur tars found in producer gases are mercaptans, thioethers, disulfides, thiophenes, benzothiophenes, and dibenzothiophenes [5].

Bulk desulfurization of H_2S can be done by wet scrubbers or sorption materials. So far, sorption materials can only be used to remove H_2S from the product gas to sub-ppm levels but not for other sulfur compounds, such as sulfur tars. Therefore sulfur species other than H_2S have to be converted to H_2S to enable desulfurization of the producer gas.

3.2.4 Halide Compounds

Halides are mainly represented by hydrogen chloride (HCl) in biomass producer gases and to a minor extent by hydrogen fluoride (HF) and hydrogen bromide (HBr). Halides are known to cause high temperature corrosion and catalyst poisoning. With other impurities, HCl can react to ammonium chloride (NH_4Cl) and sodium chloride (NaCl) which condense at lower temperature causing fouling of equipment. Halides can be removed from the producer gas by wet scrubbing or sorption materials.

3.2.5 Alkali Compounds

Biomass feedstock contains mainly potassium (K) and to a lesser extent sodium (Na). Alkali compounds in the producer gas are found in the form of chlorides (KCl, NaCl), hydroxides (KOH, NaOH), and sulfites (K_2SO_4, Na_2SO_3). Alkali metal compounds cause sintering in boilers, corrosion to turbine blades or heat exchangers, and damage to catalysts.

Alkali salts evaporate at temperatures above 800 °C and can be removed by filtration at temperatures below 600 °C. At higher temperatures, sorption materials can be used to remove alkali compounds from the producer gas.

3.2.6 Nitrogen Compounds

Nitrogen-containing compounds in biomass-derived producer gas are mainly ammonia (NH_3) and less abundant hydrogen cyanide (HCN), but also organic compounds such as, for example, pyridine (C_5H_5N) have to be considered. Thermodynamic equilibrium favors N_2 at gasification temperatures. At high temperatures, as present in gas engines and turbines, nitrogen oxides (NO_x) can be formed that have to be removed from the exhaust gas. Strict legal limits and difficulties to remove nitrogen oxides in the exhaust gas make removal prior to combustion the preferred solution. Ammonia can adsorb on catalyst active sites, reducing catalyst activity, but can also be considered as fuel in high temperature SOFC.

If ammonia needs to be removed from the producer gas, its good solubility in water makes wet scrubbers an efficient solution at low temperatures. At high temperatures, only catalytic decomposition is possible with similar catalysts used for tar decomposition such as dolomites, nickel- and iron-based catalysts.

3.2.7 Other Impurities

Other impurities such as trace elements can be found in biomass-derived producer gases. Trace elements originate from feedstock but also from components of different process units. The amount of trace elements is often limited to ppm and sub-ppm levels [24, 25]. Some impurities such as magnesium (Mg), calcium (Ca), lead (Pb), and vanadium (V) are known to cause depositions and corrosion to gas turbines. Trace elements such as arsenic (As) and phosphorus (P) are reported to cause severe damage to SOFC; cadmium (Cd) causes significant performance loss, zinc (Zn), mercury (Hg), antimony (Sb) degrade the cell power density to a lesser extent [26–30]. It is expected that trace elements will have negative long-term effects on other catalysts as well. Sorption material can be used to remove some of the trace elements, though further investigation is needed.

3.3   COLD, WARM AND HOT GAS CLEANING

Different gas cleaning technologies are operated at temperatures ranging from –60 °C to ambient and up to 1000 °C. There is no common definition about the temperature ranges and according names for gas cleaning. Similar to other definitions [31], we suggest "cold gas cleaning" for processes operated around ambient or lower temperatures, "warm gas cleaning" for processes operated higher than 100 °C up to 400 °C, and "hot gas cleaning" for processes operated at temperatures higher than 400 °C.

Hot gas cleaning (HGC) of producer gas derived from biomass gasification is a very promising technology that offers significant efficiency gains in the conversion process from biomass to electricity, synthetic natural gas (SNG) or liquid fuels. It has been shown by process modeling that HGC allows significant efficiency improvements by avoiding cooling and reheating of the producer gas [32, 33].

3.3.1   Example of B-IGFC Gas Cleaning Process Chains

Figure 3.1 shows three different cases of gas cleaning processes at different temperature levels for biomass integrated gasification fuel cell (B-IGFC) processes. The electrochemical conversion of producer gas to electricity by a fuel cell (e.g. SOFC) is shown as an example but the gas cleaning process is also valid for other conversion processes involving catalysts, such as methanation or liquid fuel synthesis.

3.3.1.1   Cold Gas Cleaning   The cold gas cleaning case represents the state of the art in gas cleaning. Exit temperatures of, for example, fluidized bed gasifiers can be as high as 850 °C. Low temperature gas cleaning requires cooling of the producer gas below 400 °C in order to fulfil the temperature limits of the filtration system. Downstream of the filter unit, tars, steam, and alkali aerosols will condense in quenching columns (wet scrubber) operated at temperatures as low as 10 °C. Sulfur species will be captured in cold absorbers such as wet scrubbers (e.g., Selexol, Rectisol) or in fixed beds (active carbon, metal oxide). Depending on the material used for the fixed bed bulk desulfurization (H_2S), the temperature has to be increased. Downstream of the desulfurization, steam has to be added again in the case of fuel cells as a conversion unit, and the temperature needs to be increased again to the required gas inlet temperature of the SOFC.

3.3.1.2   Warm Gas Cleaning   The warm gas cleaning case shows the B-IGFC process chain as implemented at Paul Scherrer Institut (PSI) at small pilot scale with a thermal input up to 12 kW [34]. The exit temperature of, for example, an updraft gasifier can have the same exit temperature as the filter operating at 450 °C.

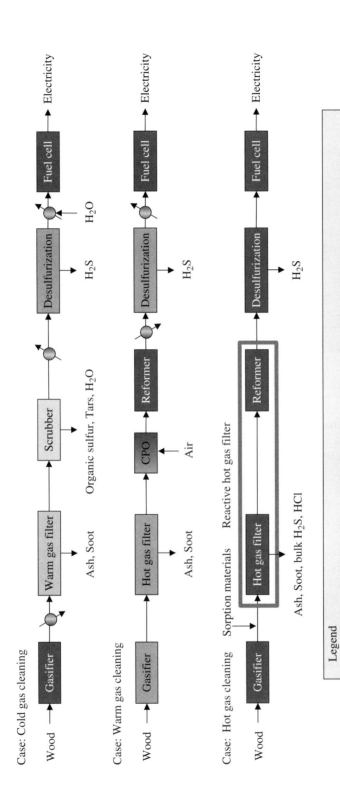

FIGURE 3.1 Cold, warm, and hot gas cleaning process chains for B-IGFC systems or other conversion processes such as methanation or liquid fuel synthesis.

Therefore, cooling of the producer gas upstream of the filter unit is not needed. Downstream of the filter unit, the producer gas is heated by catalytic partial oxidation (CPO) to reach the required temperature of the reformer catalyst. A reformer catalyst is applied to convert tars and sulfur-containing hydrocarbons to lower molecular hydrocarbons, CO, CO_2, H_2, H_2O, and H_2S. The desulfurization is done by the removal of H_2S by sorption material. If zinc oxide (ZnO) is used as the sorption material, the temperature has to be below 600 °C. No steam needs to be added as compared to the low temperature case because the process temperatures stay above the condensation temperature of water. The ability of the catalyst to decompose sulfur-containing hydrocarbons is a precondition for the proposed process chain applying HGC in combination with a metal oxide bed to remove H_2S. The conversion of sulfur-free tars has second priority.

3.3.1.3 Hot Gas Cleaning The hot gas cleaning (HGC) case is the process chain to be further investigated. The whole process chain operates at the exit temperature of the gasifier. No cooling or reheating of the producer gas is needed. The hot gas filter (HGF) operates at temperatures up to 850 °C. Concepts to integrate the reformer catalyst in the HGF unit, operating at the same temperature as the HGF, were investigated [35]. The integration of the reformer catalyst avoids heat losses due to the connection between the operating units. In addition, high temperature sorption materials can be applied upstream of the HGF to remove contaminants such as H_2S and HCl. The combination of sorption materials upstream of a HGF and a reformer catalyst is called a reactive hot gas filter.

3.4 GAS CLEANING TECHNOLOGIES

Gas cleaning technologies are presented for each of the seven groups of impurities as presented in Section 3.2.

3.4.1 Particulate Matter

Particle removal from the product gas is needed to protect downstream process units from fouling. High dust load and sticky ash particles from biomass gasification are challenging conditions for particle removal technologies. The amount of particles in the producer gas, the preferred operating temperature, and the filtration efficiency determine the preferred separation technology.

The major particle removal technologies are introduced in this chapter without going into details. The basics of most of the technologies are well known and already applied for many industrial processes. Barrier filters include bed filters, bag filters, and rigid filters. They barricade the flow of particulate matter while allowing the producer gas to pass through smaller pores. Regarding barrier filters, a more detailed introduction is given about rigid filter systems because it is a promising technology that offers significant efficiency gains in biomass conversion processes.

An overview of characteristics of particle removal technologies can be found in Table 3.1.

TABLE 3.1 Characteristics of Particle Removal Technologies [3, 31, 36, 37].

		Wet Scrubbers	Cyclones	Electrostatic Precipitators	Bed Filters	Bag Filters	Rigid Filters
Operating Temperature	[°C]	<100	<1000	<500	<870	<370	<1150
Pressure drop	[Pa]	100–1000	500–3000	30–400	1000–6000	600–2300	1000–10000
Dust concentration raw gas	[g m^{-3}]	<10	<1000	<50	<100	<100	<100
Dust concentration clean gas	[mg m^{-3}]	>10	>100	>25	<10	1–10	<1
Filtration grade		PM1	PM5 (optimized PM1)	PM5	PM3	PM1	PM0.5

3.4.1.1 Wet Scrubbers for Particle Removal Wet scrubbers are used to reduce different kinds of impurities in the producer gas. Among other impurities, they can be used to clean producer gas from particulate matter. Particles below 1 μm diameter can be removed by wash fluid droplets. Water is a common fluid used for wet scrubbers. Other liquids are described in the following sections. Wet scrubbers are robust and well known technologies usually operated around ambient temperature. Particles in the wash fluid suspension increase the risk of blockages. The commonly used venturi scrubber causes pressure drops of 3000–20 000 Pa. The disposal of the used wash fluid is expensive. The complete removal of impurities by wet scrubbers asks for big and therefore cost-intensive columns.

3.4.1.2 Cyclones Cyclones use centrifugal forces to separate particulate matter from producer gas. It is a simple, robust, and inexpensive technology to build and operate. Cyclones can remove large amounts of big particles and are therefore used as initial particle separation device and as an integral part for most fluidized bed gasification reactors to remove bed material from the producer gas. The operating temperature is only limited by the construction material and can go up to 1000 °C [38]. Several cyclones can be connected in series to improve the collection efficiency. More than 90% of particles larger than 5 μm can be removed by cyclones at a minimal pressure drop of 1000 Pa [39]. Even high performance cyclones are ineffective in removing submicron particles, such as char particles from biomass gasification, which is a major disadvantage of this technology [40].

3.4.1.3 Electrostatic Precipitators Electrostatic precipitators (ESP) separate particulate matter by applying a strong electric field. Particles get charged between the discharge electrode and the collector electrode with a very high potential difference. The ionized particles migrate to the collector electrode and are deposited on the surface. Particles are removed from the collector electrode by periodical mechanical rapping (dry ESP) or by spraying a liquid (wet ESP, usually water). More common is the dry ESP which can be operated at temperatures up to 500 °C [41, 42] and in research projects up to 1000 °C [43]. Wet ESP has a limited operating temperature range given by the boiling point of the spray liquid but is the preferred technology for explosive, corrosive, and sticky particles or particles with high resistivity [44]. The discharge electrodes are usually designed as rigid-frame wires or plate, the collector electrodes as tubes or plates. The performance of the ESP depends on different factors such as geometry of the device, applied voltage, electrical resistivity of gas and particles, and size and shape of the particles. Higher operating temperatures affect density, viscosity, and resistivity and therefore ESP operation. Higher operating pressure can counteract temperature-induced deviations.

3.4.1.4 Bed Filters Bed filters are also called packed bed or granular filters. They are depth filters which capture particulate matter in the pores of granular material. The granular material is usually sand but can also be limestone, alumina oxide, mullite, or others. Catalytic active material or sorption material can be added to the granular bed material, which is one of the advantages besides the possibility of high operating temperatures and the capability to filter sticky and tar containing dusts. Bed filters are operated as fixed or moving bed filters. The latter introduces more complexity as compared to the simple fixed bed filter but offers continuous operation. Fixed bed filters are recleaned or removed once a certain pressure drop is reached. Filter regeneration, handling, and disposal are disadvantages of bed filters.

3.4.1.5 Bag Filters   Bag filters apply flexible filter media such as textile fabric from fibers. Wire cages and spreader rings are needed as supports due to the low mechanical strength of the textile filter media. Regeneration is accomplished if a certain pressure drop over the filter medium is reached by both reverse pressure pulses and mechanical movement. Synthetic fibers (polyester, polypropylene, polypeptide) have replaced natural fibers (wool, cotton) due to better chemical, thermal, and mechanical properties. The operating temperature is limited by the melting point of the polymer fibers. Inorganic fibers (ceramic, glass) are used rarely due to their high price but enable operating temperatures above 300 °C [45]. For example,3 M's Nextel™ material is an aluminium, boron, and silica oxide composite that can withstand temperatures up to 370 °C [46]. Overviews of chemical and physical properties of fabric materials can be found elsewhere [37, 47]. Flexible filter media are the most common barrier filters used in industry for low temperature applications.

3.4.1.6 Rigid Filters   Rigid filters are made of porous metallic or ceramic materials that block particulate matter on the surface while the producer gas can pass through the pores. Grain ceramic filter candles are made of silicon carbide (SiC), alumina (Al_2O_3), or cordierite (MgO, Al_2O_3, SiO_2) [48]. Fiber ceramic filter candles are mainly made of alumosilicate fibers (Al_2O_3, SiO_2) [49]. Additional membranes on the surface of the filter candles (e.g., Mullite) can be applied to improve filtration and recleaning performance. Reactions of gas phase alkali with ceramic filter candles can reduce filter life time. Grain ceramic filter candles provide filtration grades for gases of 0.5 μm and lower and filtration efficiencies up to 99.999% [13]. Metal filter candles are made of sintered metal powders, sintered metal fibers (nonwoven) or metal fabrics. Thermal and chemical stability decide which steel grades and metal alloys have to be used to build resistant metal filter candles. Irreversible destruction of the metal filter elements by corrosion and oxidation occurs due to the higher volume of the reaction products than the pure metal.

Rigid filter elements in the form of filter candles are used in industrial applications. Particulate matter is collected at the outside of the filter candle building a filter cake. Back pressure pulses from the inside of the filter candle are used to detach the filter cake from the surface. The candle geometry enables a reliable operation by minimizing the risk of plugging. Filter candles can be as long as 3 m with diameters from 6 to 15 cm. Other geometries with higher filtration area density such as honeycomb monolith structure or cross flow filter exist but are not applied for industrial applications so far. There are also designs collecting the dust at the inside of the filter tube [43, 50, 51].

Conventional rigid filter designs install filter elements in vertical position with a closed filter end in the raw gas side. The filter elements are fixed on one side on the separation plate dividing raw gas and clean gas sector. Mechanical strength (filter fracture) can be of concern for rigid filter candles fixed at one side and operated at high temperatures. Filter elements applied in horizontal filter designs are shorter than in vertical designs. The filter candles are open at both ends, fixed on both sides of the raw gas sector, creating a clean gas sector and an additional recleaning sector. Fixing ceramic filter candles on two sides increases the mechanical stability, and filter fracture is of no concern.

A high content of ash and soot particles requires regular cleaning of the filter elements. The filter elements are cleaned from the filter cake by back pressure pulses. Cleaning the filter pores and filter surface by back pressure pulses has to guarantee stable filter operation. A stable operation of the filter equals a constant pressure drop over the HGF unit. Still, cleaning pulses should be as few as possible and at the lowest pressure difference possible in order to protect downstream equipment from pressure pulses and to save costs for energy and recleaning medium.

Different parameters such as recleaning pressure, recleaning gas volume, time interval between the cleaning pulses, and filtration velocity have to be optimized to enable a stable HGF operation. Filtration velocity ($m\,s^{-1}$) equals the gas flow of raw gas ($m^3\,s^{-1}$) divided by total filter area (m^2). It has been observed that the maximum pressure in the filter sector to be recleaned and the gradient of the pressure build-up during a back pressure pulse are the most important parameter to ensure the performance of the filter element recleaning [52, 53]. The pressure increase velocity (pressure gradient) is a function of the kind of recleaning system, type of filter elements, filtration velocity, filter design, and recleaning intensity. The recleaning intensity equals the pressure difference of the filter candle inside to the raw gas sector.

For cost saving and filter performance reasons, it is crucial to understand the filtration and recleaning mechanisms as detailed as possible. Dynamic pressure measurements help the early detection of filter failures [54]. The following filter failures reduce filter performance and increase operational costs: leakages, blockages, filter candle fracture, valve malfunctioning, filter area reduction (patchy cleaning), and filter pore size reduction (depth filtration). The main causes for filter failure are the design of filter unit, type of filter material, filter candle wall thickness and strength, thermal stress, and residual ash deposition [51]. Correct filter operating parameters such as flush tank pressure, valve opening time, and recleaning interval can avoid accelerated filter area reduction and filter pore size reduction. Knowledge about filter material resistance, ash and particle properties, dust load, filtration velocity, and filter design permits finding the correct filter operating parameters.

There are many examples of rigid filters applied for biomass gasification and IGCC processes. More information is available for ceramic filters than metal filters.

3.4.1.7 Hot Gas Filtration   Hot gas filtration (HGF) temperatures stay above the condensation temperature of tars and water. This has several advantages. Processing tars above their dew points prevents fouling of equipment due to tar condensation. Because condensation in quenching columns can be avoided by hot gas cleaning (HGC), depending on the gasification process, no steam needs to be added to the producer gas again. Steam content is needed downstream of the filter unit for steam reforming or to prevent soot formation in catalytic process units (e.g., fuel cells). In addition, contaminated liquid in quenching columns can be avoided, which improves energy efficiency and reduces environmental impact and costs.

Hot gas filtration (HGF) prevents exposing heat exchangers to particle loaded producer gas because the hot gas can be filtered at exit temperatures of gasifiers. High exit temperatures of gasifiers enable the application of high temperature sorption materials upstream or inside of a hot gas filter to reduce the sulfur and alkali content of the producer gas [55]. High temperatures are needed, for example, to catalytically convert sulfur-containing hydrocarbons (sulfur tars) and sulfur-free tars to lower molecular hydrocarbons and H_2S. Hydrogen sulfide can be adsorbed by a fixed bed of metal oxides completing the desulfurization of the producer gas. Catalysts can be used downstream of a hot gas filter protected from particulate matter. In a dust-free environment, catalyst structures, for example, monolith channels, can be smaller allowing more compact process units.

Up to now, most conventional high temperature filtration units cannot ensure long-term stable operation with dust from biomass conversion for temperatures above 450 °C. One of the main reasons is the limitation of the jet pulse recleaning system. The newly developed coupled pressure pulse (CPP) recleaning system overcomes these limitations. CPP recleaning system enables higher recleaning intensities at lower recleaning pressures as compared to conventional jet pulse systems [56, 57].

Jet pulse technology applies a high pressure, high speed gas jet directed to a bundle of filter candles. Different nozzle designs and venturi ejectors are used to optimize the gas flow of the jet. The kinetic energy of the jet has to be transformed to static pressure along the filter candle. After deceleration of the jet, static pressure is recovered and a recleaning back pressure pulse is generated. Lower recleaning intensities are measured at the entrance of the filter candle applying jet pulse technology. Not the whole recleaning sector is equally set under pressure, as compared to CPP technology.

CPP technology applies a high volume gas flow setting the whole recleaning sector under overpressure. There is no gas jet and gas velocities stay below the speed of sound ($M < 1$). The recleaning back pressure pulse is directly coupled to the recleaning compartment. The volume of the flush tank, the diameter of the high speed valves, and the connection to the recleaning sector have to be large enough to enable a fast pressure increase of the whole recleaning sector. The pressure in the flush tank can be set at 20 000–100 000 Pa above the pressure in the raw gas sector, which is enough to reach high recleaning intensities as compared to jet pulse.

The combination of HGF and CPP was successfully installed at different scales of pilot plants for several biomass conversion processes [58–60].

3.4.2 Tars

Tar removal from producer gas is needed to avoid fouling of equipment by tar condensation and soot formation. Tar conversion to lower molecular hydrocarbons is preferred, if a considerable amount of energy is contained in the form of tars in the producer gas. Depending on the kind of final energy conversion system, tars can be considered as fuel.

Physical tar removal methods are options for cold or warm gas cleaning. They are based on tar condensation which starts at temperatures below 450 °C. The resulting tar films built by tar aerosols are difficult to remove from the cleaning device. Hot gas cleaning options include catalytic and noncatalytic conversion options where tars are kept in the gas phase, avoiding tar condensation.

3.4.2.1 *Physical Tar Removal*

At temperatures below 450 °C, tars begin to condense and form aerosols. Particle removal devices such as wet scrubbers, cyclones, electrostatic precipitators, and barrier filters operated at temperatures below 450 °C are able remove tar aerosols from the producer gas. Except for wet scrubbers, the challenge is how to remove the tar films from cyclone walls, collector electrodes, and filter surfaces.

Wet scrubbers collect tar aerosols and soluble tar compounds. The lower the operating temperature of the wet scrubber, the more tar compounds condensate. The solubility of tars in oil based scrubbers is higher than in water based scrubbers. Lowest tar removal efficiencies of wet scrubbers were measured for simple spray towers. Higher efficiencies were found for venturi and vortex scrubbers and combinations with venturi and cyclones geometries [61]. Tar removal efficiencies of wet scrubbers are often not rigorous enough and the treatment of the contaminated liquid is costly.

The Energy Research Center of the Netherlands (ECN) developed a multiple stage scrubber tar removal concept called OLGA [62–64]. Heavy tars are removed by condensation with the first oil based scrubbing liquid. Light tars are removed by absorption with the second oil based scrubbing liquid. Both scrubbers operate above the condensation temperature of water. Heavy and light tars can be recovered from the scrubbing liquids and recycled to the gasifier contributing to the energy efficiency.

Electrostatic precipitators work the same way for tar droplets as they do for particulate matter. Wire and tube designs are preferred for tar collection [39]. The collector surfaces are washed continuously (wet ESP) with water or oil based washing liquids to remove the tar deposits. The lower the operating temperatures of the wet ESP are, the more efficient is the tar removal.

3.4.2.2 *Noncatalytic Tar Removal*

Thermal cracking means the decomposition of tars into lighter hydrocarbons at high temperatures. Operating temperatures are reported to range from 900 to 1300 °C [65–68]. Longer residence times are needed at 900 than at 1300 °C. For example, naphthalene was reduced by more than 80% in about 1 s at 1150 °C and it took more than 5 s at 1075 °C [65, 69]. Residence times of 0.5 s are reported at 1250 °C [66]. If the exit temperature of the gasifier is not high enough for thermal cracking, the temperature of the producer gas can be increased by heat exchangers or partial oxidation of the producer gas. Noncatalytic partial oxidation means cracking of tars by increasing the temperature by adding air or pure oxygen to the producer gas [67, 68].

The advantage of this technology is its simplicity but there are many disadvantages. Expensive alloys resisting the high temperatures have to be used for thermal cracking. The heating value of the producer gas decreases significantly by partial oxidation and CO levels could increase at the expense of conversion efficiency [70]. Process efficiency decreases if heat exchangers are needed to increase the temperature of the producer gas. Thermal cracking and according operating conditions can produce soot and polyaromatic hydrocarbons (PAH) [65, 71–73].

Plasmas are reactive atmospheres of free radicals, ions, and other excited molecules. The decomposition of tars can be initiated by reactive plasma species [74]. There are several types of electrical discharge reactors to generate the plasma. The most promising technology is the pulsed corona plasma which reduces tars around 400 °C [75,76]. Other plasma technologies are dielectric barrier discharge, DC corona discharges, RF plasma, or microwave plasma [70, 74–76]. Costs, energy demand, lifetime, and operational complexity prevent industrial scale applications [77]. The influence of other contaminants on plasma performance is not clear [3].

3.4.2.3 *Catalytic Tar Removal*

High temperature tar removal options include non-catalytic and catalytic solutions. Catalytic tar removal operates at lower temperatures than thermal cracking. Common operating temperatures are between 600 and 900 °C. Catalytic tar removal can be applied in situ or downstream of the gasifier. Catalytic material is used in situ as bed material or in addition to bed material. Downstream of the gasifier, catalytic material is specific reactors. Fixed beds and monoliths are the most common reactor designs but there are more complex designs such as chemical looping reformers (CLR) [78] or reverse flow catalytic tar converters (RFTCs) [79].

Many different classifications exist for tar removal catalyst [8, 21, 71, 80–85]. Abu El Rub et al. suggest the most straightforward classification based on the origin of the catalysts [85, 86]. Catalysts are classified as natural and synthetic. Natural catalysts include natural minerals such as dolomite, olivine, clay minerals, and ferrous metal oxides. Synthetic catalysts include chars, fluid catalytic cracking (FCC) catalysts, alkali metal carbonates, activated alumina, and transition metals.

Natural minerals are low cost catalysts and often applied in situ. Calcined rocks such as dolomites, limestone, or magnesium carbonate are used. Tar conversion up to 95% was measured with dolomite [85]. The easy deactivation and the need for a high CO_2 partial pressure to keep the active state are disadvantages of calcined minerals [87].

Olivine is a silicate mineral containing iron and magnesium. The activity of olivine is lower but the attrition resistance is higher than dolomite.

Clay minerals show catalytic activity due to their silica and alumina content [85]. Their activity is also lower than the activity of dolomite. The thermal resistance at common gasification temperatures is limited due to the porous structure.

Ferrous metal oxides, for example, iron ore, show lower activities than dolomite and are prone to coking [88]. Iron-rich mineral compounds consist of 35–70% of iron oxides. The metallic form shows higher activity than the oxide, carbonate, silicate, or sulfide forms [85, 88]. Further information can be found in the literature [81, 86, 88, 89].

Char is usually a by-product of thermochemical biomass conversion. Therefore char is cheap and abundant. Physical and chemical properties of char are not well definable because biomass feedstock and conversion process can influence the characteristics of char [90–94]. Significant tar conversions could be measured using char as a catalyst [86, 95, 96]. In addition, sorption of alkali and sulfur contaminants was observed [96]. Some char is continuously used with steam and CO_2 in gasification reactions. Barrier filters, such as ceramic filter candles, build filter cakes on the surface consisting of char and ash. This filter cake shows not only catalytic activity if operated at high temperatures but also sorption capabilities [26, 97].

Activated alumina shows relatively high activities similar to dolomite. Alumina is activated by removing hydroxyl groups (–OH) from minerals such as bauxite and aluminum oxide by heating. High mechanical and thermal stability are advantages of activated alumina [85, 98]. To improve the performance of activated alumina regarding activity, coking and poisoning, other metal oxides can be used (e.g., CoO, CuO, Cr_2O_3, Fe_2O_3, Mn_2O_3, MoO_3, NiO, V_2O_5 [99]).

Fluid catalytic cracking (FCC) catalysts are usually aluminosilicate zeolites used in the petroleum industry to convert heavy fuel oil to lighter components. Partial sulfur tolerance, low price, and greater stability compared with regular alumina based catalysts make zeolites promising tar removal catalysts [85, 100]. Moderate to low tar conversions were measured for FCC catalysts used in situ or in a sand bed [95, 101]. Parallel water–gas shift reactions of zeolites reduce tar conversion in gasification environments. Zeolites combined with nickel (Ni) showed higher activity as compared to pure zeolites [100, 102].

Alkali metal carbonates are represented by sodium carbonate (Na_2CO_3) and potassium carbonate (K_2CO_3). Biomass feedstock contains naturally alkali but tar conversion can be increased by adding alkali minerals (trona, borax) or ash or by wet impregnation with alkali carbonates [71, 72, 89, 103–107]. Adding alkali to the biomass increases ash content.

Transition metals such as cobalt (Co), copper (Cu), iron (Fe) nickel (Ni), platinum (Pt), rhodium (Rh), ruthenium (Ru), and zirconium (Zr) are used in combination with catalyst supporting and promoting materials. The activity regarding tar conversion decreases in the following order: Rh > Pd > Pt > Ni, Ru [70].

Several studies cover the performance of catalytic conversion of tars in biomass gasification fuel gases [108–120]. Ni-based catalysts show high activities but are prone to sulfur poisoning below 900 °C. Therefore, the performance of noble metal catalysts below 900 °C in the presence of sulfur is of special interest. Lower operating temperatures will better suit exit temperatures of biomass gasifiers. Additional heating of the producer gas to reach temperature around 900 °C can be avoided.

Noble metal catalysts show promising results due to superior coking resistance and sulfur tolerance. Significantly higher tar conversions could be measured for an $Rh/CeO_2/SiO_2$ catalyst used for catalytic partial oxidation (CPO) of tars as compared to a typical Ni-based catalyst [77, 121]. Rhyner et al. showed high activities for a 400 cpsi noble metal catalyst regarding the conversion of tars (toluene, naphthalene, phenanthrene, pyrene) and sulfur-containing hydrocarbons (thiophene, benzothiophene, dibenzothiophene) [34, 35, 122, 123].

Rönkkönen et al. investigated ZrO_2 based catalysts in the temperature range of 600–900 °C and report naphthalene conversion rates around 80% with high O_2 concentration in the gas and temperatures of 900 °C, indicating that the main reactions in naphthalene decomposition on ZrO_2 are oxidations [124]. The performance of Rh, Ru, Pt, and Pd on modified commercial zirconia support (m-ZrO_2) regarding naphthalene, toluene and ammonia decomposition was compared to a benchmark Ni/m-ZrO$_2$ catalyst in the presence of H$_2$S in recent studies. Rh/m-ZrO$_2$ was found to be the most promising catalyst.

Furusawa from the University of Tokyo compared Co/MgO and Ni/MgO catalysts for the steam reforming of naphthalene as a model compound of tar derived from biomass gasification [125]. Although the catalytic performance showed that the Co/MgO catalyst had higher activity (conversion: 23%) than any kind of Ni/MgO catalyst tested in that study, the conversion rate is very low. In the latest studies from Furusawa et al., they studied the influence of support on the catalytic performances of Pt and Ni based catalysts for the steam reforming of naphthalene and benzene as model tar compounds of biomass gasification. They concluded that Pt/Al$_2$O$_3$ showed the highest and most stable activity of the tested catalyst supports at 800 °C [126].

Cui et al. [127] looked at permanent gas species, tar compounds, sulfur compounds, and ammonia produced from a bench scale (1 kg h^{-1}) fluidized bed biomass gasifier. Two commercial Ni based catalysts and one commercial ZnO sorbent were evaluated under varying conditions by quantifying contaminants from the gasification reactor inlet and outlet.

There are several groups working with catalytic active filter elements. Nacken et al. worked with silicon carbide based filter elements catalytically activated with Ni [111, 112, 128]. Naphthalene conversion up to 66% was reported in an environment with 100 ppm H$_2$S. Rapagnà et al. applied activated filter elements in the free board of a laboratory scale fluidized bed gasification reactor [113, 129, 130]. At temperatures up to 840 °C, tar conversion of 58% was reported. Simeone et al. [59, 60] tested ceramic hot gas filter elements with mullite membrane coating and integrated Ni-based catalyst in a dust-free model gas. Naphthalene conversion of 99.4% was measured at 850 °C and 30 vol% H$_2$O with 2.5 g m$_n^{-3}$ of naphthalene. Experiments with higher steam content showed higher conversion rates.

Professor José Corella's group from the University Complutense of Madrid published four articles about high dust catalytic hot gas cleaning with monoliths in biomass gasification in fluidized beds, reporting: (a) their effectiveness for tar elimination [131], (b) modeling of the monolithic reactor [132], (c) their effectiveness for ammonia elimination [133], and (d) the performance of an advanced, second generation, two-layer based monolithic reactor [134]. As the conclusion of the fourth paper, Toledo et al. wrote that Ni-based catalysts are not the definitive answer to the problem and that non-nickel based monoliths working at even lower temperatures would be welcomed.

At Chalmers University, a chemical looping reformer (CLR) concept was tested using ilmenite (FeTiO$_3$), Mn$_2$O$_3$ and NiO as catalyst materials [78, 135–140]. CLR allows catalytic tar cleaning with simultaneous regeneration of the catalyst from carbon deposits. The producer gas is cleaned by the catalyst in one reactor (fuel reactor), while the catalyst is continuously regenerated in another reactor (air reactor). Tar conversion rates of 35 and 44% were reported.

Steam, tars and sulfur species have to be considered to simulate realistic producer gas. Sulfur in the form of H$_2$S was often considered in the studies mentioned before because sulfur-free producer gases will hardly exist. It is important to consider steam content when evaluating the performances of tar reforming catalysts, since gasifiers using steam as the gasification agent create producer gases with steam contents up to 50 vol% [10]. Most of the studies used merely toluene and naphthalene as model compounds.

The disadvantages of catalytic tar removal options are catalyst deactivation and costs, especially for synthetic catalysts. In particular, noble metals such as Ir, Rh, Ru, Re, and Pd are two to three orders of magnitude more expensive than Mo, W, Ni, and Co. Catalyst deactivation is caused by fouling (soot and coke formation), poisoning, sintering, evaporation of catalyst components, and attrition. Catalyst poisoning means strong chemisorption of contaminants such as sulfur, ammonia, and trace elements on the active catalyst sites. Catalyst inhibition is a milder form of catalyst poisoning, which means a reversible adsorption of impurities or compounds.

3.4.3 Sulfur Compounds

Sulfur compounds in the producer gas are mainly hydrogen sulfide (H_2S), carbonyl sulfide (COS), carbon disulfide (CS_2), and sulfur-containing hydrocarbons. After combustion, sulfur exists in the form of sulfur dioxide (SO_2), which is a regulated pollutant. Sulfur compounds have to be removed from the producer gas in order to protect downstream catalysts used for gas upgrading and conversion processes. There are basically three options to remove sulfur from producer gas: wet scrubbing, hydrodesulfurization (HDS), and sorption materials [39, 141, 142]. HDS catalysts convert sulfur-containing hydrocarbons to H_2S which can be removed by wet scrubbing or sorption materials. Wet scrubbing is a cold gas cleaning method, whereas HDS is a hot gas cleaning method. Sorption materials can be used in a temperature range of 300–1000 °C. Upstream of a hot gas filter unit, sorption materials can be added for bulk sulfur removal [2, 34, 55]. Fixed bed reactors are used downstream of a HDS reactor.

Sulfur-containing hydrocarbons are often neglected in gas cleaning studies of biomass gas. The lack of appropriate analytical equipment and methods could be one reason for neglecting sulfur tars. The variety of sulfur-containing hydrocarbons found in biomass gasifier gas was shown, for example, by Rechulski et al. and Cui et al. [5, 127]. Up to 41 different sulfur tars could be detected. The most abundant were thiophenes, followed by benzothiophenes and dibenzothiophenes. The amount of sulfur-containing hydrocarbons in biomass producer gases can be above the acceptable tolerance levels of the catalytic process.

3.4.3.1 *Wet Scrubber for Sulfur Removal* Liquid solvents are used to remove H_2S from producer gas by physical or chemical sorption. The process is also called acid gas removal. Depending on the solvent, COS, CO_2, and hydrocarbons are absorbed in addition to H_2S. Regeneration processes (e.g., strippers) are integrated to enable a continuous operation. Elemental sulfur recovery processes can be integrated as well (e.g., Claus process). Energy intense cooling equipment is needed due to operating temperatures in the range of –60 to +20 °C. Wet scrubbers for sulfur removal are efficient but operation and equipment are expensive.

Amine based solvents build weak chemical bonds between the amine component and H_2S or CO_2. Primary, secondary, or tertiary amines are commonly used for absorption processes. Widely used amines are monoethanolamine (MEA), diethanolamine (DEA), and methyldiethanolamine (MDEA). COS is not efficiently removed by amine scrubbers and can even degrade the solvent. Therefore, hydrogenation of COS to H_2S is necessary. Amine loss takes place continuously in the process and fresh solvent has to be added constantly.

Solvents used for physical absorption of H_2S are often preferred as compared to chemical solvents. Advantages are minimal solvent loss, high loadings, and stripping of impurities by reducing pressure without the addition of heat. A selection of common solvents can be found in Table 3.2.

Liquid redox processes are used for direct H_2S removal as well as sulfur recovery. Dissolved vanadium catalyst (Stretford, Sulfinol, Unisulf processes) or chelated iron slurries (LO-CAT process) are used for liquid redox processes. The low severity process is a third kind of liquid redox process where H_2S is absorbed into a polar solvent and hydroquinone and elemental sulfur are formed [143].

There are also bacteria that are able to remove sulfur compounds [144], though reactions conditions are limited to the comfort zone of the living bacteria. For separated biological sulfur recovery processes (Thiopaq, Biopuric, Bio-SR), the operating conditions can be optimized more easily for bacteria [145]. Reductions in operating costs (energy savings) are the motivation for chemo-biological processes.

3.4.3.2 Sorption Materials for Sulfur Removal

Solid sorption materials can be used to remove mainly H_2S from the producer gas down to sub-ppm levels. Sulfur-containing hydrocarbons are usually not captured efficiently enough by sorption materials [146] and have to be converted to H_2S first. Depending on reversible or irreversible sorption reactions, sorption materials can be regenerated and reused. Cheap sorption materials, usually natural minerals, are used for once through processes where irreversible sorption occurs. More expensive sorption materials (synthetic materials) can ideally be regenerated and be used for several reversible sorption cycles.

Sorption materials for bulk removal of sulfur can be added to the producer gas directly in the gasifier or upstream of the particle removal unit. Fixed beds are installed downstream of the particle removal to remove H_2S. Reactions can take

TABLE 3.2   Solvents for Sulphur Removal with Wet Scrubbers.

Compound	Abbreviation	Product	Supplier	Operating Temperature (°C)
Dimethyl ether of polyethylene glycol	DEPG	Selexol	DOW, UOP	> -18
Dialkyl ethers of polyethylene glycol		Genosorb	Clariant	
Methanol	MeOH	Rectisol	Lurgi	-60 to -40
N-methyl-2-pyrrolidone	NMP	Purisol	Lurgi	-15

place in the gasifier, during the flight of the particle in the direction of the particle removal unit, and in the filter cake once the sorption particle reaches the filter element. Adding sorption materials to the producer gas stream will decrease the temperature of the producer gas, increase the amount of particulate matter, and change the characteristics of the filter cake. Sorption materials can be used to remove not only H_2S but also alkali, halide, and trace elements. Optimization is needed regarding costs, removal efficiency, and increased lifetime of downstream equipment due to the absence of impurities. Costs include sorption materials and feeding system, heat reduction, and increased filtration intensity.

Bulk removal of H_2S can be done for example with calcium based sorbents or trona. Calcium based sorbents include naturally available dolomite and limestone but also calcium acetate or calcium magnesium acetate [Equations (3.1) and (3.2)]. Trona can be used to remove H_2S and HCl. H_2S capture was investigated at 600 and 800 °C, resulting in 1.8 and 1.0 ppmV with corresponding H_2S in the raw syngas of 100 and 200 ppmV, respectively [147, 148].

$$CaCO_3 + H_2S \rightarrow CaS + H_2O + CO_2 \qquad (3.1)$$

$$CaO + H_2S \rightarrow CaS + H_2O \qquad (3.2)$$

Fixed bed sorption materials for H_2S removal to sub-ppm levels commonly use metal oxides [Equation (3.3)]. ZnO is a very good sorption material for H_2S removal because it shows the most favorable sulfidation thermodynamics. However, vaporization of elemental zinc at temperatures above 600 °C is a drawback. For that reason, zinc ferrite ($ZnFe_2O_4$) and zinc titanate ($ZnTiO_3$) are considered as alternatives where zinc titanate shows better removal efficiencies at temperature up to 900 °C [Equation (3.4)] [3]. There are also copper, iron, manganese, and cerium based metal oxide sorption materials for H_2S removal. Besides calcium based sorbents, zinc titanate shows the highest operating temperatures. An overview of sorption materials for H_2S removal can be found in Meng et al. [2].

$$Me_xO_y + xH_2S + (y-x)H_2 \rightarrow xMeS + yH_2O \qquad (3.3)$$

$$Zn_2TiO_4 + 2H_2S \rightarrow 2ZnS + TiO_2 + 2H_2O \qquad (3.4)$$

3.4.4 Hydrodesulfurization

To convert sulfur-containing hydrocarbons to H_2S, a hydrodesulfurization (HDS) catalyst is needed. H_2S can be captured efficiently by sorption materials down to sub-ppm levels whereas sulfur tars such as mercaptans, disulfides, and thiophenes cannot. HDS catalysts are traditionally used in the hydrotreatment units of oil refining and coal liquefaction industries. Usually supported molybdenum or tungsten sulfide catalysts promoted by nickel or cobalt (CoMoS, NiMoS, CoWS, or NiWS) are used in hydrotreatment units where HDS activity is desired [149, 150]. Their high activity and low cost as compared to transition metals are reasons for their wide application. Hydrogenation of hydrocarbons, methanation, and water–gas shift reactions can also be promoted by

CoMoS and NiMoS [151]. Catalysts with transition metal sulfides (e.g., Mo or W sulfides), carbides (e.g., Mo_2C, NbC), and phosphides (e.g., MoP, NiP_2) are also used as HDS catalysts [149, 151–155]. Furimsky [149] published a review on the selection of catalysts for the hydrodesulfurization of petroleum refinery feeds. The activity of transition metal sulfides can be related to their position in the periodic table, as shown in several studies [156–173]. Rh, Re, and Ru present the highest activity for HDS, often comparable or higher than the commercial Co(Ni)Mo based catalysts.

Since HDS catalysts are mainly used in petroleum industries, only a few studies are available regarding their performance under biomass gasification conditions. Temperature, pressure, and gas composition differ between traditional oil refinery application and the more recent biomass gasification application [5].

Rechulski tested the performance of four different catalysts (CoMoS/Al_2O_3, NiMoS/Al_2O_3, RuS_2/Al_2O_3 type I, and RuS_2/Al_2O_3 type II) regarding hydrogenolysis capabilities under biomass producer gas conditions [5]. In the presence of sulfur, the catalytic hydrogenolysis reactor should: (a) hydrogenate sulfur-containing compounds to H_2S, (b) hydrogenate or steam reform tars, (c) promote methanation, and (d) promote the water–gas shift reaction. None of the catalysts promoted sulfur-resistant methanation. The highest activities for all reactions observed at operating temperatures of 500 and 600 °C showed the RuS_2/Al_2O_3 type II catalyst. However, the activity of the Ru-based catalysts could not justify the price of the noble metal, which is about two orders of magnitude higher than the price of Mo. A commercial Mo based hydrogenolysis reactor operated at 400 °C is suggested to be economically more feasible than the self-made Ru based catalysts.

Rhyner et al. used a 400 cpsi noble metal catalyst to test the conversion of tars and sulfur-containing hydrocarbons in the presence of steam, hydrogen sulfide, and ethene [34, 35, 122]. In order to simulate producer gas from biomass gasification, higher molecular hydrocarbons (toluene, naphthalene, phenanthrene, pyrene) and sulfur-containing hydrocarbons (thiophene, benzothiophene, dibenzothiophene) were added to a syngas. The catalyst was operated at temperatures between 620 and 750 °C. Conversions of sulfur-containing hydrocarbons (41–99.6%) were on average higher than conversions of sulfur-free tars (0–47%). High temperature, low GHSV, low steam, and sulfur content favored high conversions of tars and sulfur tars. As the catalyst was able to decompose sulfur tars under operating conditions close to a real wood gasification plant, it is possible to use it for hot gas cleaning in any process that includes sulfur-sensitive catalysts such as fuel cells, liquid fuel synthesis, or methanation processes. In such processes, H_2S produced by the reforming catalyst from sulfur tars can be captured downstream of the reformer in a metal oxide bed such as ZnO.

3.4.5 Chlorine (Halides)

Halides are mainly represented by hydrogen chloride (HCl) in biomass producer gas. Chlorine exists as HCl in producer gas and reacts with ammonia (NH_3) also present in producer gas at temperatures below 300 °C to give ammonium chloride (NH_4Cl). Wet scrubbers are applied for cold gas cleaning and sorption materials for hot gas cleaning processes. Wet scrubbers can be used to collect chloride salts or to adsorb HCl vapor.

Calcined limestone and dolomites are the most often used sorption materials for HCl capture in flue gas from coal combustion plants [2]. CaO and MgO react easily with HCl forming $CaCl_2$ and $MgCl_2$ with melting points of 774 and 695 °C, respectively [174, 175]. Sodium and potassium compounds are also capable of reducing HCl concentration, see Equation (3.5). Nahcolite is another naturally available option for HCl removal, see Equation (3.6). Nahcolite is a cheap, natural mineral, having a cost of about US\$ $50\,t^{-1}$, and sorbent regeneration is not necessary. This approach to halide removal has been demonstrated at a pilot scale by Siemens [97]. Trona can be used to remove H_2S and HCl. With trona sorbents, HCl was reduced to a 40 ppbV level at 600 °C from initially 20 ppmV [147, 148].

$$2NaCl + Al_2O_3 + H_2O \rightarrow 2NaAlO_2 + 2HCl \qquad (3.5)$$

$$NaHCO_3 + HCl \rightarrow NaCl + H_2O + CO_2 \qquad (3.6)$$

3.4.6 Alkali

Alkali can be removed from producer gas by condensation or sorption materials. Alkali vapors condense at temperatures lower than 600 °C. Therefore, alkali can be captured in wet scrubbers, filter cakes, and any other surface offered below 600 °C. Sorption materials are the preferred solution for high temperature gas cleaning process chains. Sorbent in alkali removal is generally termed an "alkali getter". Natural minerals such as diatomaceous earth (silica), clays, or kaolinite, but also synthetic materials such as activated alumina from bauxite minerals are used as sorption materials [43, 176, 177].

Dou et al. [1, 174] tested the alkali metal removal capabilities of second grade alumina, bauxite, kaoline, acidic white clay, and activated alumina in a fixed bed reactor at 840 °C with coal-derived gas. Al_2O_3 showed the highest adsorption efficiency. Tran et al. conducted tests with kaoline in a fixed bed reactor and reported good efficiencies for the removal of KCl [178]. Turn et al. report good physical adsorption and chemisorption of Na and K by bauxite but no effect for Cl [179].

There are also processes to remove alkali content in biomass prior to thermo-chemical conversion by leaching, though costs are incurred for the washing, drying, waste treatment, and mechanical processes involved [180–182].

3.4.7 Nitrogen-containing Compounds

Nitrogen-containing compounds in biomass-derived producer gas are mainly ammonia (NH_3) and less abundant hydrogen cyanide (HCN) but also organic compounds such as, for example, pyridine (C_5H_5N).

If NH_3 and HCN need to be removed from the producer gas, its good solubility in water makes wet scrubbers an efficient solution at low temperatures. Even condensed water from vapor in the producer gas can be enough for a substantial removal of nitrogen compounds in amine or oil based scrubbers or in a chilled condenser [183–185].

At high temperatures, only catalytic decomposition to N_2 and H_2 or selective oxidation to NO_x is possible. Oxidizing nitrogen compounds selectively with oxygen molecules while not affecting other gas species such as CH_4, CO, or H_2 is difficult though [77]. The decomposition to N_2 and H_2 seems easier. Common tar cracking or hydrocarbon reforming catalysts such as dolomites or nickel and iron based catalysts have shown promising NH_3 reductions [39, 186]. Catalytic materials based on Ru, W, or other noble metals are more expensive but showed good activities for the decomposition of tars and ammonia [39, 77, 102]. Further research is needed investigating the activity of these catalysts used for tar decomposition regarding their capabilities for ammonia decomposition. In particular the resistance to steam and sulfur is of interest besides the ammonia decomposition.

The cleaning of NO_x from exhaust gases by selective catalytic reduction (SCR) or selective noncatalytic reduction (SNCR) are widely used in industry to cope with emission levels but are not discussed here.

3.4.8   Other Impurities

Trace elements are usually removed by sorption materials. Common sorption materials used to remove sulfur, alkali, and chloride compounds can also be used as sorbents for trace elements. Often limited information is available about sorption capacities of different trace elements and interactions of contaminants. Most information comes from coal gasification and little is known regarding the performance under biomass gasification conditions. Silica, bauxite, kaolinite, zeolite, lime, activated carbon, fly ash, alumina, metal oxides, and others are applied as sorption materials for trace element removal [24, 26].

High As and Se removal rates were measured for fly ash, limestone, and metal oxides. Fly ash is also effective in removing Cd and Zn [24]. The sorption capacity of the filter cake consisting of very fine fly ash generated during gasification is not to be underestimated, providing a large amount of high surface area material. It appears to be an excellent sorbent on which metal and contaminant compound vapors can condense and then be captured by particulate collection devices [26, 97]. It is also found that Cd, Se, As, Pb, and Zn may be removed to some extent by fly ash sorption [24, 26].

Celatom is a granular, calcined, diatomaceous earth and represents a low cost, high temperature sorbent often used for a variety of applications requiring high stability and sorption capacity. It is an alumino-silicate material and a rather effective polishing sorbent for As, Se, and Zn [26]. CaO and CuO on activated carbon are other sorbents suitable for the capture of arsenic [26, 187].

Ideally sorption materials can be used to capture several different impurities. Table 3.3 lists a selection of sorption materials, including tested temperature, pressure, and contaminants.

3.5   REACTIVE HOT GAS FILTER

Integrated concepts of hot gas filtration (HGF) and reformer catalyst promise cost and energy savings due to the reduction of operating units, the decrease of temperature losses between operating units, the compact design, and straightforward process units.

TABLE 3.3   Sorption Materials Overview.

Sorbent	Tested Temperature [°C]	Tested Pressure [bar]	References	Observation
CaO	650	1, 5	[26]	H_2S, As
Calcium carbonate ($CaCO_3$)	600–800	1	[147, 188]	H_2S, HCl
Celatom (SiO_2)	650	1, 5	[26]	As, Se, Zn
Nahcolite ($NaHCO_3$)	430–600	1	[97]	H_2S, HCl
Trona (Na_2CO_3, $NaHCO_3$ $2H_2O$)	430–600	1	[55, 97, 148, 188]	H_2S, HCl
Na_2CO_3	450–500	1	[55]	H_2S, HCl
Na_2O	450–500	1	[55]	H_2S, HCl
K_2CO_3	450–500	1	[55]	H_2S, HCl
CuO/C	30–140	52	[55]	As
Dolomite [$CaMg(CO_3)_2$]	650–1050	1	[1,2]	H_2S
Limestone ($CaCO_3$)	500–1050	1–20	[2]	H_2S
Calcium acetate $Ca(CH_3COO)_2$	600–1050	1	[2]	H_2S
Calcium magnesium acetate $Ca_xMg_y(CH_3COO)_{2(x+y)}$	800–1000	1	[2]	H_2S
Al_2O_3	840	1	[174]	NaCl, KCl

The performance of HGF and reformer catalyst has to be well understood to permit the combination of the two process units.

So far the limiting factor has been the maximum temperature of the filter units (400 °C), hence the reformer or catalytic partial oxidation (CPO) unit was operated upstream of the filter with high dust load. Monoliths with low cell densities (cpsi) were selected in order to lower the risk of particle and soot deposition. With the recent development of hot gas filter designs and ceramic filter elements that can be used up to 850 °C, it is possible to install reformer/CPO units in a low dust environment at high temperatures.

There are two examples in Europe where catalytic tar reforming is applied at industrial scale combined heat and power (CHP) plants: Kokemäki in Finland and Skive in Denmark. Both reformers are operated at high dust load. At the time of basic engineering of these plants (around 2003), no hot gas filtration technology in the range of 400–800 °C was available. For this reason, the reformers were built on a massive scale to be be able to cope with the high particle load. The reformers were almost as big as the gasifiers themselves. The two plants differ in the gasifier and reformer technology, as shown in Table 3.4.

Only a little information about the experiences of the two plants has been published. Kurkela et al. published the final activity report of the BIGPower project, which is the Kokemäki plant [189]. Rönkkönen et al. was involved in the same project [124, 190, 191]. Only workshop presentations are available about the Skive plant [192, 193]. Reference [192] stated that noble metal reformers show very promising results compared to Ni based catalysts, with tar conversion rates higher than 93% at temperatures of 850 °C. It also stated that the overall costs of such processes

TABLE 3.4   Gasifier and Reformer Technology of the Skive and Kokemäki Plants.

Site of Installation	Gasifier Technology	Reformer Technology
Skive (Denmark)	Autothermal fluidized bed	Steam reformer
Kokemäki (Finland)	Updraft fixed bed	Catalytic partial oxidation

need to be reduced by simplification and new innovations [192]. The other presentation [193] states that the monolith tar removal capabilities of 50–70% at operating temperatures between 850 and 930 °C are not satisfactory. A photograph showing the monolith with most of the channels covered heavily with dust explains the problem of tar reforming in a high dust environment [193].

The company Pall Filtersystems (M. Nacken and S. Heidenreich) has been involved with in situ catalytic ceramic filters for tar reforming over many years, together with the Universities of Teramo and L'Aquila in Italy, as well as with the Vrije Universiteit Brussel in Belgium and the Delft University of Technology in the Netherlands [111, 112, 128, 194].

Rapagnà et al. applied catalytic active filter elements placed in the free board of a laboratory scale fluidized bed gasification reactor [113, 129, 130]. These investigations are part of the terminated EU project UNIQUE [195] aimed at integrating the fluidized bed steam gasification of biomass and the hot gas cleaning and conditioning system into one single gasification reactor vessel.

In the framework of the terminated EU project CHRISGAS [196], ceramic hot gas filter elements with a mullite membrane coating and integrated Ni-based catalyst were tested with a dust-free model gas by Simeone et al. [197, 198].

Rhyner et al. investigated a reactive HGF system consisting of HGF and catalytic reformer in order to implement a hot gas cleaning (HGC) process in thermochemical conversion processes from biomass to electricity, bio-SNG, or liquid fuels. Stable operation for more than 1000 h could be shown for a HGF system with a coupled pressure pulse (CPP) recleaning system. The filter unit was operated at 450 °C filtering particulate matter from an updraft wood gasifier. A noble metal catalyst was found which is able to form H_2S out of sulfur bound in hydrocarbons at temperatures of 600–850 °C.

The ability of the catalyst to decompose sulfur-containing hydrocarbons is a precondition for the proposed process chain applying HGC in combination with a metal oxide bed to remove H_2S. The conversion to H_2S is needed, because so far, high temperature desulfurization by sorption materials is efficient for H_2S only, but not for sulfur tars. The conversion of sulfur-free tars has second priority. The catalyst activity regarding water gas shift (WGS) reaction and steam reforming of methane (SRM) has minor relevance regarding the desulfurization step. Tars and methane in the producer gas are not problematic to some extent when solid oxide fuel cells (SOFC) are used for electricity production. Tars can be considered as fuel and methane is used for internal cooling of the fuel cell [14].

Calculations based on the simulation of the tested reformer catalyst showed that, besides other possibilities, the integration of a catalytic monolith at the exit of the HGF vessel is feasible. Assuming that the presented HGF design works at

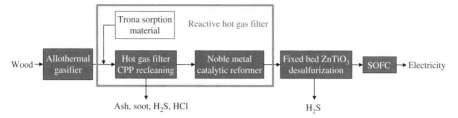

FIGURE 3.2 Suggested HGC at 700–850 °C, including reactive HGF for B-IGFC process chain.

temperatures of 850 °C, a reactive HGF system combining filter and catalyst, operating at exit temperatures of 850 °C, is possible. To complete the reactive HGF system, Trona is suggested for use as a high temperature sorption material upstream of the HGF. Bulk H_2S and HCl can be removed by Trona. The complete HGC process includes a final desulfurization polishing step. $ZnTiO_3$ can be applied for H_2S removal at high temperatures. Figure 3.2 shows the suggested HGC including HGF for a B-IGFC process chain.

REFERENCES

[1] Dou B, Wang C, Chen H, Song Y, Xie B, Xu Y, et al. Research progress of hot gas filtration, desulphurization and HCl removal in coal-derived fuel gas: A review. *Chemical Engineering Research and Design* **90**(11):1901–1917; 2012.

[2] Meng X, de Jong W, Pal R, Verkooijen AHM. In bed and downstream hot gas desulphurization during solid fuel gasification: A review. *Fuel Processing Technology* **91**(8): 964–981; 2010.

[3] Aravind PV, de Jong W. Evaluation of high temperature gas cleaning options for biomass gasification product gas for solid oxide fuel cells. *Progress in Energy and Combustion Science* **38**: 737–764; 2012.

[4] Judex JW. *Grass for Power Generation – Extending the Fuel Flexibility for IGCC Power Plants*. Report 18865, ETH Zürich, Zurich, p. 199; 2010.

[5] Kaufman-Rechulski MD. *Catalysts for High Temperature Gas Cleaning in the Production of Synthetic Natural Gas from Biomass*. Report 5484, EPFL, Lausanne, p. 270; 2012.

[6] Vassilev SV, Baxter D, Andersen LK, Vassileva CG, Morgan TJ. An overview of the organic and inorganic phase composition of biomass. *Fuel* **94**: 1–33; 2012.

[7] Nagel FP. *Electricity from Wood Through the Combination of Gasification and Solid Oxide Fuel Cells, Systems Analysis and Proof-of-Concept*. Report 17856, ETH Zürich, Zurich, p. 328; 2008.

[8] Milne TA, Evans RJ, Abatzoglou N. *Biomass Gasifier "Tars": Their Nature, Formation, and Conversion*. NREL Report 1998, National Renewable Energy Laboratory, Colorado; 1998.

[9] Berger B, Bacq A, Jeanmart H, Bourgois F. Experimental and Numerical Investigation of the Air Ratio on the Tar Content in the Syngas of a Two-Stage Gasifier. In: Florence ETA, 18th European Biomass Conference. ETA, Lyon; 2010.

[10] Gallmetzer G, Ackermann P, Schweiger A, et al. The agnion heatpipe-reformer – operating experiences and evaluation of fuel conversion and syngas composition. *Biomass Conversion and Biorefinery* **2**(3): 207–215; 2012.

[11] Kinoshita CM, Wang Y, Zhou J. Tar formation under different biomass gasification conditions. *Journal of Analytical and Applied Pyrolysis* **29**: 169–181; 1994.

[12] Meng X, de Jong W, Fu N, Verkooijen AHM. Biomass gasification in a 100 kWth steam–oxygen blown circulating fluidized bed gasifier: Effects of operational conditions on product gas distribution and tar formation. *Biomass and Bioenergy* **35**: 2910–2924; 2011.

[13] Martin RA, Gardner B, Guan X, Hendrix H. *Power Systems Development Facility: High Temperature, High Pressure Filtration in Gasification Operation.* Fifth International Symposium on Gas Cleaning at High Temperatures. Morgantown, USA, pp. 1–14; 2002.

[14] Nagel FP, Ghosh S, Pitta C, Schildhauer TJ, Biollaz SMA. Biomass integrated gasification fuel cell systems – Concept development and experimental results. *Biomass and Bioenergy* **35**: 354–362; 2011.

[15] Good J, L V, Knoef H, Zielke U, Hansen P, van de Kamp W. Sampling and analysis of tar and particles in biomass producer gases. *European Communication EFTA* **2005**: 1–44; 2005.

[16] CEN. *Biomass Gasification – Tar and Particles in Product Gases – Sampling and Analysis.* European Commission and the European Free Trade Association; 2005.

[17] Maniatis K, Beenackers A. Tar protocols. IEA bioenergy gasification task. *Biomass and Bioenergy* **18**(1): 1–4; 2000.

[18] Kaufman-Rechulski MD, Schneebeli J, Geiger S, Schildhauer TJ, Biollaz SMA, Ludwig C. Liquid-quench sampling system for the analysis of gas streams from biomass gasification processes. Part 1: Sampling noncondensable compounds. *Energy and Fuels* **26**: 7308–7315; 2012.

[19] Kaufman-Rechulski MD, Schneebeli J, Geiger S, Schildhauer TJ, Biollaz SMA, Ludwig C. Liquid-quench sampling system for the analysis of gas streams from biomass gasification processes. Part 2: Sampling condensable compounds. *Energy and Fuels* **26**: 6358–6365; 2012.

[20] ECN. *Thersites – The ECN Tar Dew Point Site.* Energy Research Center of the Netherlands (ECN), Rotterdam; 2009.

[21] Devi L, Ptasinski KJ, Janssen FJJG. Decomposition of naphthalene as a biomass tar over pretreated olivine: effect of gas composition, kinetic approach, and reaction scheme. *Industrial and Engineering Chemistry Research* **44**: 9096–9104; 2005.

[22] Van Paasen S, Kiel J. *Tar Formation in a Fluidised-Bed Gasifier – Impact of Fuel Properties and Operating Conditions.* Energy Research Center of the Netherlands, Rotterdam; 2004.

[23] Lovell R, Dylewski S, Peterson C. *Control of Sulfur Emissions from Oil Shale Retorts.* Environmental Protection Agency, Cincinnati, Ohio, p. 190; 1981.

[24] Diaz-Somoano M, Lopez-Anton M, Martinez-Tarazona M. Solid sorbents for trace element removal at high temperatures in coal gasification. *Global NEST Journal* **8**(2): 137–145; 2006.

[25] Salo K, Mojtahedi W. Fate of alkali and trace metals in biomass gasification. *Biomass and Bioenergy* **15**(3): 263–267; 1998.

[26] Pigeaud A, Maru H, Wilemski G, Helble J. *Trace Element Emissions, Semi-Annual Report.* U.S. Department of Energy, Office of Fossile Energy, Morgantown, USA; 1995.

[27] Bao J, Krishnan GN, Jayaweera P, Lau K-H, Sanjurjo A. Effect of various coal contaminants on the performance of solid oxide fuel cells: Part II. ppm and sub-ppm level testing. *Journal of Power Sources* **193**: 617–624; 2009.

[28] Bao J, Krishnan GN, Jayaweera P, Perez-Mariano J, Sanjurjo A. Effect of various coal contaminants on the performance of solid oxide fuel cells: Part I. Accelerated testing. *Journal of Power Sources* **193**: 607–616; 2009.

[29] Bao J, Krishnan GN, Jayaweera P, Sanjurjo A. Effect of various coal gas contaminants on the performance of solid oxide fuel cells: Part III. Synergistic effects. *Journal of Power Sources* **195**: 1316–1324; 2010.

[30] Marina OA, Pederson L, Gemmen R, Gerdes K, Finklea H, Celik I. Overview of SOFC Anode Interactions with Coal Gas Impurities. *ECS Transactions* **26**(1): 363–370; 2010.

[31] Woolcock PJ, Brown RC. A review of cleaning technologies for biomass-derived syngas. *Biomass and Bioenergy* **52**: 54–84; 2013.

[32] Nagel FP, Schildhauer TJ, Biollaz SMA. Biomass-integrated gasification fuel cell systems – Part 1: Definition of systems and technical analysis. *International Journal of Hydrogen Energy* **34**: 6809–6825; 2009.

[33] Gassner M, Marechal F. Thermo-economic optimisation of the polygeneration of synthetic natural gas (SNG), power and heat from lignocellulosic biomass by gasification and methanation. *Energy and Environmental Science* **5**(2): 5768–5789; 2012.

[34] Rhyner U. *Reactive Hot Gas Filter for Biomass Gasification*. Report 21102. ETH Zürich, Zurich, p. 160; 2013.

[35] Rhyner U, Edinger P, Schildhauer TJ, Biollaz SMA. Applied kinetics for modeling of reactive hot gas filters. *Applied Energy* **113**: 766–780; 2014.

[36] VDI. Filtering separators – High temperature gas filtration. VDI 3677. *VDI/DIN-Handbuch Reinhaltung der Luft* **6**: 3; 2010.

[37] VDI. Filtering separators – Surface Filters. VDI 3677. *VDI/DIN-Handbuch Reinhaltung der Luft* **6**: 1; 2010.

[38] Brouwers JJH. Phase separation in centrifugal fields with emphasis on the rotational particle separator. *Experimental Thermal and Fluid Science* **26**(2/4): 325–334; 2002.

[39] Stevens DJ. *Hot Gas Conditioning: Recent Progress With Larger-Scale Biomass Gasification Systems*. Pacific Northwest National Laboratory for US DOE, NETL, Richland, Washington; 2001.

[40] Cortés C, Gil A. Modeling the gas and particle flow inside cyclone separators. *Progress in Energy and Combustion Science* **33**(5): 409–452; 2007.

[41] Probstein R, Hicks R. *Synthetic Fuels*. Dover Pubn Inc.; 2006.

[42] McDonald J, Dean A. *Electrostatic Precipitator Manual*. Noyes Data Corporation, Park Ridge, N.; 1982.

[43] Seville JPK. *Gas Cleaning in Demanding Applications, 1st Edition*. Blackie Academic and Professional, London; 1997.

[44] EPA. *Electrostatic Precipitator Operation. APTI Virtual Classroom*. U.S. Environmental Protection Agency, Washington, D.C.; 1998.

[45] Peukert W. High temperature filtration in the process industry. *Filtration and Separation* **35**(5): 461–464; 1998.

[46] 3M Deutschland GmbH. *Ceramic Textiles and Composites Europe*. 3M Deutschland GmbH, Neuss, Germany; 2012. www.3m.com/market/industrial/ceramics/pdfs/3M_Filter_Bags.pdf (accessed 15 December 2015).

[47] Löffler F, Dietrich H, Flatt W. *Staubabscheidung mit Schlauchfiltern und Taschenfiltern.* Vieweg, Braunschweig; 1991.

[48] Pall Filtersystems GmbH. *Homepage.* Pall Filtersystems GmbH, Crailsheim, Germany; 2012. www.pall.com/main/fuels-and-chemicals (accessed 15 December 2015).

[49] TENMAT Ltd. *Homepage.* TENMAT Ltd, Manchester, England; 2012. www.tenmat. com/Content/Hot20Gas20Filtration (accessed 15 December 2015).

[50] Sharma SD, Dolan M, Ilyushechkin AY, McLennan KG, Nguyen T, Chase D. Recent developments in dry hot syngas cleaning processes. *Fuel* **89**: 817–826; 2010.

[51] Sharma SD, Dolan M, Park D, et al. A critical review of syngas cleaning technologies – fundamental limitations and practical problems. *Powder Technology* **180**(1/2): 115–121; 2008.

[52] Heidenreich S, Haag W, Mai R, Leibold H, Seifert H. Untersuchungen zur Abreinigungsleistung verschiedener Rückpulssysteme für Oberflächenfilter aus starren Filtermedien. *Chemie Ingenieur Technik* **75**(9): 1280–1283; 2003.

[53] Leubner H, Riebel U. Pulse jet cleaning of textile and rigid filter media – characteristic parameters. *Chemical Engineering and Technology* **27**(6): 652–661; 2004.

[54] Rhyner U, Mai R, Leibold H, Biollaz SMA. Dynamic pressure measurements of a hot gas filter as a diagnostic tool to assess the time dependent performance. *Biomass and Bioenergy* **53**: 72–80; 2013.

[55] Leibold H, Hornung A, Seifert H. HTHP syngas cleaning concept of two stage biomass gasification for FT synthesis. *Powder Technology* **180**: 265–270; 2008.

[56] Mai R, Kreft D, Leibold H, Seifert H. *Coupled Pressure Pulse (CPP) Recleaning of Ceramic Filter Candles Components and System Performance.* Fifth European Conference on Industrial Furnaces and Boilers (INFUB). Porto, Portugal; 2000.

[57] Mai R, Leibold H, Seifert H, Heidenreich S, Walch A. Coupled pressure pulse (CPP) recleaning system for ceramic hot-gas filters with an integrated safety filter. *Chemical Engineering and Technology* **26**(5): 577–579; 2003.

[58] Leibold H. *Trockene HT Synthesegasreinigung.* First Nürnberger Fach-Kolloquium, Methanisierung und Second Generation Fuels. Nürnberg, Germany; 2012.

[59] Simeone E, Nacken M, Haag W, Heidenreich S, de Jong W. Filtration performance at high temperatures and analysis of ceramic filter elements during biomass gasification. *Biomass and Bioenergy* **35**: 87–104; 2011.

[60] Simeone E, Siedlecki M, Nacken M, Heidenreich S, De Jong W. High temperature gas filtration with ceramic candles and ashes characterisation during steam – oxygen blown gasification of biomass. *Fuel* **108**: 99–111; 2013.

[61] Heidenreich S. Heissgasfiltration. *Chemie Ingenieur Technik* **84**(6): 795–807; 2012.

[62] Zwart R, Bos A, Kuipers J. *Principle of OLGA Tar Removal System.* Energy Research Centre of the Netherlands, Rotterdam, p. 2; 2010.

[63] Zwart RWR, Van der Drift A, Bos A, Visser HJM, Cieplik MK, Könemann HWJ. Oil-based gas washing – Flexible tar removal for high-efficient production of clean heat and power as well as sustainable fuels and chemicals. *Environmental Progress and Sustainable Energy* **28**(3): 324–335; 2009.

[64] Boerrigter H, Bergman P. *Method and System for Gasifying Biomass.* Energy Research Centre of the Netherlands, Rotterdam; 2002.

[65] Houben MP, de Lange HC, van Steenhoven AA. Tar reduction through partial combustion of fuel gas. *Fuel* **84**(7/8): 817–824; 2005.

[66] Brandt P, Henriksen U. *Decomposition of Tar in Gas from Updraft Gasifier by Thermal Cracking*. First World Conference on Biomass for Energy and Industry. ETA Florence, Seville, Spain, pp. 1756–1758; 2000.

[67] Jess A. Reaktionskinetische Untersuchungen zur thermischen Zersetzung von Modellkohlenwasserstoffen. *Erdöl, Erdgas und Kohle* **111**(11): 479–483; 1995.

[68] Jess A, Depner H. Thermische und katalytische Aufarbeitung von Rohgasen der Vergasung und Verkokung fester Brennstoffe. *Chemie Ingenieur Technik* **69**(7): 970–973; 1997.

[69] Fjellerup J, Ahrenfeldt J, Henriksen U, Gobel B. *Formation, Decomposition, and Cracking of Biomass Tars in Gasification*. Technical University of Denmark, Copenhagen, p. 60; 2005.

[70] Han J, Kim H. The reduction and control technology of tar during biomass gasification/pyrolysis: An overview. *Renewable and Sustainable Energy Reviews* **12**(2): 397–416; 2008.

[71] Sutton D, Kelleher B, Ross JRH. Review of literature on catalysts for biomass gasification. *Fuel Processing Technology* **73**(3): 155–173; 2001.

[72] Sutton D, Kelleher B, Ross JRH. Catalytic conditioning of organic volatile products produced by peat pyrolysis. *Biomass and Bioenergy* **23**(3): 209–216; 2002.

[73] Houben MP, Verschuur K, de Lange R, Neeft J, Daey Ouwens C. *An analysis and Experimental Investigation of the Cracking and Polymerisation of Tar*. 12th European Conference on Biomass for Energy and Industry and Climate Protection. Amsterdam, Netherlands, pp. 581–584; 2002.

[74] Pemen AJM, Nair SA, Yan K, van Heesch EJM, Ptasinski KJ, Drinkenburg AAH. Pulsed Corona Discharges for Tar Removal from Biomass Derived Fuel Gas. *Plasmas and Polymers* **8**(3): 209–224; 2003.

[75] Nair SA, Yan K, Safitri A, et al. Streamer corona plasma for fuel gas cleaning: comparison of energization techniques. *Journal of Electrostatics* **63**(12): 1105–1114; 2005.

[76] Nair SA, Pemen AJM, Yan K, et al. Tar removal from biomass-derived fuel gas by pulsed corona discharges. *Fuel Processing Technology* **84**(1/3): 161–173; 2003.

[77] Torres W, Pansare SS, Goodwin JG. Hot gas removal of tars, ammonia, and hydrogen sulfide from biomass gasification gas. *Catalysis Reviews* **49**(4): 407–456; 2007.

[78] Lind F, Seemann M, Thunman H. Continuous catalytic tar reforming of biomass derived raw gas with simultaneous catalyst regeneration. *Industrial and Engineering Chemistry Research* **50**(20): 11553–11562; 2011.

[79] Beld L, Wagenaar BM, Prins W. *Cleaning of Hot Producer Gas in a Catalytic Adiabatic Packed Bed Reactor with Periodic Flow Reversal*. In: Bridgwater A. V., Boocock D. G. B. (eds) Developments in Thermochemical Biomass Conversion. Springer,Dordrecht, pp. 907–920; 1997.

[80] Li C, Suzuki K. Tar property, analysis, reforming mechanism and model for biomass gasification – An overview. *Renewable and Sustainable Energy Reviews* **13**: 594–604; 2009.

[81] Mastellone ML, Arena U. Olivine as a tar removal catalyst during fluidized bed gasification of plastic waste. *AIChE Journal* **54**(6): 1656–1667; 2008.

[82] Dou B, Gao J, Sha X, Baek SW. Catalytic cracking of tar component from high-temperature fuel gas. *Applied Thermal Engineering* **23**(17): 2229–2239; 2003.

[83] Yung MM, Jablonski WS, Magrini-Bair KA. Review of catalytic conditioning of biomass-derived syngas. *Energy and Fuels* **23**(4): 1874–1887; 2009.

[84] Xu C, Donald J, Byambajav E, Ohtsuka Y. Recent advances in catalysts for hot-gas removal of tar and NH3 from biomass gasification. *Fuel* **89**(8): 1784–1795; 2010.

[85] Abu El-Rub Z, Bramer EA, Brem G. Review of catalysts for tar elimination in biomass gasification processes. *Industrial and Engineering Chemistry Research* **43**(22): 6911–6919; 2004.

[86] El-rub ZA. *Biomass Char as an In-situ Catalyst for Tar Removal in Gasification Systems*. Dissertation, Twente University, Enschede, The Netherlands; 2008.

[87] Simell PA, Hirvensalo EK, Smolander VT, Krause AOI. Steam reforming of gasification gas tar over dolomite with benzene as a model compound. *Industrial and Engineering Chemistry Research* **38**(4): 1250–1257; 1999.

[88] Tamhankar SS, Tsuchiya K, Riggs JB. Catalytic cracking of benzene on iron oxide-silica: catalyst activity and reaction mechanism. *Applied Catalysis* **16**(1): 103–121; 1985.

[89] Dayton DC. *A Review of the Literature on Catalytic Biomass Tar Destruction Milestone Completion Report*. Technical University of Denmark, Copenhagen; 2002.

[90] Brandt P, Larsen E, Henriksen U. High tar reduction in a two-stage gasifier. *Energy and Fuels* **14**(4): 816–819; 2000.

[91] Zanzi R, Sjöström K, Björnbom E. Rapid high-temperature pyrolysis of biomass in a free-fall reactor. *Fuel* **75**(5): 545–550; 1996.

[92] Chembukulam SK, Dandge AS, Rao NLK, Seshagiri K, Vaidyeswaran R. Smokeless fuel from carbonized sawdust. *Industrial and Engineering Chemistry Product Research and Development* **20**(4): 714–719; 1981.

[93] Zhang T, Walawender WP, Fan LT, Fan M, Daugaard D, Brown RC. Preparation of activated carbon from forest and agricultural residues through CO2 activation. *Chemical Engineering Journal* **105**(1/2): 53–59; 2004.

[94] Brown RA, Kercher AK, Nguyen TH, Nagle DC, Ball WP. Production and characterization of synthetic wood chars for use as surrogates for natural sorbents. *Organic Geochemistry* **37**(3): 321–333; 2006.

[95] Abu El-Rub Z, Bramer EA, Brem G. Experimental comparison of biomass chars with other catalysts for tar reduction. *Fuel* **87**(10/11): 2243–2252; 2008.

[96] Hosokai S, Kumabe K, Ohshita M, Norinaga K, Li C-Z, Hayashi J-i. Mechanism of decomposition of aromatics over charcoal and necessary condition for maintaining its activity. *Fuel* **87**(13/14): 2914–2922; 2008.

[97] Gerdes K, Grol E, Keairns D, Newby R. *Integrated Gasification Fuel Cell Performance and Cost Assessment*. National Energy Technology Laboratory, U.S. Department of Energy; 2009.

[98] Ma L, Baron GV. Mixed zirconia–alumina supports for Ni/MgO based catalytic filters for biomass fuel gas cleaning. *Powder Technology* **180**(1/2): 21–29; 2008.

[99] Devi L, Ptasinski KJ, Janssen FJJG. A review of the primary measures for tar elimination in biomass gasification processes. *Biomass and Bioenergy* **24**(2): 125–140; 2003.

[100] Luengnaruemitchai A, Kaengsilalai A. Activity of different zeolite-supported Ni catalysts for methane reforming with carbon dioxide. *Chemical Engineering Journal* **144**(1): 96–102; 2008.

[101] Corella J, Toledo JM, Molina G. A review on dual fluidized-bed biomass gasifiers. *Industrial and Engineering Chemistry Research* **46**(21): 6831–6839; 2007.

[102] Pansare SS, Goodwin JG, Gangwal S. Simultaneous ammonia and toluene decomposition on tungsten-based catalysts for hot gas cleanup. *Industrial and Engineering Chemistry Research* **47**: 8602–8611; 2008.

[103] Dayton DC, French RJ, Milne TA. Direct observation of alkali vapor release during biomass combustion and gasification. 1. Application of molecular beam/mass spectrometry to switchgrass combustion. *Energy and Fuels* **9**(5): 855–865; 1995.

[104] Cui H, Turn SQ, Keffer V, Evans D, Tran T, Foley M. Study on the fate of metal elements from biomass in a bench-scale fluidized bed gasifier. *Fuel* **89**: 74–81; 2011.

[105] Turn SQ, Kinoshita CM, Ishimura DM, Zhou J. The fate of inorganic constituents of biomass in fluidized bed gasification. *Fuel* **77**(3): 135–146; 1998.

[106] Hauserman WB. High-yield hydrogen production by catalytic gasification of coal or biomass. *International Journal of Hydrogen Energy* **19**(5): 413–419; 1994.

[107] Raveendran K, Ganesh A, Khilar KC. Influence of mineral matter on biomass pyrolysis characteristics. *Fuel* **74**(12): 1812–1822; 1995.

[108] Draelants DJ, Zhao H, Baron GV. Preparation of catalytic filters by the urea method and its application for benzene cracking in H2S-containing biomass gasification gas. *Industrial and Engineering Chemistry Research* **40**: 3309–3316; 2001.

[109] Engelen K, Zhang Y, Draelants DJ, Baron GV. A novel catalytic filter for tar removal from biomass gasification gas: Improvement of the catalytic activity in presence of H2S. *Chemical Engineering Science* **58**: 665–670; 2003.

[110] Ma L, Verelst H, Baron GV. Integrated high temperature gas cleaning: Tar removal in biomass gasification with a catalytic filter. *Catal Today* 2005;**105**: 729–734.

[111] Nacken M, Ma L, Heidenreich S, Baron GV. Catalytic activity in naphthalene reforming of two types of catalytic filters for hot gas cleaning of biomass-derived syngas. *Industrial and Engineering Chemistry Research* **49**: 5536–5542; 2010.

[112] Nacken M, Ma L, Heidenreich S, Verpoort F, Baron GV. Development of a catalytic ceramic foam for efficient tar reforming of a catalytic filter for hot gas cleaning of biomass-derived syngas. *Applied Catalysis B: Environmental* **125**: 111–119; 2012.

[113] Rapagnà S, Gallucci K, Di Marcello M, et al. First Al2O3 based catalytic filter candles operating in the fluidized bed gasifier freeboard. *Fuel* **97**: 718–724; 2012.

[114] Rönkkönen H, Simell P, Niemelä M, Krause O. Precious metal catalysts in the clean-up of biomass gasification gas part 2: Performance and sulfur tolerance of rhodium based catalysts. *Fuel Processing Technology* **92**: 1881–1889; 2011.

[115] Rönkkönen H, Simell P, Reinikainen M, Krause O, Niemelä MV. Catalytic clean-up of gasification gas with precious metal catalysts – A novel catalytic reformer development. *Fuel* **89**: 3272–3277; 2010.

[116] Rönkkönen H, Simell P, Reinikainen M, Niemelä M, Krause O. Precious metal catalysts in the clean-up of biomass gasification gas. Part 1: Monometallic catalysts and their impact on gasification gas composition. *Fuel Processing Technology* **92**: 1457–1465; 2011.

[117] Toledo JM, Corella J, Molina G. Catalytic hot gas cleaning with monoliths in biomass gasification in fluidized beds. 4. Performance of an advanced, second-generation, two-layers-based monolithic reactor. *Industrial and Engineering Chemistry Research* **45**: 1389–1396; 2006.

[118] Tomishige K, Miyazawa T, Asadullah M, Ito S-i, Kunimori K. Catalyst performance in reforming of tar derived from biomass over noble metal catalysts. *Green Chemistry* **5**: 399; 2003.

[119] Zhao H, Draelants DJ, Baron GV. Performance of a nickel-activated candle filter for naphthalene cracking in synthetic biomass gasification gas. *Industrial and Engineering Chemistry Research* **39**: 3195–3201; 2000.

[120] Pfeifer C, Hofbauer H. Development of catalytic tar decomposition downstream from a dual fluidized bed biomass steam gasifier. *Powder Technology* **180**: 9–16; 2008.

[121] Miyazawa T, Kimura T, Nishikawa J, Kunimori K, Tomishige K. Catalytic properties of Rh/CeO2/SiO2 for synthesis gas production from biomass by catalytic partial oxidation of tar. *Science and Technology of Advanced Materials* **6**(6): 604–614; 2005.

[122] Rhyner U, Edinger P, Schildhauer TJ, Biollaz SMA. Experimental study on high temperature catalytic conversion of tars and organic sulfur compounds. *Hydrogen Energy* **80**: 13–19; 2014.

[123] Rhyner U, Schildhauer TJ, Biollaz SMA. *Catalytic Conversion of Tars in a Monolith in the Presence of H2S at 750 °C.* In: Florence ETA. 19th European Biomass Conference and Exhibition. Berlin, Germany; 2011.

[124] Rönkkönen H, Rikkinen E, Linnekoski J, Simell P, Reinikainen M, Krause O. Effect of gasification gas components on naphthalene decomposition over ZrO2. *Catalysis Today* **147**: 230–236; 2009.

[125] Furusawa T, Tsutsumi A. Comparison of Co/MgO and Ni/MgO catalysts for the steam reforming of naphthalene as a model compound of tar derived from biomass gasification. *Applied Catalysis A: General* **278**(2): 207–212; 2005.

[126] Furusawa T, Saito K, Kori Y, Miura Y, Sato M, Suzuki N. Steam reforming of naphthalene/benzene with various types of Pt- and Ni-based catalysts for hydrogen production. *Fuel* **103**: 111–121; 2013.

[127] Cui H, Turn SQ, Keffer V, Evans D, Tran T, Foley M. Contaminant estimates and removal in product gas from biomass steam gasification. *Energy and Fuels* **24**: 1222–1233; 2010.

[128] Nacken M, Ma L, Heidenreich S, Baron GV. Performance of a catalytically activated ceramic hot gas filter for catalytic tar removal from biomass gasification gas. *Applied Catalysis B: Environmental* **88**: 292–298; 2009.

[129] Rapagna S, Gallucci K, Di Marcello M, Foscolo PU, Nacken M, Heidenreich S. In situ catalytic ceramic candle filtration for tar reforming and particulate abatement in a fluidized-bed biomass gasifier. *Energy and Fuels* **23**(7): 3804–3809; 2009.

[130] Rapagna S, Gallucci K, Di Marcello M, Matt M, Nacken M, Heidenreich S, et al. Gas cleaning, gas conditioning and tar abatement by means of a catalytic filter candle in a biomass fluidized-bed gasifier. *Bioresource Technology* **101**(18): 7134–7141; 2010.

[131] Corella J, Toledo JM, Padilla R. Catalytic hot gas cleaning with monoliths in biomass gasification in fluidized beds. 1. Their effectiveness for tar elimination. *Industrial and Engineering Chemistry Research* **43**(10): 2433–2445; 2004.

[132] Corella J, Toledo JM, Padilla R. Catalytic hot gas cleaning with monoliths in biomass gasification in fluidized beds. 2. Modeling of the monolithic reactor. *Industrial and Engineering Chemistry Research* **43**(26): 8207–8216; 2004.

[133] Corella J, Toledo JM, Padilla R. Catalytic hot gas cleaning with monoliths in biomass gasification in fluidized beds. 3. Their effectiveness for ammonia elimination. *Industrial and Engineering Chemistry Research* **44**(7): 2036–2045; 2005.

[134] Toledo JM, Corella J, Molina G. Catalytic hot gas cleaning with monoliths in biomass gasification in fluidized beds. 4. Performance of an advanced, second-generation,

two-layers-based monolithic reactor. *Industrial and Engineering Chemistry Research* **45**(4): 1389–1396; 2006.

[135] Lind F, Berguerand N, Seemann M, Henrik T. *Comparing Three Materials for Secondary Catalytic Tar Reforming of Biomass Derived Gas.* Fourth International Symposium on Hydrogen from Renewable Sources and Refinery Pennsylvania. American Chemical Society, New York; 2012.

[136] Berguerand N, Lind F, Israelsson M, Seemann M, Biollaz S, Thunman H. Use of Nickel Oxide as a Catalyst for Tar Elimination in a Chemical-Looping Reforming Reactor Operated with Biomass Producer Gas. *Industrial and Engineering Chemistry Research* **51**(51): 16610–16616; 2012.

[137] Berguerand N, Lind F, Seemann M, Thunman H. Producer gas cleaning in a dual fluidized bed reformer – a comparative study of performance with ilmenite and a manganese oxide as catalyst. *Biomass Conversion and Biorefinery* **2**(3): 245–252; 2012.

[138] Berguerand N, Lind F, Seemann M, Thunman H. Manganese oxide as catalyst for tar cleaning of biomass-derived gas. *Biomass Conversion and Biorefinery* **2**(2): 133–140; 2012.

[139] König CFJ, Nachtegaal M, Seemann M, Clemens F, van Garderen N, Biollaz SMA, et al. Mechanistic studies of chemical looping desulfurization of Mn-based oxides using in situ X-ray absorption spectroscopy. *Applied Energy* **113**: 1895–1901; 2014.

[140] Lind F, Berguerand N, Seemann M, Thunman H. Ilmenite and nickel as catalysts for upgrading of raw gas derived from biomass gasification. *Energy and Fuels* **27**(2): 997–1007; 2013.

[141] EPA. *Flue Gas Desulfurization (Acid Gas Removal) Systems. APTI Virtual Classroom.* U.S. Environmental Protection Agency, Washington D.C.; 1998.

[142] Liu Y, Bisson TM, Yang H, Xu Z. Recent developments in novel sorbents for flue gas clean up. *Fuel Processing Technology* **91**: 1175–1197; 2010.

[143] Plummer MA. Sulfur and hydrogen from H2S. *Hydrocarbon Processing* **66**(4): 38–40; 1987.

[144] Jensen AB, Webb C. Treatment of H2S-containing gases: A review of microbiological alternatives. *Enzyme and Microbial Technology* **17**(1): 2–10; 1995.

[145] Fortuny M, Baeza JA, Gamisans X, Casas C, Lafuente J, Deshusses MA, et al. Biological sweetening of energy gases mimics in biotrickling filters. *Chemosphere* **71**(1): 10–17; 2008.

[146] Babich IV, Moulijn JA. Science and technology of novel processes for deep desulfurization of oil refinery streams: a review. *Fuel* **82**(6): 607–631; 2003.

[147] Leibold H. *Trockene Synthesegasreinigung bei hohen Temperaturen.* Kolloquium Sustainable BioEconomy. Forschungszentrum Karlsruhe, Germany; 2008.

[148] Seifert H, Kolb T, Leibold H. *Syngas aus Biomasse – Flugstromvergasung und Gasreinigung.* 41st Kraftwerkstechnisches Kolloquium. Dresden, Germany; 2009.

[149] Furimsky E. Selection of catalysts and reactors for hydroprocessing. *Applied Catalysis A: General* **171**(2): 177–206; 1998.

[150] Topsøe H, Clausen B, Massoth F. *Hydrotreating Catalysis.* In: Anderson JohnR, Boudart Michel (eds.) Catalysis. Springer, Heidelberg, pp. 1–269; 1996.

[151] Twigg MV. *Catalyst Handbook. 2nd Edn.* Manson Publishing, London, UK; 1996.

[152] Deutschmann O, Knözinger H, Kochloefl K, Turek T. *Heterogeneous Catalysis and Solid Catalysts, 1. Fundamentals*. In: Ullmann's Encyclopedia of Industrial Chemistry. Wiley-VCH Verlag, Heidelberg, pp. 16–25; 2000.

[153] Burns AW, Gaudette AF, Bussell ME. Hydrodesulfurization properties of cobalt–nickel phosphide catalysts: Ni-rich materials are highly active. *Journal of Catalysis* **260**(2): 262–269; 2008.

[154] Duan X, Teng Y, Wang A, Kogan VM, Li X, Wang Y. Role of sulfur in hydrotreating catalysis over nickel phosphide. *Journal of Catalysis* **261**(2): 232–240; 2009.

[155] Liu P, Rodriguez JA, Muckerman JT. Sulfur adsorption and sulfidation of transition metal carbides as hydrotreating catalysts. *Journal of Molecular Catalysis A: Chemical* **239**(1/2): 116–124; 2005.

[156] Pecoraro TA, Chianelli RR. Hydrodesulfurization catalysis by transition metal sulfides. *Journal of Catalysis* **67**(2): 430–445; 1981.

[157] Chianelli RR, Berhault G, Raybaud P, Kasztelan S, Hafner J, Toulhoat H. Periodic trends in hydrodesulfurization: in support of the Sabatier principle. *Applied Catalysis A: General* **227**(1/2): 83–96; 2002.

[158] Topsøe H, Clausen BS, Topsøe NY, Nørskov JK, Ovesen CV, Jacobsen CJH. The Bond Energy Model for Hydrotreating Reactions: Theoretical and Experimental Aspects. *Bulletin des Sociétés Chimiques Belges* **104**(4/5): 283–291; 1995.

[159] Hermann N, Brorson M, Topsøe H. Activities of unsupported second transition series metal sulfides for hydrodesulfurization of sterically hindered 4,6-dimethyldibenzothiophene and of unsubstituted dibenzothiophene. *Catalysis Letters* **65**(4): 169–174; 2000.

[160] Breysse M, Portefaix JL, Vrinat M. Support effects on hydrotreating catalysts. *Catalysis Today* **10**(4): 489–505; 1991.

[161] Daudin A, Lamic AF, Pérot G, Brunet S, Raybaud P, Bouchy C. Microkinetic interpretation of HDS/HYDO selectivity of the transformation of a model FCC gasoline over transition metal sulfides. *Catalysis Today* **130**(1): 221–230; 2008.

[162] Raje AP, Liaw S-J, Srinivasan R, Davis BH. Second row transition metal sulfides for the hydrotreatment of coal-derived naphtha I. Catalyst preparation, characterization and comparison of rate of simultaneous removal of total sulfur, nitrogen and oxygen. *Applied Catalysis A: General* **150**(2): 297–318; 1997.

[163] Kuo Y-J, Cocco RA, Tatarchuk BJ. Hydrogenation and hydrodesulfurization over sulfided ruthenium catalysts: II. Impact of surface phase behavior on activity and selectivity. *Journal of Catalysis* **112**(1): 250–266; 1988.

[164] Hensen EJM, Brans HJA, Lardinois GMHJ, de Beer VHJ, van Veen JAR, van Santen RA. Periodic Trends in Hydrotreating Catalysis: Thiophene Hydrodesulfurization over Carbon-Supported 4d Transition Metal Sulfides. *Journal of Catalysis* **192**(1): 98–107; 2000.

[165] Lacroix M, Boutarfa N, Guillard C, Vrinat M, Breysse M. Hydrogenating properties of unsupported transition metal sulphides. *Journal of Catalysis* **120**(2): 473–477; 1989.

[166] Dhar GM, Srinivas BN, Rana MS, Kumar M, Maity SK. Mixed oxide supported hydrodesulfurization catalysts – a review. *Catalysis Today* **86**(1/4): 45–60; 2003.

[167] Breysse M, Djega-Mariadassou G, Pessayre S, et al. Deep desulfurization: reactions, catalysts and technological challenges. *Catalysis Today* **84**(3/4): 129–138; 2003.

[168] Pérez-Martínez D, Giraldo SA, Centeno A. Effects of the H2S partial pressure on the performance of bimetallic noble-metal molybdenum catalysts in simultaneous

hydrogenation and hydrodesulfurization reactions. *Applied Catalysis A: General* **315**: 35–43; 2006.

[169] Vít Z, Gulková D, Kaluža L, Zdražil M. Synergetic effects of Pt and Ru added to Mo/Al2O3 sulfide catalyst in simultaneous hydrodesulfurization of thiophene and hydrogenation of cyclohexene. *Journal of Catalysis* **232**(2): 447–455; 2005.

[170] Pinzón MH, Meriño LI, Centeno A, Giraldo SA. *Performance of Noble Metal-Mo/γ-Al2O3 Catalysts: Effect of Preparation Parameters.* In: B. Delmon G. F. Froment, Grange P. (eds.) Studies in Surface Science and Catalysis. Elsevier, New York, pp. 97–104; 1999.

[171] Meriño LI, Centeno A, Giraldo SA. Influence of the activation conditions of bimetallic catalysts NM–Mo/γ-Al2O3 (NM=Pt, Pd and Ru) on the activity in HDT reactions. *Applied Catalysis A: General* **197**(1): 61–68; 2000.

[172] Ishihara A, Dumeignil F, Lee J, Mitsuhashi K, Qian EW, Kabe T. Hydrodesulfurization of sulfur-containing polyaromatic compounds in light gas oil using noble metal catalysts. *Applied Catalysis A: General* **289**(2): 163–173; 2005.

[173] Wang J, Li WZ, Perot G, Lemberton JL, Yu CY, Thomas C, et al. *Study on the Role of Platinum in PtMo/Al2O3 for Hydrodesulfurization of Dibenzothiophene.* In: Can Li, Qin Xin (eds.) Studies in Surface Science and Catalysis. Elsevier, New York, pp. 171–178; 1997.

[174] Dou B, Shen W, Gao J, Sha X. Adsorption of alkali metal vapor from high-temperature coal-derived gas by solid sorbents. *Fuel Processing Technology* **82**: 51–60; 2003.

[175] Corella J, Toledo JM, Molina G. Performance of CaO and MgO for the hot gas clean up in gasification of a chlorine-containing (RDF) feedstock. *Bioresource Technology* **99**(16): 7539–7544; 2008.

[176] Dou B, Pan W, Ren J, Chen B, Hwang J, Yu T-U. Single and combined removal of hcl and alkali metal vapor from high-temperature gas by solid sorbents. *Energy and Fuels* **21**(2): 1019–1023; 2007.

[177] Mulik PR, Alvin MA, Bachovchin DM. *Simultaneous High-Temperature Removal of Alkali and Particulates in a Pressurized Gasification System. Final technical progress report, April 1981–July 1983.* U.S. Department of Energy, National Energy Technology Laboratory, Morgantown, USA. pp. 337; 1983.

[178] Tran K-Q, Iisa K, Steenari B-M, Lindqvist O. A kinetic study of gaseous alkali capture by kaolin in the fixed bed reactor equipped with an alkali detector. *Fuel* **84**(2/3): 169–175; 2005.

[179] Turn SQ, Kinoshita CM, Ishimura DM, Hiraki TT, Zhou J, Masutani SM. An experimental investigation of alkali removal from biomass producer gas using a fixed bed of solid sorbent. *Industrial and Engineering Chemistry Research* **40**(8): 1960–1967; 2001.

[180] Cummer KR, Brown RC. Ancillary equipment for biomass gasification. *Biomass and Bioenergy* **23**(2): 113–128; 2002.

[181] Turn SQ, Kinoshita CM, Ishimura DM. Removal of inorganic constituents of biomass feedstocks by mechanical dewatering and leaching. *Biomass and Bioenergy* **12**(4): 241–252; 1997.

[182] Davidsson KO, Korsgren JG, Pettersson JBC, Jäglid U. The effects of fuel washing techniques on alkali release from biomass. *Fuel* **81**(2): 137–142; 2002.

[183] Pröll T, Siefert IG, Friedl A, Hofbauer H. Removal of NH3 from biomass gasification producer gas by water condensing in an organic solvent scrubber. *Industrial and Engineering Chemistry Research* **44**(5): 1576–1584; 2005.

[184] Pinto F, Lopes H, André RN, Dias M, Gulyurtlu I, Cabrita I. Effect of experimental conditions on gas quality and solids produced by sewage sludge cogasification. 1. Sewage sludge mixed with coal. *Energy and Fuels* **21**(5): 2737–2745; 2007.

[185] Koveal RJJ, Alexion DG. Gas conversion with rejuvenation ammonia removal. *Fuel Processing Technology* **49**: 18–35; 1999.

[186] Mojtahedi W, Ylitalo M, Maunula T, Abbasian J. Catalytic decomposition of ammonia in fuel gas produced in pilot-scale pressurized fluidized-bed gasifier. *Fuel Processing Technology* **45**(3): 221–236; 1995.

[187] Air Products and Chemicals, Inc., Eastman Chemical Company. *Removal of Trace Contaminants from Coal-Derived Synthesis Gas, Topical Report.* U.S. Department of Energy, National Energy Technology Laboratory, Morgantown, USA; 2003.

[188] Leibold H, Mai R, Linek A, Zimmerlin B, Seifert H. *Dry High Temperature Sorption of HCl and H2S with Natural Carbonates.* Seventh International Symposium and Exhibition, Gas Cleaning at High Temperatures (GCHT-7). Newcastle, Australia; 2008.

[189] Kurkela E, Kurkela M. *Advanced Biomass Gasification for High-Efficiency Power. Final Activity Report of BiGPower Project.* VTT Research Notes 2511, VTT; 2009.

[190] Rönkkönen H, Simell P, Reinikainen M, Krause O. The effect of sulfur on ZrO2-based biomass gasification gas clean-up catalysts. *Topics in Catalysis* **52**(8): 1070–1078; 2009.

[191] Viinikainen T, Rönkkönen H, Bradshaw H, et al. Acidic and basic surface sites of zirconia-based biomass gasification gas clean-up catalysts. *Applied Catalysis A: General* **362**: 169–177; 2009.

[192] Leppin D, Basu A. *Novel Bio-Syngas Cleanup Process.* Presentation to ICPS09. Gas Technology Institute, Chicago, USA; 2009.

[193] Horvath A. *Operating Experience with Biomass Gasifiers, Needs to Improve Gasification Plant Operation.* IEA Task 33 Workshop. Breda, Netherlands; 2009.

[194] Heidenreich S, Nacken M, Hackel M, Schaub G. Catalytic filter elements for combined particle separation and nitrogen oxides removal from gas streams. *Powder Technology* **180**: 86–90; 2008.

[195] EC. *UNIQUE – Theme: Energy*, Seventh Framework Program, EC, Brussels; 2010.

[196] EC. *CHRISGAS – Fuels from Biomass*, Sixth Framework Program, EC, Brussels; 2009.

[197] Simeone E, Hölsken E, Nacken M, Heidenreich S, De Jong W. Study of the behaviour of a catalytic ceramic candle filter in a lab-scale unit at high temperatures. *International Journal of Chemical Reaction Engineering* **8**: 16; 2010.

[198] Simeone E, Pal R, Nacken M, Heidenreich S, Verkooijen AHM. Tar removal in a catalytic ceramic candle filter unit at high temperatures. *Florence ETA. 18th European Biomass Conference and Exhibition.* Lyon, France, pp. 1–20; 2010.

4

METHANATION FOR SYNTHETIC NATURAL GAS PRODUCTION – CHEMICAL REACTION ENGINEERING ASPECTS

TILMAN J. SCHILDHAUER

4.1 METHANATION – THE SYNTHESIS STEP IN THE PRODUCTION OF SYNTHETIC NATURAL GAS

The conversion of chemical energy carriers is in many cases either endothermic (e.g., steam reforming, gasification) or exothermic (e.g., combustions, hydrogenations). Especially, when this step is heterogeneously catalyzed, this brings special challenges for the reactor containing the solid catalyst, because heat management has to be considered in the reactor design. A detailed understanding of the interaction of heat transfer, hydrodynamic phenomena (such as mixing, residence time distribution, mass transfer), and the kinetics of the involved reactions is necessary to properly design and optimize such reactors and to limit the risks during up-scaling from laboratory over pilot to commercial scale. This applies also for the methanation reaction which is the synthesis step in the production of synthetic natural gas (SNG).

The task of the synthesis step is to maximize the conversion of carbon-containing molecules to species which can easily be injected into the natural gas grid. According to the specifications (e.g., [1]), these are especially methane (more than 96%) and, to a lower extent, ethane. For species with low or even no volumetric heating values, such as hydrogen, nitrogen, and carbon dioxide, upper limits in the range of a small

Synthetic Natural Gas from Coal, Dry Biomass, and Power-to-Gas Applications, First Edition.
Edited by Tilman J. Schildhauer and Serge M.A. Biollaz.
© 2016 John Wiley & Sons, Inc. Published 2016 by John Wiley & Sons, Inc.

percentage exist. Even stricter are the rules concerning carbon monoxide; due to its toxicity, usually a maximum of 0.5% [2] is permitted.

The main carbon containing compounds in producer gas from gasification of solid feedstocks are carbon monoxide, CO, carbon dioxide, CO_2, methane, CH_4, and C_2 species such as ethylene, C_2H_4, ethane, C_2H_6, acetylene, C_2H_2 (see also Chapter 2 in this book). In traces, aromatic species such as benzene, C_6H_6, toluene, C_7H_8, naphthalene, $C_{10}H_8$, and even larger poly-aromatic hydrocarbons [3] are found, but are largely removed in the gas cleaning upstream of the methanation (see chapter 3 in this book). Due to this complex gas mixture, besides the name-giving methanation, also a number of other reactions have to be considered in the synthesis step of SNG production.

The main reactions are CO- and CO_2-methanation, also known as the Sabatier reaction [4]:

$$3H_2 + CO \leftrightarrow CH_4 + H_2O \quad \Delta H_R^0 = -206.28\,kJ\,mol^{-1} \tag{4.1}$$

$$4H_2 + CO_2 \leftrightarrow CH_4 + 2H_2O \quad \Delta H_R^0 = -165.12\,kJ\,mol^{-1} \tag{4.2}$$

As steam, $H_2O(g)$, and hydrogen, H_2, are always present in gasification producer gas, the methanation reaction is always accompanied by the reversible homogeneous water gas shift reaction:

$$CO + H_2O \leftrightarrow CO_2 + H_2 \quad \Delta H_R^0 = -41.16\,kJ\,mol^{-1} \tag{4.3}$$

Further, the C_2 species undergo serial or even direct hydrogenation to methane:

$$C_2H_2 + H_2 \rightarrow C_2H_4 \quad \Delta H_R^0 = -175.6\,kJ\,mol^{-1} \tag{4.4}$$

$$C_2H_4 + H_2 \rightarrow C_2H_6 \quad \Delta H_R^0 = -136.9\,kJ\,mol^{-1} \tag{4.5}$$

$$C_2H_x + (4 - x/2)H_2 \rightarrow 2CH_4 \tag{4.6}$$

All hydrocarbons, but especially CO, are known to form carbon on the catalyst surface; the latter by the so-called Boudouard reaction:

$$2CO \leftrightarrow C(s) + CO_2 \quad \Delta H_R^0 = -172.54\,kJ\,mol^{-1} \tag{4.7}$$

The surface carbon then can further polymerize to form carbon depositions which may deactivate the catalyst; but it can also react with H_2 or steam in the hydrogasification or the heterogeneous water gas shift reaction to form again gaseous compounds:

$$2H_2 + C(s) \rightarrow CH_4 \tag{4.8}$$

$$H_2O(g) + C(s) \rightarrow CO + H_2 \tag{4.9}$$

Which reactions occur to which extent depends on the actual gas mixture, the chosen catalyst, the reactor type or rather its design and the applied operation conditions.

Therefore, the next sections in the first part of this chapter present the most important feed gas compositions, the compositions to be expected after conversion according the thermodynamic equilibrium, the kinetics and reaction mechanisms of the main reactions, and important aspects of catalyst deactivation. The following part of this chapter discusses methanation reactor types and their operation conditions. The last part of this chapter focusses on the modeling and simulation of methanation reactors.

4.1.1 Feed Gas Mixtures for Methanation Reactors

Depending on the upstream processes, for example, the choice of the gasification process, its operating conditions, and the subsequent gas cleaning and conditioning steps, the gas composition of the feed gas shows a significant variability. In the next paragraphs, typical feed gases and the corresponding process chains upstream of the methanation reactor are presented.

4.1.1.1 Stoichiometric $H_2/CO = 3.1$ The stoichiometric mixture of three hydrogen molecules and one carbon monoxide is the most simple gas composition possible. A slight excess of hydrogen is applied, both to limit carbon deposition and to achieve high selectivity to methane. This feed gas composition to the methanation reactor is the state of the art solution for coal to SNG plants and has the important advantage that, besides methane and traces of unreacted hydrogen, only steam is formed in the reaction. The steam can be separated by condensation and drying, which simplifies the gas up-grading before injection into the gas grid.

One disadvantage of the state of the art stoichiometric feed gas composition is that all higher hydrocarbons such as ethylene, which may be contained in the gasification producer gas, have either to be removed or converted beforehand. This leads to an increased complexity of the gas cleaning or gas conditioning steps upstream of the methanation reactor and eventually to a loss of chemical efficiency.

In state of the art coal to SNG plants such as the 1.5 GW$_{SNG}$ Dakota Gas Company plant in Great Plains, United States (built by Lurgi), the coal is gasified in a steam–oxygen blown gasifier. The producer gas is de-dusted and then partly fed over a molybdenum sulfide based catalyst where the water gas shift reaction is catalyzed in the presence of hydrogen sulfide (H_2S) to achieve a H_2/CO ratio of about 3.1. The gas mixture then is cleaned in a so-called Rectisol© wash with methanol as solvent under 25–70 bar and –40 °C. In this step, all contaminants, but also steam, CO_2, and ethylene are removed, leading to a purified, slightly over-stoichiometric mixture of H_2 and CO.

4.1.1.2 Non-stoichiometric Feed Gas Compositions from Gasification Non-stoichiometric feed gas mixtures are obtained when the upstream water gas shift step (WGS) or the CO_2 removal are omitted, further when methane, C_2 species, traces of aromatic compounds or steam are contained in the cleaned producer gas.

Omission of Water Gas Shift Step or CO_2 Removal   As the hydrogen to carbon monoxide ratio in most producer gases lies significantly below the stoichiometric value (three) for the methanation, omitting the upstream water gas shift inherently means that the water content has to be adjusted (eventually by adding steam) to allow for a simultaneous water gas shift reaction inside the methanation reactor. The fact that the methanation reaction produces steam inside the reactor allows decreasing the overall steam consumption compared to the above-discussed case of upstream WGS. Still, a downstream CO_2 removal in the gas upgrading step becomes necessary. In such cases, the upstream CO_2 removal can be omitted and also the CO_2 already contained in the gasification producer gas might be fed to the methanation reactor. Although this dilutes the reactants, CO_2 may suppress carbon formation, dampening the temperature increase and influencing the WGS equilibrium such that it increases the selectivity of CO to methane.

Influence of Gasifier Type and Gas Cleaning Steps   As discussed in Chapter 2 in this book, producer gas of direct gasification contains significantly more CO_2 than that of allothermal gasifiers, because the CO_2 originating from combustion processes is contained, while it leaves the gasifier with the flue gas in the case of allothermal gasifiers. While low temperature gasification generally leads to higher amounts of methane and C_2 species, their concentration and the water content in the feed gas to the methanation reactor depend on the gas cleaning. Steam reforming units will lead to conversion of C_2 species and methane to carbon oxides [5]. Hydro-desulfurization catalysts such as molybdenum sulfide may hydrogenate ethylene to ethane [6]. Applying cold gas cleaning at temperatures below the dew point of water leads to dry gas mixtures, to which steam has to be added to avoid carbon deposition. On the other hand, warm gas cleaning such as the oil wash "OLGA" developed by ECN lead to a high steam content of up to 35% (see Chapter 9 in this book). Table 4.1 shows the molar fractions (dry gas) of the main species for typical gasification producer gases or methanation feed gas compositions.

4.1.1.3   Input Gases for Methanation Reactors in Power to Gas Applications   In recent years, the power to gas concept has been developed to store electricity in the natural gas grid [7]. Within the concept, excess electricity, for example, from stochastic sources such as wind or solar power that cannot be stored in pumped-storage hydropower plants, is used to generate hydrogen by electrolysis. The hydrogen may further be converted in methanation plants, which allow converting CO_2 to methane, the main compound in natural gas. This way, it is possible to convert CO_2, either in form of biogas (CO_2/CH_4 mixtures from biomass fermentation), gasification derived producer gas or as pure stream separated from flue gas of a combustion process, to SNG. The SNG can then either be used for, for example, mobility application in compressed natural gas (CNG) cars or, in times of peak electricity consumption, can be converted back to electricity in combined cycle or combined heat and power plants (CHP). Chapter 7 in this book presents one example for this process in more detail.

TABLE 4.1 Molar Fractions (dry gas) of the Main Species for Typical Gasification Producer Gases or Methanation Feed Gas Compositions.

%	Stoichiometric $H_2/CO = 3.1$	Allothermal Gasification	Direct Gasification	Stoichiometric Power to Gas	Power to Gas with Biogas	Power to Gas with Producer Gas
H_2	75.6	30–40	25–30	80.4	62.1–67.2	70–75
CO	24.4	20–30	16–57	0	0	10–12
CO_2	0	20	2–36	19.6	15.2–16.4	10–12
CH_4	0	10	5–12	0	16.4–22.7	4–5
C_2H_2	0	0–1	0–0.5	0	0	0–0.5
C_2H_4	0	1–4	0–3	0	0	1.0–2.5
C_2H_6	0	0–1	0–0.5	0	0	0–0.5

In the first power to gas pilot and demonstration plant in Werlte (Germany), CO_2 from biogas is used to convert hydrogen to methane. The CO_2 is separated from the biogas by means of an amine wash and mixed with hydrogen, leading to a stoichiometric mixture of H_2 and CO_2 with a ratio of around four. Similarly, it is possible to obtain such a gas mixture by removing CO_2 from flue gases, for example, from combustion processes or cement kilns. If a biogas plant is used as CO_2 source, and the methanation step is able to handle the methane content in the mixture (50–60% CH_4 in biogas), one could omit the CO_2 separation step and convert the biogas nearly completely to methane and (easily condensable) steam. This results in a feed stream for methanation reactors which consist of around one-sixth of each methane and carbon dioxide, while the remaining two-thirds are hydrogen, enabling a H_2/CO_2 ratio of 4.1.

In principle, it is also possible to add hydrogen to producer gas from gasification to increase the H_2/CO ratio. This way, the necessity of simultaneous water gas shift can be decreased, and more carbon monoxide is converted to methane instead to carbon dioxide resulting in higher carbon efficiency (i.e., the fraction of CO converted to CH_4). Further, the CO_2 already contained in the producer gas could be converted with H_2 as well. Due to the full use of carbon oxides as carbon sources for methane, the SNG output of a wood to SNG plant can be doubled by the addition of hydrogen from excess electricity. Alongside the water consumption, the heat duty for water evaporation and the CO_2 production in the WGS can be decreased.

Besides these technical benefits, the integration of both processes also offers further advantages for the economic feasibility of the power to gas concept. First, the investment for the methanation step can be used during the whole year, not only in times of peak electricity. Further, the capacity of wood to SNG plants (20–200 MW) is comparatively larger than that of biogas plants therefore facilitating the integration with electrolysis plants of suitable capacity. Adding H_2 to producer gas from more or less constant running gasification asks for some flexibility in the reactor concepts as, with hydrogen addition, the overall flow rate will also be increased. In consequence, the cooling system and the heat exchanger network have to cope not only with a higher amount of reaction heat (due to additional CO_2 methanation), but also with higher throughputs and linear velocities. Furthermore, dynamically integrating additional hydrogen streams within short response times is desired to avoid the necessity of intermediate hydrogen storage. Table 4.1 shows the gas compositions to be expected at the inlet of methanation reactors for the different cases of power to gas applications.

4.1.2 Thermodynamic Equilibrium

The previous section discussed the different gas mixtures that can be expected at the inlet of a methanation reactor depending on the chosen feedstock, gasification technology, and gas cleaning steps. As the main task of the methanation is to convert as much carbon as possible to methane to optimise the chemical efficiency, the thermodynamic analysis will focus on the fate of carbon. In the following, the distribution of carbon atoms within the molecules to be expected according the thermodynamic equilibrium will be presented for several gas mixtures and the important temperature and pressure range (200–500 °C, 1–10 bar).

The equilibrium constants for methanation and water gas shift reaction can be calculated from thermodynamic data such as heat of formation (taken, e.g., from the DIPPR Project 801 database [8]) as function of the temperature by the van't Hoff equation:

$$\frac{\partial \ln\left(K_P\right)}{\partial T} = \frac{\Delta H_R^0}{\Re \cdot T^2} \tag{4.10}$$

As the reaction enthalpy changes with temperature, it has to be computed for each temperature by integrating the change of the heat capacities (using a reference temperature T_0 at standard condition of 298.15 K):

$$\Delta H_R\left(T\right) = \Delta H_R^0\left(T_0\right) + \int_{T_0}^{T} v_i \cdot c_{p,i} \cdot dT \tag{4.11}$$

For the results shown in the next sections, the software package HSC© [9] was used.

4.1.2.1 Stoichiometric Mixtures for CO and CO₂ Methanation

4.1.2.1 Stoichiometric Mixtures for CO and CO$_2$ Methanation Figure 4.1 shows how the carbon atoms are distributed within the different molecules as predicted by the thermodynamic equilibrium. Figure 4.1a, b compares the situation for stoichiometric mixtures of hydrogen and carbon monoxide ($H_2/CO=3$), and of hydrogen and carbon dioxide ($H_2/CO_2=4$), respectively. It can be observed that the exothermic nature of the methanation favors high methane yields at lower temperatures, while higher pressure have a positive influence due to the volume contraction in the reaction. Still, the step from 1 to 5 bar has a significantly higher impact then the step from 5 to 10 bar. For all pressures, more than 95% methane yield can be achieved at temperature between 350 and 400 °C in the case of CO methanation, and between 300 and 350 °C in the case of CO_2 methanation. This would allow simplifying the subsequent upgrading steps before injection of the SNG into the natural gas grid: a CO_2 separation could be omitted.

While the carbon conversion would be high (and the fraction of carbon in carbon monoxide is therefore very low) according to thermodynamics, the coupled equilibrium for methanation and water gas shift reaction predicts a small percentage of CO_2, especially in the case of CO_2 methanation. The reason is that the CO_2 methanation produces two water molecules per methane. Because of this higher water content, equilibrium requires lower temperatures in the case of CO_2 methanation to reach 95% conversion, and the pressure influence for CO_2 methanation is stronger than for CO methanation, but is still mild.

Figure 4.1c shows the atomic carbon distribution in the case of a mixture of hydrogen and biogas, where the biogas content is 50% of methane and 50% of carbon dioxide and enough hydrogen is added to convert the CO_2 to CH_4. Such a mixture relates to a power to gas application in which biogas is converted with hydrogen from

(a)

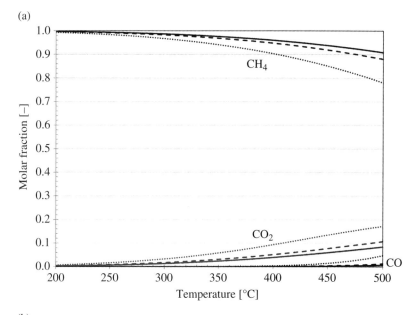

(b)

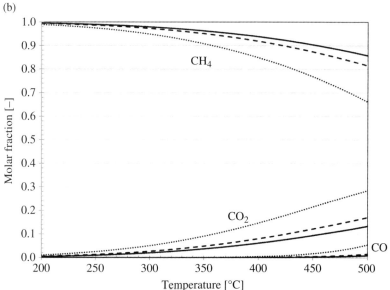

FIGURE 4.1 Distribution of carbon atoms within the different molecules as predicted by the thermodynamic equilibrium (HSC© [9]) for different gas mixtures in the range from 200 to 500 °C and different pressures: 1 bar (dotted lines), 5 bar (dashed lines) and 10 bar (full line). (a) Stoichiometric mixture of hydrogen and carbon monoxide ($H_2/CO = 3$). (b) Stoichiometric mixture of hydrogen and carbon dioxide ($H_2/CO_2 = 4$). (c) Mixture of hydrogen and biogas, where the biogas consists 50% of methane and 50% of carbon dioxide ($H_2/CO_2 = 4$).

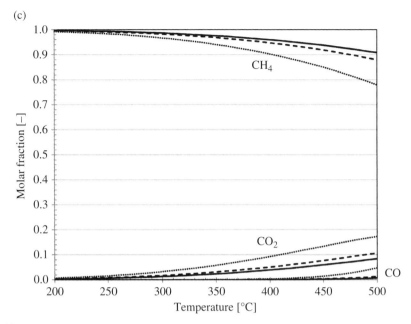

FIGURE 4.1 (*Continued*)

excess electricity without separating the CO_2 from CH_4. As can be seen from Figure 4.1c, according to thermodynamics the dilution with CH_4 facilitates reaching 95% methane content in the reactor outlet; this is now possible at temperatures around 350–400 °C; the situation is very close to that of stoichiometric CO methanation.

The residual hydrogen content can be calculated by multiplying the CO_2-content by four and the CO content by three. Therefore, to obtain 95% methane content even without H_2 removal in the gas upgrading, a CO_2 content below 1% has to be reached, which is the case around 250 °C.

4.1.2.2 Purified Product Gas from Auto- and Allothermal Gasifiers

In the purified producer gas from coal or biomass gasification, the ratio between hydrogen and carbon monoxide is significantly lower than the stoichiometric value of three; moreover methane, carbon dioxide, C_2 species (such as acetylene, ethylene, and ethane), and even benzene are found. Depending on the different gas cleaning steps, steam contents up to 40% (in the case of non-condensing warm gas cleaning) can be expected. Especially, in the case of autothermal gasification, relatively high concentrations of CO_2 and steam can be observed as the combustion products are contained in the product gas, while they leave the gasification system with the flue gas in the case of allothermal gasification.

As a result (see Figure 4.2a), far more carbon dioxide can be expected according to thermodynamics than in case of stoichiometric mixtures. While the carbon monoxide content at equilibrium is negligible at temperatures below 400–450 °C, the carbon dioxide content is slightly higher than the methane content. This is caused, on

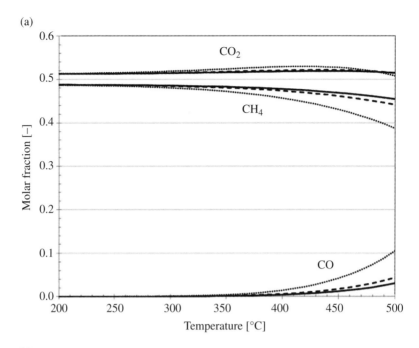

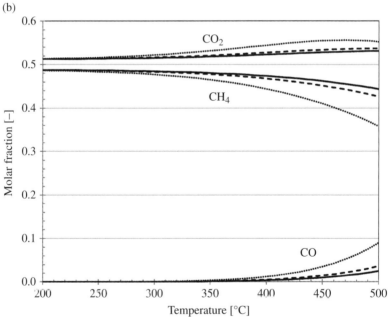

FIGURE 4.2 Distribution of carbon atoms within the different molecules as predicted by the thermodynamic equilibrium (HSC© [9]) for different gas mixtures in the range from 200 to 500 °C and different pressures: 1 bar (dotted lines), 5 bar (dashed lines) and 10 bar (full line). (a) Purified product gas from allothermal gasification (40% H_2, 25% CO, 22% CO_2, 10% CH_4, 3% C_2H_4) at moderate steam contents. (b) Purified product gas from allothermal gasification at 40% steam (in case of hot gas cleaning). (c) Purified product gas from allothermal gasification with stoichiometric hydrogen addition (power to gas application: 73.8% H_2, 10.9% CO, 9.6% CO_2, 4.4% CH_4, 1.3% C_2H_4), no steam addition.

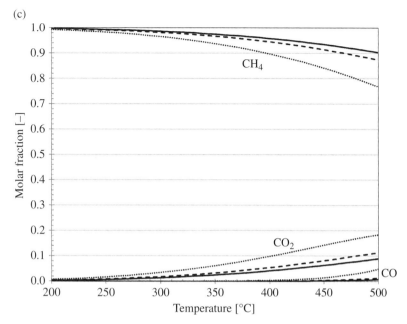

FIGURE 4.2   (*Continued*)

the one hand, by the carbon dioxide already contained in the producer gas; on the other hand, under-stoichiometric amounts of hydrogen and high water contents favor the water gas shift reaction. As already observed for stoichiometric mixtures, for higher pressures and lower temperatures, thermodynamics predict higher methane yields whereby the impact of the temperature is stronger. A very remarkable second observation is that according to thermodynamics, no higher hydrocarbons (C_2 species, benzene) can be expected at the outlet of a methanation reactor. Figure 4.2b shows the equilibrium predictions for converting producer gas from allothermal gasification combined with warm gas cleaning, which results in 40% steam content. At low pressure, the higher steam content shifts methane into CO_2, however at 10 bars, this effect is relatively small.

If the product gas is enriched with hydrogen (e.g., in the case of combining SNG production via gasification of dry feedstock with power to gas application), again all equilibria are shifted such that the methane yield becomes very high. These equilibrium predictions are as expected very similar to the stoichiometric ones, because so much hydrogen and no steam is added, that all carbon could be converted to CH_4. In consequence, only a small percentage of carbon atoms can be found in carbon dioxide. Still, temperatures as low as 325 °C are necessary according to thermodynamics to reach 95% methane yield and to allow omission of a downstream carbon dioxide separation. The residual hydrogen content can be calculated by multiplying the CO_2 content by four. Therefore, to obtain 95% methane content even without H_2 removal in the gas upgrading, CO_2 contents below 1% had to be reached which is the case around 250 °C.

4.1.3 Methanation Catalysts: Kinetics and Reaction Mechanisms

Sabatier and Senderens [4] found in 1902 that a number of metals catalyse the methanation reaction: rhodium, ruthenium, iridium, cobalt, iron, and nickel. Due to its high activity and the relative low price, nickel is by far the most applied catalyst. Ruthenium is discussed for several special applications, such as low temperature and selective methanation in gas cleaning [10]. As will be discussed in the second part of this chapter, in most cases the methanation reactor performance for SNG production is not limited by the catalyst activity but rather by heat removal, thermodynamic equilibrium or mass transfer. In consequence, there is no urgent research need to further improve the activity of nickel based catalysts.

For the methanation reaction, a number of reliable industrial catalysts on different supports, especially alumina and silica based, have been developed for different conditions: for isothermal operation, highly porous γ-alumina may be used as support, which however is not stable under adiabatic methanation conditions. For these high temperature conditions, α-alumina is often used as support, eventually under addition of a small percentage of magnesia for stabilization [11]. As nickel catalyst are handled in an oxidized form, methanation catalysts should also allow for reduction at reasonably low temperatures (300–500 °C). Further, they should show also high mechanic and temperature stability. A detailed overview on methanation catalysts has been given by Ross [11]. Many catalyst manufacturers can supply methanation catalysts, of which BASF AG, Haldor-Topsoe A/S, and Johnson–Matthey plc are the best known ones. As will be discussed in the second part of this chapter, available commercial methanation catalysts are very active and stable. In consequence, the reactor performances are limited by mass or heat transfer rather than by catalyst activity, which in turn strongly decreases the necessity of developing new methanation catalysts.

4.1.3.1 Kinetics and Reaction Mechanism of Main Reactants
Although the methanation reaction has been investigated by many groups, still a variety of proposed reaction mechanisms, surface intermediates, and rate-determining steps is presented. Partly, this can be explained by the wide range of catalysts, operation conditions, and experimental methods applied for the determination of kinetic parameters and for the elucidation of the mechanisms. Section 4.3.1 of this chapter discusses in more detail the challenges in determining kinetic parameters of real catalysts for the highly exothermic methanation reaction.

As shown in Table 4.2, the literature suggests different kinetic approaches ranging from simple power law to a more complex Langmuir–Hinshelwood (L-H) type of models. While power laws are suited for interpolation within the measured range only, L-H type models aim at representing the physico-chemical aspects in more detail by considering the rate determining step. They can therefore be used with more confidence to extrapolate the predictions to a wider range of operating conditions.

Most published kinetic rate equations assume one of the elementary steps in two main models as the rate-determining step for the reaction mechanisms proposed.

TABLE 4.2 Kinetic Approaches for CO Methanation on Nickel Catalysts [12].

Reactor	Catalyst/d_p [mm]	Temp. [K]	Pressure [bar]	Proposed Kinetic Model Equation	E_A, ΔH [kJ/mol]	Comments	Ref.
Berty reactor	18 wt% Ni/ Al$_2$O$_3$ 1.0 × 1.4	453–557	p_{tot}: 1–25 p_{H_2}: 1–25 p_{CO}: 0.001–0.6	$r_{CH_4} = \dfrac{k_{CH_2} \cdot K_C \cdot K_H^2 \cdot p_{CO}^{0.5} \cdot p_{H_2}}{\left(1 + K_C \cdot p_{CO}^{0.5} + K_H \cdot p_{H_2}^{0.5}\right)^3}$	E_A: 106 ± 1.7 ΔH_{H2}: −42 ± 2.6 ΔH_{CO}: −16 ± 3.3	RDS: A4 or A5 no pore diffusion	[13]
Berty reactor	18 wt% Ni/ Al$_2$O$_3$ 1.0 × 1.4	453–557	p_{tot}: 1–3 p_{H_2}: 0.2–3 p_{CO}: 0.005–0.5	$r_{C_2H_4} = \dfrac{k_{CH_2} \cdot K_C^2 \cdot p_{CO}}{\left(1 + K_C \cdot p_{CO}^{0.5} + K_H \cdot p_{H_2}^{0.5}\right)^2}$	E_A: 103 ± 1.7		[13]
Flow Ø 7–12 mm	5 wt% Ni/ SiO$_2$ 0.3 × 0.6	463–843	p_{tot}: 1 p_{H_2}: 0.007–1.0 p_{CO}: 0.001–0.87	$r_{CH_4} = \dfrac{Z_1 \cdot p_{CO}^{0.5}}{\left(1 + Z_2 \cdot p_{CO}^{0.5} \cdot p_{H_2}^{-0.5}\right)^2}$		RDS: A5 no influence of CH$_4$, H$_2$O T < 350 °C no WGS T > 450 °C WGS	[14]
Fluidised bed Ø 8 mm	12 wt% Ni/ Al$_2$O$_3$; 20 wt% Ni/ Mg/Al$_2$O$_3$ 0.3 × 0.5	443–573	p_{tot}: 1–15 p_{H_2}: 1–15 p_{CO}: 0.0002–1.5	$r_{CH_4} = \dfrac{k_1 \cdot p_{CO}^{0.15}}{\left(1 + K_1 \cdot p_{CO} \cdot p_{H_2}^{-1}\right)^{0.5}}$	E_A: 75–117 ΔH_{H2}: −69.7	RDS: A1 H$_2$ + 2* → 2 H* no influence of CH$_4$, H$_2$O, and CO$_2$ no pore diffusion	[15, 16]
Flow Reactor	Ni/SiO$_2$ 3.2	573–623	p_{tot}: 1 p_{H_2}: 0.55–0.8 p_{CO}: 0.2–0.45	$r_{CH_4} = \dfrac{p_{CO} \cdot p_{H_2}^3}{\left(A + B \cdot p_{CO} + C \cdot p_{CO_2} + D \cdot p_{CH_4}\right)^4}$		RDS: C* + 2H* → no pore diffusion	[17]
Flow Ø 10 mm	G65; Ni/ Al$_2$O$_3$ 0.35 × 0.42	443–483	p_{tot}: 1 p_{H_2}: 1 − p_{CO} p_{CO}: <0.02	$r_{CO} = \dfrac{k_1 \cdot p_{CO}}{\left(1 + K_{CO} \cdot p_{CO}\right)^2}$	E_A: 42 ΔH_{CO}: −53	RDS: (H$_2$...CO)* + H* → no influence of CH$_4$, H$_2$O	[18]

(Continued)

TABLE 4.2 (Continued)

Reactor	Catalyst/d_P [mm]	Temp. [K]	Pressure [bar]	Proposed Kinetic Model Equation	E_A, ΔH [kJ/mol]	Comments	Ref.
Flow Ø10 mm	G65; Ni/Al$_2$O$_3$ 0.35 × 0.42	443–483	p_{tot}: 1 p_{H_2}: 1 − p_{CO} p_{CO}: <0.02	$r_{CO_2} = \dfrac{k_2 \cdot p_{CO_2}}{1 + 1270 \cdot p_{CO_2}}$	E_A: 106	RDS: no influence of CH$_4$, H$_2$O CO poisons CO$_2$ methanation	[18]
Diff. reactor	2; 10 wt% Ni/SiO$_2$ 0.3 × 0.6	473–673	p_{tot}: 1–3 p_{H_2}: 0.2–3 p_{CO}: 0.005–0.5	$r_{CH_4} = \dfrac{k_1 \cdot K_{CO} \cdot K_{H_2} \cdot p_{CO} \cdot p_{H_2}}{\left(1 + K_{CO} \cdot p_{CO} + K_{H_2} \cdot p_{H_2}\right)^2}$	E_A: 84–103	RDS: B3 CO* + 2H* → C* + H$_2$O + *	[19]
Tube wall reactor Ø 6 mm	Ni	533–573	p_{tot}: 1 p_{H_2}: 1 − p_{CO} p_{CO}: 0.047–0.65	$r_{CO} = \dfrac{k_1 \cdot p_{CO}^{0.5} \cdot p_{H_2}}{\left(1 + K_{CO} \cdot p_{CO}\right)}$	E_A: 59 ΔH_{CO}: −25	RDS: A4 C* + H$_2$ → CH$_2$* no H$_2$ adsorption no WGS	[20]
Tube wall reactor Ø 6 m	Ni	523–623	p_{tot}: 1 p_{H_2}: 1 − p_{CO} p_{CO}: 0.03–0.62	$r_{CO_2} = \dfrac{k_2 \cdot p_{CO_2}^{1/3} \cdot p_{H_2}}{\left(1 + K_{CO_2} \cdot p_{CO_2} + K_{H_2} \cdot p_{H_2} + K_{H_2O} \cdot p_{H_2O}\right)}$			[20]
Berty reactor	33.8 wt% Ni/CaO/SiO$_2$ 0.5 × 1.0	453–505	p_{tot}: 1 p_{H_2}: 0.22–0.96 p_{CO}: 0.0008–0.14	$r_{CH_4} = \dfrac{k_1 \cdot p_{CO}^{0.5} \cdot p_{H_2}}{\left(p_{H_2}^{0.5} + K_2 \cdot p_{CO} + K_3 \cdot p_{CO}^{0.5} \cdot p_{H_2}^{0.5}\right)}$	E_A: 81.1 ΔH_2: −33 ΔH_3: −23.4	RDS: CH* + H	[21]
Berty reactor	5 wt% Ni/Al$_2$O$_3$ 12.5 25.5 (Monolith)	473–623	p_{tot}: 6.9 p_{H_2}: 0.12–0.3 p_{CO}: 0.005–0.12	$r_{CH_4} = \dfrac{k_1 \cdot k_2 \cdot p_{H_2}}{k_1 \cdot \left(1 + K_{CO} \cdot p_{CO} + K_{H_2}^{0.5} \cdot p_{H_2}^{0.5}\right)^2 + k_2 \cdot \left(1 + K_{CO} \cdot p_{CO}\right)^2}$	E_{A1}: 143.7 E_{A2}: 70 ΔH_{H_2}: −92 ΔH_{CO}: −68.5	RDS: Change from A1 to A4 H$_2$O retarded the reaction rate no influence of CH$_4$	[22]

Reactor	Catalyst	Temperature	Partial pressures	Rate equation		Comments	Reference
Fixed bed	27 wt% Ni/Al$_2$O$_3$ 0.06	573	p_{tot}: 1 H$_2$/CO : 1.2–16	$$r_{CH_4} = \frac{k_1 \cdot p_{H_2}}{1 + K_{CO} \cdot p_{CO} + K_{H_2} \cdot p_{H_2} + K_{CO_2} \cdot p_{CO_2} + K_{CH_2} \cdot p_{CH_4}}$$		high N$_2$ dilution, CO-conversion <1%	[23]
Fixed bed	27 wt% Ni/Al$_2$O$_3$ 0.06	573	p_{tot}: 1 H$_2$/CO : 1.2–16	$$r_{CO_2} = \frac{k_2 \cdot p_{H_2O}}{\left(p_{CO} \cdot p_{H_2}\right)^{0.5}} - \frac{k_2 \cdot p_{CO_2}}{k_{eq} \cdot p_{CO}} \left(\frac{p_{H_2}}{p_{CO}}\right)^{0.5}$$		Water gas shift	[23]
Fixed bed	29 wt% Ni/Al$_2$O$_3$ 0.2–0.4	533–573	p_{tot}: 1–5 p_{H2}: 0.0.96–0.8 p_{CO}: 0.032–0.16	$$r = k_1 \cdot p_{CO}^{-0.87} \cdot p_{H_2}^{1.27} \cdot p_{H_2O}^{-0.13}$$	E_A: 78	0.2–0.5 g catalyst N$_2$ dilution, no pore diffusion	[24]

Source: Kopyscinski J. Production of Synthetic Natural Gas in a Fluidized Bed Reactor – Understanding the Hydrodynamic, Mass Transfer, and Kinetic Effects. Dissertation ETH Zürich Nr. 18800, 2009.

TABLE 4.3 Reaction pathways for mechanism A with rate-determining steps (RDS) suggested in the literature [12].

H_2	+	2*	↔	2H*				Dissociative adsorption of H_2	A 1
CO	+	*	↔	CO*			RDS	CO adsorption	A 2
CO*	+	*	↔	C*	+	O*	RDS	Dissociation of CO	A 3
C*	+	H*	↔	CH*	+	*	RDS	Hydrogenation of C	A 4
CH*	+	H*	↔	CH_2*	+	*	RDS	Hydrogenation of CH	A 5
CH_2*	+	H*	↔	CH_3*	+	*	RDS	Hydrogenation of CH_2	A 6
CH_3*	+	H*	↔	CH_4*	+	*		Hydrogenation of CH_3	A 7
CH_4*			↔	CH_4	+	*		CH_4 desorption	A 8
CO*	+	O*	↔	CO_2	+	*		CO_2 formation	A 9
CO_2*			↔	CO_2	+	*		CO_2 desorption	A 10
O*	+	H*	↔	OH*	+	*		OH formation	A 11
OH*	+	H*	↔	H_2O*	+	*		H_2O formation	A 12
H_2O*			↔	H_2O	+	*		H_2O desorption	A 13
CO*	+	OH*	↔	CO_2*	+	H*		CO_2 formation via OH	A 14
CO*	+	H_2O*	↔	CO_2*	+	2H*		CO_2 formation via H_2O	A 15

Empty active site. C Adsorbed species (e.g., adsorbed carbon).

Source: Kopyscinski J. Production of synthetic natural gas in a fluidized bed reactor – Understanding the hydrodynamic, mass transfer, and kinetic effects. Dissertation ETH Zürich Nr. 18800, 2009.

Mechanism A (see Table 4.3) assumes the methanation to proceed via molecular adsorption and subsequent dissociation of CO. This leads to an adsorbed carbon (C_{ads}) as intermediate on the catalyst surface, which is stepwise hydrogenated to methane. Mechanism B (see Table 4.4) proposes that a hydrogenation of the adsorbed CO is necessary to facilitate splitting of the carbon–oxygen bond. The intermediate assumed is an oxygenated compound, i.e. a COH_x complex.

Mechanism A (assuming an adsorbed carbon atom as intermediate on the catalyst surface) was proposed by Araki and Ponec in 1976 [25]. The hypothesis was supported by isotopic labeling experiments where ^{13}CO was fed over nickel catalyst followed by H_2 and ^{12}CO. Formation of $^{13}CH_4$ occurred before $^{12}CH_4$ and $^{12}CO_2$ were formed, while no $^{13}CO_2$ was detected [26]. Both the adsorbed carbon and oxygen atoms then reacted with hydrogen. Oxygen and hydrogen formed water, the second reaction product of the methanation reaction and adsorbed OH* onto the surface. Both water and the OH* could react with adsorbed CO* to form CO_2.

The adsorbed carbon atom underwent stepwise hydrogenation to finally form methane. An adsorbed *CH_x species on Ni/γ-Al_2O_3 was observed by Galuszka et al. [27] at 100 °C by using infrared spectroscopy. Many further authors [14, 28–34] accepted and adopted the stepwise hydrogenation of surface carbon on nickel to methane. Several authors tested L-H type models, which were based on mechanism A, for the evaluation of the experimental kinetic data and found that they could explain their experimental data [13, 14, 19, 20, 28, 31, 34–37], however assuming different rate-determining steps.

Reaction mechanism B assumes non-dissociative adsorption of CO which then reacts with adsorbed hydrogen atoms to *COH_x intermediates. Due to the lower

TABLE 4.4 Reaction pathways for mechanism B with rate-determining steps (RDS) suggested in the literature [12].

						RDS		
H_2	+	$2*$	$\leftrightarrow 2H*$				Dissociative adsorption of H_2	B 1
CO	+	$*$	$\leftrightarrow CO*$			RDS	CO adsorption	B 2
$CO*$	+	$H*$	$\leftrightarrow COH*$	+	$*$	RDS	COH formation	B 3
$COH*$	+	$*$	$\leftrightarrow CH*$	+	$O*$	RDS	Dissociation of COH complex	B 4
$COH*$	+	$*$	$\leftrightarrow C*$	+	$OH*$	RDS	Dissociation of COH complex	B 5
$COH*$	+	$H*$	$\leftrightarrow COH_2*$	+	$*$	RDS	COH_2 formation	B 6
$COH*$	+	$H*$	$\leftrightarrow CH*$	+	$OH*$	RDS	Dissociation of COH with H	B 7
COH_2*	+	$H*$	$\leftrightarrow CH*$	+	H_2O*	RDS	Dissociation of COH_2 with H	B 8
COH_2*	+	$H*$	$\leftrightarrow COH_3*$	+	$*$	RDS	COH_3 formation	B 9
COH_3*	+	$H*$	$\leftrightarrow CH_2*$	+	H_2O*	RDS	Dissociation of COH_3 with H	B 10
$CO*$	+	$OH*$	$\leftrightarrow CO_2* +$	$H*$			CO_2 formation	B 11
$CO*$	+	H_2O*	$\leftrightarrow CO_2* +$	$2H*$			CO_2 formation	B 12
CO_2*			$\leftrightarrow CO_2$	+	$*$		CO_2 desorption	B 13
$O*$	+	$H*$	$\leftrightarrow OH*$	+	$*$		OH formation	B 14
$OH*$	+	$H*$	$\leftrightarrow H_2O*$	+	$*$		H_2O formation	B 15
H_2O*			$\leftrightarrow H_2O$	+	$*$		H_2O desorption	B 16
$C*$	+	$H*$	$\leftrightarrow CH*$	+	$*$	RDS	Hydrogenation of C	B 17
$CH*$	+	$H*$	$\leftrightarrow CH_2*$	+	$*$	RDS	Hydrogenation of CH	B 18
CH_2*	+	$H*$	$\leftrightarrow CH_3*$	+	$*$	RDS	Hydrogenation of CH_2	B 19
CH_3*	+	$H*$	$\leftrightarrow CH_4*$	+	$*$		Hydrogenation of CH_3	B 20
CH_4*			$\leftrightarrow CH_4$	+	$*$		CH_4 desorption	B 21

Empty active site. C Adsorbed species (e.g., adsorbed carbon).
Source: Kopyscinski J. Production of synthetic natural gas in a fluidized bed reactor – Understanding the hydrodynamic, mass transfer, and kinetic effects. Dissertation ETH Zürich Nr. 18800, 2009.

activation barrier energy for the C-O bond dissociation [38], these are splitting off *OH or water molecules leading to adsorbed *C or *CH$_x$ on the surface. While these undergo, like in mechanism A, stepwise further hydrogenation to methane, *OH may react further to form CO_2 or water. Coenen et al. [39] found no isotope exchange when they conducted methanation with equimolar amounts of $^{13}C^{16}O$ and $^{12}C^{18}O$. Again, different authors assumed different rate-determining steps and therefore different intermediates to evaluate their data [18, 21, 37–45].

Very recent work has applied temperature programmed desorption and hydrogenation as well as modulation excitation infrared spectroscopy (diffuse reflectance infrared spectroscopy, DRIFTS) and mass spectrometry (MS) on an industrial nickel based catalyst, whereby the modulation–excitation technique allows enhanced sensitivity to observe even small changes in the IR signal [46]. It was found that linear CO is less strongly bonded to nickel than bridge CO and therefore more reactive for CO methanation. While linearly adsorbed CO dissociates on nickel defect sites, it seems to be less reactive and accumulates on well ordered sites. Isotope labeling (regular addition of $^{12}CH_4$ to a constant flow of ^{13}CO and deuterium D$_2$) showed that, besides the expected main product $^{13}CD_4$, in phases of $^{12}CH_4$ addition also $^{13}CHD_3$ and $^{12}CD_4$ are formed. This proves that CH_4 can completely be dissociated at 300 °C to form atomic carbon on the surface which reacts with the excess deuterium D_2. The formation of $^{13}CHD_3$ shows on the other hand that $^{13}CD_x*$ can be considered as an important intermediate in the methanation of ^{13}CO with deuterium D_2 as it evidently is present on the catalyst surface long enough to react with H* atoms from $^{12}CH_4$ dissociation.

4.1.3.2   Kinetics and Reaction Mechanism of Side Reactions (C_2 species)   As discussed above, in the gasification of carbonaceous feedstock, besides typical synthesis gas components such as CO, H_2, CO_2, and H_2O, also hydrocarbons such as methane, ethene/ethylene, ethane, ethyne/acetylene, propene, and aromatic species are formed. The formation of methane in the gasification is advantageous for the cold gas efficiency of the overall process chain because less CO has to be converted by exothermic methanation in the synthesis step. Unfortunately, the necessary low temperature in the gasification favors also the formation of other hydrocarbons, especially C_2 species.

The producer gas of most allothermal wood gasification processes, for example, the commercial dual fluidized bed gasifier in Güssing (Austria), contains around 10% methane, but also more than 2% of ethene and nearly 0.5% of both ethane and ethyne. Especially ethene and ethyne are known to be harmful by carbon deposition on the nickel catalyst and even due to the formation of carbon fibres or so-called whiskers [47], if the catalyst is used in fixed bed operation. Therefore, upstream of fixed bed methanation reactors applying a nickel catalyst, ethylene in the feed gas has to be either removed (e.g., by Rectisol® wash [48]) or be converted. Catalytic conversion of ethene can be achieved by reforming over nickel or noble metal catalysts [49] or in a hydrodesulfurization step over molybdenum sulfide based catalysts [6, 50]. To simplify the gas cleaning, it would be favorable to process ethene and the other C_2 species in the methanation step without harming the catalyst. Besides the omission of the gas cleaning steps which are so far necessary to remove or convert ethene, it would also allow to convert the ethene in the gas into an injectable energy carrier, thus improving the overall chemical efficiency of the process.

If the nickel catalyst is used under fluidized bed operating conditions, ethylene seems to be converted to methane [51] and long-term stability up to 1000 h could be demonstrated in a laboratory-scale fluidized bed reactor which was connected to a slip stream of the commercial wood gasifier in Güssing (Austria) [48]. More recent investigation in a micro-fluidized bed reactor [52] showed that, at low temperature operation conditions, more ethane, and at higher temperatures, more methane can be formed when ethene is fed with a synthetic gasification producer gas to a nickel based methanation catalyst (see Figure 4.3).

The reaction pathways or mechanisms of the different C_2 species under methanation conditions are not completely understood, although significant work has been conducted on the reactions of these species with nickel based catalysts.

As mentioned in the introduction, the saturated C_2 species ethane can form methane in a hydrogenolysis reaction. References [53, 54] suggest that ethane adsorbs on nickel surfaces where first C–H bonds and then the C–C bond of the hydrogen-deficient molecule are broken, followed by a hydrogenation to form

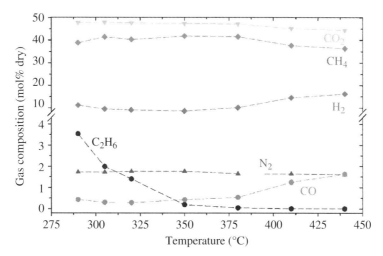

FIGURE 4.3 Measured dry outlet gas composition (averaged over 18 h of stable reaction) as a function of temperature for the methanation. Dry gas feed: $H_2 = 39$ vol%, $CO = 27$ vol%, $CO_2 = 19$ vol%, $CH_4 = 10$ vol%, $C_2H_4 = 4$ vol%, $N_2 = 1$ vol%. The symbols denote experimental data and the dashed lines are for guidance only; from [52]. *Source*: Kopyscinski 2013 [52]. Reproduced with permission of Elsevier.

methane. Single crystal studies [55, 56] reported for Ni(100) and Ni(111) surfaces different activation energies for ethane hydrogenolysis, but the same value for ethane hydrogenolysis and CO methanation on the Ni(100) surface. This indicates that on Ni(100) both reactions include similar steps, that is, the hydrogenation of a surface carbon species. Isotope labeling experiments, [57] and recently [58, 59] showed the formation of CD_4 when C_2H_6 and D_2 were fed over a nickel catalyst.

Ethene under conditions typical for methanation reactors has been shown to form ethane [52, 60], surface carbon [47, 61–63], and partly methane [52, 61–63]; see the results of experiments in a catalytic plate reactor with axially moveable gas sampling probe (Figure 4.4).

Similar to ethene, also ethyne is reported to form surface carbon on nickel catalysts at elevated temperatures [64, 65], a fact that is exploited for the formation of carbon nanotube materials for several applications by feeding ethyne over a nickel catalyst above 600 °C [66].

Sheppard et al. [67] showed using DRIFTS that hydrogen has to be pre-adsorbed to enable ethyne hydrogenation. Otherwise, hydrogenation is slowed down and carbon species such as surface alkyl groups are formed by hydrogenation and polymerization.

The selective hydrogenation of alkynes to alkenes in the presence of CO (several hundred ppm) is an important cleaning step upstream of the olefin polymerization over Pd and Ni based catalysts and has been investigated widely [68]. Under the conditions for this application (100–200 °C), hydrogen can form a subsurface or bulk hydrides which play a role for the (undesired) full conversion of alkenes to alkanes. Following the Horiuti–Polanyi mechanism [69], ethylene is adsorbed on the surface

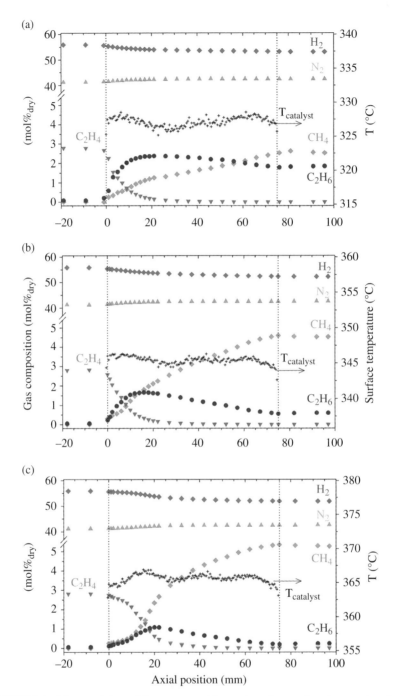

FIGURE 4.4 Measured dry axial gas composition and catalyst surface temperature for the ethylene hydrogenation at: (a) 325, (b) 345 and (c) 365 °C. Gas feed: $H_2 = 200$ ml$_N$ min^{-1}, $N_2 = 150$ ml$_N$ min^{-1}, $C_2H_4 = 10$ ml$_N$ min^{-1} and $H_2O = 40$ ml$_N$ min^{-1} ($H_2/C_2H_4 = 20$). The dotted lines indicate where the catalyst area begins and ends (75 mm length, 70 mg catalyst); from [52]. *Source*: Kopyscinski 2013 [52]. Reproduced with permission of Elsevier.

forming σ bonds with the surface. Atomic hydrogen may form another σ bond with one of the two carbon atoms leading to an ethyl group on the surface. The hydrogenation of the ethyl group seems to be the rate determining step [70] while several studies have observed the formation of ethylidyne species on Pd surfaces (*C-CH$_3$) [71].

Similarly, ethyne can adsorb on the surface and stepwise add atomic hydrogen to form adsorbed vinyl groups (*HC=CH$_2$) and then adsorbed ethylidene groups (*HC-CH$_3$). The latter can form ethylene or be further hydrogenated to ethyl groups which both can form (undesired) ethane. The presence of CO significantly changes this reaction network, as the activity for alkyne hydrogenation drops, but the selectivity to desired alkene is increased while the formation of undesired alkane is decreased [72]. This may be explained by the CO adsorption on the surface which lowers the number of sites for hydrogen adsorption and the formation of hydrides [68]. Further, formation of oligomers was found [72] which could be decreased by CO adsorption [68] or addition of Zn to nickel catalysts [73].

A very recent study [58] further elucidated the reaction pathways of the C$_2$ species (ethane, ethene, ethyne) on a commercial nickel catalyst under methanation reactions (i.e. CO hydrogenation at 200 and 300 °C). The use of modulation excitation DRIFTS improved the sensitivity with respect to trace and intermediate species, while isotope labeling helped to understand in more detail some of the elementary steps.

In these very systematic experiments, it was found that, at 300 °C, all three species are decomposed completely to atomic surface carbon that is stepwise further hydrogenated to methane or can, according water gas shift equilibrium, form CO$_2$. The decomposition rate order increases from ethane over ethene to ethyne where the latter leads to fast accumulation of unreactive carbon deposits on the catalyst surface. Further, formation of ethane was observed upon ethene addition during CO hydrogenation. Besides by the carbon deposition and the slight change in hydrogen concentration, which influences the equilibrium situation with respect to water gas shift reaction, addition of C$_2$ species seemed not to influence the methanation reaction. At 200 °C, the decomposition rate was clearly decreased, which favors the sequential hydrogenation of ethyne to ethene and further ethane, while the latter is practically unreactive at such low temperatures (see Figure 4.5).

4.1.4 Catalyst Deactivation

Generally, catalyst deactivation can be ascribed to one or a combination of the following phenomena (refer to the very useful review on catalyst deactivation by Bartholomew [74]):

- Fouling/blockage (i.e., the physical access of the reactants to the catalyst surface is not given).
- Poisoning or solid reaction (i.e., the active sites of the catalyst are changed).
- Loss of active surface (i.e., sintering of nickel crystallites).
- Loss of active phase (i.e., the active phase is physically or chemically separated from the main part of the catalyst and transported out of the reactor).

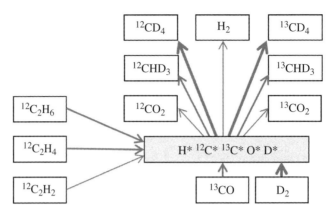

FIGURE 4.5 Species evolution observed by modulation excitation-DRIFTS and mass spectrometry after addition of 0.2 mol% $^{12}C_2$ species (acetylene only 0.05 mol%) to an isotope labeled methanation of 2 mol% ^{13}CO with 10 mol% deuterium at 300 °C [58]. The surface species were not directly observed, but are assumed to be present.

In industrial application, none of these processes can completely be avoided, therefore the aim is to understand and control them such that a catalyst lifetime of at least a year is achievable to avoid unplanned plant shut-down and production interruption. Exceptions to this are SNG process concepts that accept a certain degree of catalyst deactivation and comprise a suitable way of more frequent catalyst exchange; see the second part of this book. But also there, control of catalyst deactivation is economically favorable.

Proper design of the gas cleaning section (see Chapter 3 in this book), should avoid *fouling* of the catalyst due to condensation and/or polymerization of poly-aromatic hydrocarbons stemming from the gasification by removing these in scrubbers or filters or convert them in catalytic reforming units.

Thermal sintering of the nickel catalyst and the support or sintering by redox cycles should be limited if industrial catalysts are used according to the specifications of the manufacturers and therefore play a minor role within the guaranteed lifetime (see the discussion of the TREMP® process in Section 4.2.1.2). Similarly, *catalyst loss by attrition* in fluidized bed applications can be controled to a very low level by choice of the proper catalyst (see the discussion of the COMFLUX process in Section 4.2.2.2).

An inherent challenge for nickel based methanation catalysts is the formation of nickel tetra-carbonyl in the presence of high CO partial pressures. $Ni(CO)_4$ is highly toxic and volatile and can therefore lead to *chemical sintering* (i.e., the formation of large crystallites due to *gas phase transport of nickel* as tetra carbonyl from the small crystallites to the bigger ones) or even to loss out of the reactor. In practice, both can be limited by choosing sufficiently high temperatures (higher than 200–250 °C, depending on CO partial pressure) and hydrogen to CO ratios [11].

In the following two sections, the phenomena leading to catalyst deactivation due to compounds within the gas feed to the methanation reactor (sulfur species, carbon species) are discussed, while the third section presents some means to identify and quantify carbon depositions on catalyst samples taken from methanation reactors.

4.1.4.1   Sulfur Poisoning   As discussed in detail in Chapter 3 of this book, the feedstocks for gasification such as coal and biomass also contain (besides carbon, hydrogen, and oxygen) a number of hetero-atoms (sulfur and nitrogen) and further elements such as phosphor, chlorine, and alkali metals. While the removal of most of these usually is sufficiently achieved with state of the art gas cleaning technologies, the removal of nitrogen and sulfur is a challenge. Nitrogen is mostly present as ammonia, NH_3, and in organic compounds such as pyridine. Sulfur can form hydrogen sulfide H_2S, carbonyl sulfide COS, carbon disulfide CS_2, mercaptans (also referred to as thio-alcohols and thio-ethers with alkyl-groups such as methyl-mercaptan CH_3SH), and a high number of thiophenic species (thiophene, benzo-thiophene and dibenzo-thiophene and their derivatives), see e.g., [75].

In large scale coal-to-SNG plants, these species are removed reliably by the Rectisol® scrubbing leading to total sulfur content below 100 ppb. In small and medium scale biomass-to-SNG plants however, economics do not allow the application of such low temperature/high pressure physical washing units. Rather, slightly chilled atmospheric scrubbers, chemical conversion or, in future, hot gas cleaning are applied. The organic sulfur and nitrogen species are largely removed in the chilled scrubber or converted to ammonia and H_2S, respectively, in hydro-treating [6, 50] and reforming units [5, 76]. H_2S is adsorbed by active carbon or metal oxide based sorbents such as zinc oxide. Still, the sum of all unconverted sulfur species can be significant; e.g., [77] showed that thiophenic species deactivated a methanation catalyst that converted wood gasification derived producer gas. Sulfur adsorbs and dissociates on nickel surfaces, leading to a highly stable and hardly reversible adsorbate [74] that over time will cover the complete surface when sulfur traces are added continuously. Sulfur blocks the adsorption of CO and hydrogen on the nickel sites and has to be considered as a very selective poison, that is, one sulfur atom can easily block 10 nickel surface atoms [74]. Still the tolerable level of sulfur in a syngas feed to a nickel catalyst depends on the hydrogen content, the sulfur partial pressure and the temperature. Reference [78] gives a good indication based on substantial industrial experience (see Figure 4.6).

Due to its strong adsorption, the removal of sulfur from a nickel catalyst is very challenging and can be achieved only by a complex redox cycle procedure using very low oxygen partial pressures and subsequent reduction. For more details, refer to Chapter 12 of this book. Too high oxygen partial pressures lead immediately to the formation of nickel sulfate. The sulfate can be removed by oxidative treatment above 830 °C [77]. Although this method will destroy the catalyst, it can be used to determine the total sulfur content by proper gravimetric analysis or measurement of the released sulfur oxides. For temperature programmed oxidation with gravimetric analysis, however, the weight increase by oxidation of the nickel as well as the weight loss by combustion of carbon deposits has to be taken into account [77].

4.1.4.2   Carbon and Coke Deposition   The conversion of carbon monoxide is inherently connected to the presence of carbon atoms on the catalyst surface which in turn can react with each other and the nickel. This way, they form a number of chemically different species that, depending on their nature, can deactivate or destroy

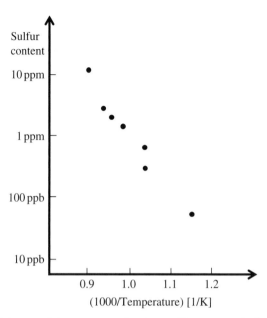

FIGURE 4.6 Minimum sulphur concentration that could cause practical poisoning of nickel based catalyst in hydrocarbon steam reforming; re-drawn from [78].

the catalyst. Due to the importance of the topic for the operation of nickel based catalysts in steam reforming and methanation, an enormous number of investigations has been published, especially by Bartholomew (e.g., [79]), Rostrup-Nielsen (e.g., [80]), Trimm (e.g., [81]), McCarty (e.g., [82]), Figueiredo (e.g., [83]) and their coworkers, and many other authors.

While, in steam reforming, larger coke formation due to higher hydrocarbon decomposition also plays an important role, in this section we will focus on the species to be expected in methanation reactors, that is, carbon depositions due to CO dissociation as well as coke formation mainly due to the presence of C_2 species (ethene, ethane, ethyne), and to a lesser extent, light aromatics. The differentiation between carbon (stemming from CO) and coke (caused by hydrocarbons) should not distract from the fact that the several carbon depositions can be caused by both CO and hydrocarbons. The conditions of formation of the different carbon depositions mainly depend on temperature, on the partial pressures of steam, hydrogen, CO, and hydrocarbons, on the nickel crystallite size and on the catalyst support and may show some overlaps. It should be noted that thermodynamic calculations using graphite as a representation of the carbon deposition are helpful in the case of steam reforming at high temperatures where the carbon depositions have a chemical similarity to graphite; under typical methanation conditions however, and in the presence of olefins and hydrocarbons, significant deviations from thermodynamics have to be expected [79].

As was discussed in the previous section, both CO and C_2 species may dissociate to form single carbon atoms on the catalyst surface. These *adsorbed carbon atoms*, the intermediates for methane formation, are generally referred to as C_α, but the expression *surface carbide* has also been used. Under circumstances where the carbon deposition rate on the surface exceeds the conversion with adsorbed steam, hydrogen and/or oxygen atoms, the carbon atoms may diffuse into the metal bulk forming *bulk nickel carbide*, usually referred to as C_γ. As bulk nickel carbide is very reactive with both hydrogen and steam under typical methanation reactions (see next section), it is stable only below 350 °C and its catalyst deactivation potential can be considered as small.

Further, in the temperature range up to 500 °C, the adsorbed carbon atoms C_α may polymerize forming an *amorphous carbon* film C_β that can deactivate the catalyst by covering active sites, sometimes also referred to as *gum*. With ageing, this amorphous carbon structure is reported to alter to an even more stable form [79]. As shown by [85], amorphous carbon exhibiting a significant fraction of sp^3-hybridized carbon changes with time to a mostly graphitic structure with more stable sp^2-hybridization. It is referred to by Bartholomew [79] as *graphitic carbon* C_C. Based on the formation conditions and the reactivity with hydrogen, this graphitic carbon can be considered identical to the *encapsulating carbon* C_δ found by McCarty and co-authors in a very broad systematic study using ethene and ethyne as sources for carbon deposition at temperatures from 300 to 1000 °C [84]. C_C/C_δ can form graphene layer type films encapsulating the nickel crystallites and deactivating the catalyst. McCarty et al. [84] found for ethene and ethyne addition to nickel catalysts at higher temperatures between 500 and 800 °C so-called *platelet carbon* C_ε [84] that sometimes was not easily differentiated from C_δ. In consequence, the definition of Bartholomew [79] of C_C with respect to formation conditions and reactivity covers the properties of both C_δ and C_ε found by McCarty et al. [84].

Under certain circumstances, carbon can form so-called *whiskers* or carbon nanofibres referred to as *vermicular carbon* C_V by Bartholomew [79] or *filamentous carbon* $C_{\delta'}$ by McCarty et al. [84]. In this special form, the carbon diffuses, most probably over the surface [86], to the rear end of the nickel crystallites where they precipitate at nickel step sites and form the carbon fibre that lifts the crystallite from the support. As the nickel crystallite stays active, the process can continue as long as the carbon deposition on the front end of the catalyst is faster than the reaction to methane, but not fast enough to form an encapsulating film. Therefore, sufficient, but not too much hydrogen is necessary, while water decreases the growth rates. The deposition rate of carbon and therefore the growth rate of the fibres increases from alkanes over alkenes to alkynes [87]. These whiskers were also observed on methanation catalysts applied in the presence of ethene under fixed bed conditions [47]. Although the nickel crystallite stays active for methanation and other reactions, the growth of these fibres will lead to blockage of pores and can destroy the catalyst particles mechanically, such that blockage of the complete reactor tube can be caused within hours or days. It is reported that the fibre growth, once it is started (e.g., by operation error leading to local hotspots or hydrogen deficiency), will continue even when hydrogen-rich conditions are re-established, and should therefore carefully be avoided [74].

Finally, hydrocarbons may also form in hot spots (above 600 °C), so-called *pyrolytic* or *non-catalyst carbon G* [84] that is not connected to the nickel crystallites, but

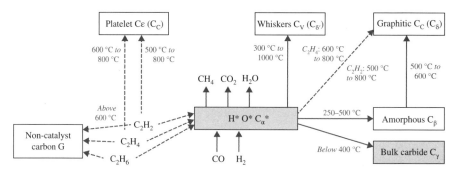

FIGURE 4.7 Overview over the formation of carbon deposits observed on nickel based catalysts for fixed bed methanation and steam reforming (including the specific formation temperatures); based on [79] and [84].

may also cause catalyst deactivation by blockage. Figure 4.7 gives an overview over the formation of carbon deposits observed on nickel based catalysts for fixed bed methanation and steam reforming and over the specific formation temperatures.

For comparison, Figure 4.8 presents the carbon depositions found at different temperatures when ethene or ethyne was added to a nickel catalyst [84]. Given the fact that encapsulating carbon C_δ, whiskers $C_{\delta'}$ and platelet carbon C_ε are formed by ethyne at significant lower temperatures (about 100 K lower than for ethene), one may assume that the formation mechanism is not exactly identical for all hydrocarbons and that the formation temperatures may vary for different nickel catalysts.

As already discussed, lower temperatures and higher hydrogen to carbon and steam to carbon ratios can decrease the rate of carbon depositions. There seems even to exist a window (ca. 330–400 °C) where the formation of C_α exceeds its conversion with hydrogen leading to accumulation, while the polymerization to C_β exceeds its gasification [79]. Operation in this window would lead to accumulation of amorphous carbon and stable operation should be possible at higher or lower temperatures. As the conditions for carbon deposition may vary for different nickel catalysts, the optimal operation conditions (temperature, hydrogen to carbon ratio, steam to carbon ratio, etc.) to avoid such carbon accumulation windows have to be determined for each case.

Further, the presence of sulfur traces may play a positive [88] or ambiguous [89] role. It was observed that, at lower temperatures, sulfur may help to avoid carbon polymerization by blocking some sites and limiting the necessary space on the surface for polymerization, while at higher temperatures, the sulfur hinders hydrogen adsorption which would be necessary for re-gasification of carbon deposits [89].

4.1.4.3 Methods to Quantify and Identify Carbon Depositions   An important aspect of carbon depositions is their chemical stability, as this decides both the regeneration of the catalyst and the options to differentiate, to identify and to quantify them on catalyst samples, which in turn is a prerequisite to optimise operation conditions with respect to catalyst stability.

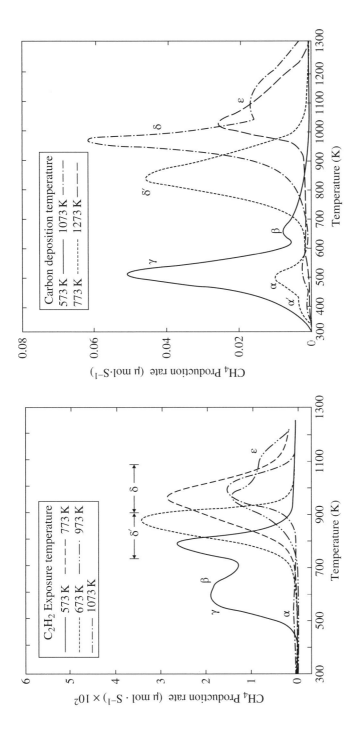

FIGURE 4.8 Temperature programmed reduction (TPR) for determination of carbon depositions formed at different temperatures when ethyne (left) or ethene (right) were added to nickel catalyst [84].

Unfortunately, none of the applied methods combines all necessary features. X-ray diffraction for example can only detect crystalline phases whose domain is sufficiently big. From the carbon depositions discussed above, this applies only to bulk nickel carbide; further, graphite (sometimes used as a binder in catalyst pellets) would be detected. Raman spectroscopy is able to detect the graphitic types of carbon depositions, but is not easily applied for detection of sp^3 hybridized carbon and fails for adsorbed carbon. Further, the energy input from the used lasers may change or deteriorate the sample within a short time. Similarly, electron microscopy (SEM and especially HRTEM) is perfect to identify whiskers and larger carbon structures, but fails for adsorbed carbon and quickly changes the catalyst sample. A general challenge for all methods is that an active nickel catalyst is reduced, therefore the catalyst sample will oxidize, at least superficially in contact with air, during transfer to a characterization method or the catalyst needs to be passivated carefully before removing a sample from the reactor. In both cases, information on the very active carbon species, that is, the adsorbed carbon atoms, might easily be lost.

Due to the availability of the set-up and robustness of the method, temperature programmed reaction has gained importance in the determination of carbon deposits on catalysts. Oxygen, hydrogen and steam could be used as reactive gases. Temperature programmed oxidation (TPO) can be useful to determine the sulfur and graphite (binder) content as well as the total carbon hold-up in a spent nickel catalyst [77], but TPO is limited to differentiate bulk nickel carbide, amorphous and graphitic carbon. The reasons are the temperature peaks caused by the bulk oxidation of the nickel catalyst (starting around 250–300 °C) and the oxidation of the respective carbon depositions, which may lead to local hot spots and mislead the ascription of a CO_2 evolution peak to the correct temperature.

The most common method is temperature programmed reduction (TPR) with hydrogen, which proved to be able to differentiate most of the carbon depositions [79, 84]. Figure 4.8 shows the results of TPR for the determination of carbon depositions formed at different temperatures when ethene or ethyne was added to a nickel catalyst [84]. The resolution, especially for the very reactive carbon depositions is very good, but one has to keep in mind that these samples were analyzed by TPR in situ, that is, without leaving the reactor and therefore without contact to air. If samples are taken from larger scale reactors, contact in air is hard to avoid. As a result, not only the state of the active nickel and the carbon depositions, but also the support may change, for example, by the formation of hydroxyl groups on high surface alumina supports. These in turn may split off water during TPR, leading to a mixture of steam and hydrogen reacting with the carbon on the catalyst surface [90]. As the reducing conditions turn the catalyst active again, besides hydro-gasification of carbon deposits, also formation of CO and CO_2 can be expected which complicates analytics and data evaluation. A further drawback of TPR is the high temperature necessary to convert the very stable carbon species such as non-catalyst carbon or graphite that sometimes is used as a binder. As mentioned in [79], depending on the conditions, the thermodynamic equilibrium limits complete the conversion of this carbon into methane.

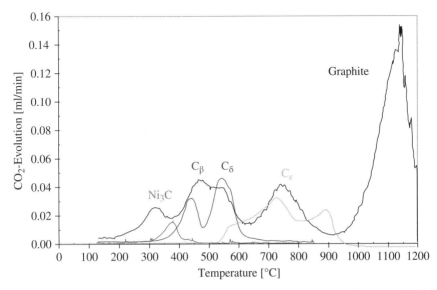

FIGURE 4.9 CO$_2$ evolution during temperature programmed reaction with steam (TPSt) in a micro-fluidized bed of used methanation catalyst (thick line) and several reference samples (bulk nickel carbide, amorphous and graphitic carbon, platelet and non-catalyst carbon) [90].

Therefore water was successfully applied as a soft oxidation mean [84, 90] which, like hydrogen, if present in larger amounts in methanation reactors might be useful for at least partial catalyst regeneration in combination with hydrogen. McCarty and coworkers [84] systematically produced carbon depositions from ethene and ethyne on a steam reforming catalyst and analyzed the samples in situ with both hydrogen (100%) and steam (3% in He). In a more recent study [90], a micro-fluidized bed was used to analyse catalyst and reference samples with 24% steam in argon (see Figure 4.9). The reference sample for bulk nickel carbide was produced by converting nickel nano-powder with ethylene (33 mol%) and hydrogen (67 mol%) at 300 °C and identified by XRD. Amorphous and graphitic carbon were produced by converting nickel nano-powder with ethylene (33 mol%) and hydrogen (67 mol%) at 400 °C and analyzed by Raman and EELS. Reference material for platelet and non-catalyst carbon was produced by converting nickel nano-powder with methane at 875 °C and analyzed by HRTEM-EDX.

Table 4.5 gives an overview of the temperature where the respective carbon depositions react with hydrogen or steam. It can be observed that, generally, steam is more reactive with respect to splitting carbon–carbon bonds, allowing better resolution of the polymerized carbon depositions at lower temperatures. For the reaction with single carbon atoms, adsorbed carbon and bulk nickel carbide, TPR is advantageous as it turns the catalyst active and therefore converts the carbon atoms at lower temperature than possible with steam.

TABLE 4.5 Overview of the Temperatures of Maximum Reaction Rates of the Respective Carbon Depositions with Hydrogen or Steam.

Carbon Type	Identification	Temperature at Maximum Reaction Rate with Hydrogen, °C [79]	Temperature at Maximum Reaction Rate with Hydrogen, °C [84]	Temperature at Maximum Reaction Rate with 3% Steam, °C [84]	Temperature at Maximum Reaction rate with 24% Steam, °C [90]	Temperature at Maximum Reaction Rate with 10% Oxygen, °C [90]
$C\alpha$	Adsorbed carbon	200	200	320–350	–	–
$C\beta$	Amorphous	400	380–390	320–380	450	400–500
$C\gamma$	Bulk nickel carbide	275	275–310	320–380	300–350[a]	350–400[a]
C_ν, $C_{\delta'}$	Whiskers	400–600	500–600	520–570	–	–
C_C, C_δ	Encapsulating	550–850	700	570	550	450–550
$C\varepsilon$ (C_C)	Platelet		830	625–650	750	–
G	Noncatalyst carbon	–	–	>950	900	–
	Graphite (binder)	–	–	–	>1100	>730

[a] Depends on size of crystallite/bulk nickel carbide domain.

4.2   METHANATION REACTOR TYPES

As described in Section 4.1.2, a broad variety of input gas mixtures for methanation reactors has to be considered. Different gas sources (e.g., producer gas from varying gasification steps or hydrogen-rich mixtures for power to gas applications) and different process chains (e.g., cold versus hot gas cleaning) lead to different gas compositions. As will be shown, already this has a strong influence on the choice of the reactor type. Moreover, also the amount of available feedstock and therefore plant capacity will have an impact on reactor choice. On the one hand, each reactor type is connected to a different size dependence of the ratio between capital costs (CAPEX) and operating costs (OPEX) and, on the other hand, offers different options for obtaining coupled products or by-products. While for example, in small-scale biomass based SNG processes the usability of the produced heat may be of lesser importance, in the big coal gasification based plants, energy and heat integration and therefore overall efficiency plays a dominant role.

Due to the strongly exothermic character of the methanation reaction and the equilibrium limitation at higher temperatures, controlling the temperature inside a catalyst bed is a serious challenge, but of utmost importance. In fact, the methanation reactor types developed so far can be differentiated according to the heat removal concepts applied.

When the concentration of carbon (mon-)oxide is high, the conversion and therefore the local heat production will be high and easily higher than the possible local heat removal. In such cases, the temperature will rise, which leads to even higher reaction rates and heat production (temperature run-away). This is shown in Figure 4.10a where the heat production increases exponentially with rising temperature while the cooling rises only linearly. This results in two stationary operation points where cooling and heat production are equal. However, only the lower operation point is stable, as small deviations from it are corrected by the system. A coincident increase of the temperature favors the cooling which leads the system back to the stable operation point at lower temperature and vice versa. The upper operation point is however unstable as small deviations in temperature cause the system to either move into the lower operation point or to turn into temperature run-away.

As recently shown in a modeling study [91] (see also Section 4.3.2 of this chapter), this temperature run-away cannot be avoided even if means such as modest catalyst dilution or specialised catalyst supports with very high convective heat transfer performance (so-called closed cross-flow structures; see [92, 93]) are applied. The temperature and the conversion rise then in a short length of the reactor until thermodynamic equilibrium conversion is reached. For stoichiometric mixtures of hydrogen and carbon monoxide, temperatures significantly above 700 °C are easily reached. For typical fixed bed nickel based methanation catalysts, such high temperatures endanger catalyst stability and have therefore to be avoided. This is usually achieved by ballast gas (often by recirculation of reaction products) which dampens the temperature increase due to a higher product of mass flow and heat capacity (see Figure 4.10b).

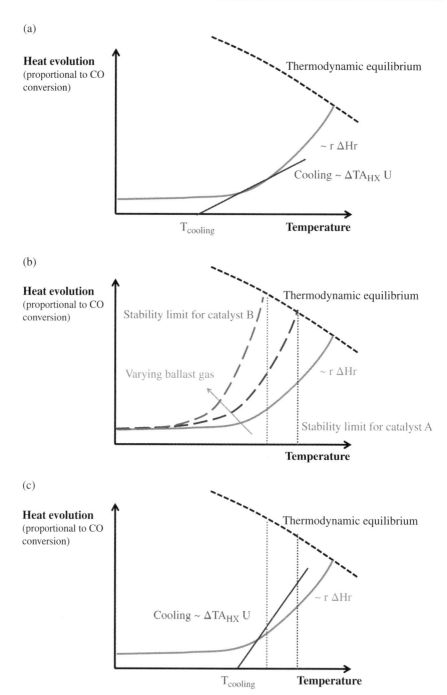

FIGURE 4.10   Heat evolution and heat removal in methanation reactors: (a) fixed bed reactor, (b) fixed bed reactor with ballast gas, (c) fluidized bed, bubble column or metal monoliths.

A further concept to limit the reactor temperature is to increase the useable cooling surface drastically. The nearby option to decrease reactor tube diameter is limited in practical application, as too small reactor tubes (and therefore a higher number of them) would lead to higher pressure drop, higher capital costs and longer shut down times during catalyst exchange. It is however possible to achieve a relatively large effective heat transfer area by dispersing the heat beyond the local volume where the exothermic reaction is taking place. This can be achieved by introducing highly conductive parts into the reactor (e.g., metal monoliths) or by moving the hot catalyst particles throughout the reactor (e.g., in fluidized beds or three-phase bubble columns) to exploit the complete available heat transfer area inside the reactor (Figure 4.10c).

4.2.1 Adiabatic Fixed Bed Reactors

Adiabatic fixed bed reactors are relatively simple with respect to construction and control. This is one reason why this concept has been adopted for all large scale coal to SNG plants so far. As mentioned above, in these reactors, the temperature and the conversion rise until thermodynamic equilibrium conversion is reached. To overcome the equilibrium limitation, the gas stream leaving the reactor has to be cooled again, before the next reactor is entered where again conversion and temperature rise until equilibrium is reached. As shown in Figure 4.11 [94], this way full conversion can be reached with a series of adiabatic reactors with intermittent and recirculation cooling.

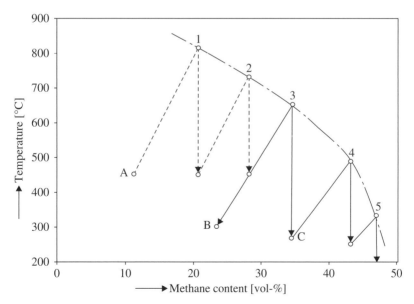

FIGURE 4.11 Temperature profiles in series of adiabatic fixed bed reactors. Dashed lines: without product gas recycle. Full lines: with product gas recycle. Dash-dotted line: equilibrium limitation. Adapted from [94]. *Source*: Harms 1980 [94]. Reproduced with permission of Wiley.

Especially for large scale plants, the high temperatures of the gas stream entering the cooling between the reactors is a second advantage, because very good steam parameters can be reached which facilitates efficient energy integration and electricity production. Still, [95] showed by modeling that, due to thermal inertia, locally transient temperature overshoot could occur during load changes.

Although there are a couple of differences in the details, large scale coal to SNG plants generally follow the same scheme [48] (see Figure 4.12): coal is gasified close to the mine mouth, then de-dusted, cooled and further compressed. Part of the raw gas (still containing hydrogen sulfide and other impurities) enters a sour gas shift section, where steam addition and a molybdenum sulfide based catalyst convert part of the carbon monoxide to carbon dioxide, producing hydrogen to adjust the hydrogen to carbon monoxide ratio in the reactor feed to a slightly over-stoichiometric ratio of about 3.0–3.1. In the next step, all the gas is cleaned in a physical washing step (such as Rectisol®) where at low temperatures (–40 °C) and high pressures (20–65 bar) practically all species besides methane, hydrogen, and carbon dioxide are removed by the solvent (e.g., methanol in Rectisol®, dimethyl ethers of polyethylene glycol in Selexol®). In one step, the CO_2 from the gasifier and the sour water gas shift section, olefins such as ethylene, ammonia, hydrogen sulfide, all higher hydrocarbons, and the remaining humidity are removed, leaving a nearly stoichiometric mixture of hydrogen and carbon monoxide which then is fed to the series of methanation reactors. As mentioned in [96], the impurity concentrations downstream of the Rectisol® scrubber are 180 ppmv C_2H_4, 750 ppmv C_2H_6, 10 ppmv C_3H_6, and 14 ppmv C_3H_8, respectively, while the total sulfur content is 0.08 mg m$^{-3}_N$ (to which H_2S contributes with 0.04 mg m$^{-3}_N$).

As shown in [48], processes from coal to SNG have been developed during the 1970s at many different places. From these more than 12 different developments, so far three concepts have been realized on a commercial scale. Lurgi built a lignite to SNG plant in the United States during the 1980s, Davy Process Technology and Haldor Topsoe A/S recently erected and commissioned plants in the only emerging market for coal to SNG plants right now, in China. Further, Foster–Wheeler and Clariant recently developed a slightly modified coal to SNG process, for which a demonstration plant in China is under construction [97]. Finally, Haldor–Topsoe adapted their adiabatic fixed bed reactor process for the specific conditions in the biomass to SNG field. In the following, the main reaction engineering aspects of some of these developments are briefly discussed.

4.2.1.1 Lurgi In the 1960s and 1970s, Lurgi developed a coal to SNG process based on their successful updraft fixed bed coal gasification (Lurgi mark IV). The methanation unit includes two adiabatic fixed bed reactors with internal recycle (and a final trim methanation step when necessary). One pilot plant was designed and erected by Lurgi and SASOL in Sasolburg (South Africa) using a syngas side-stream from the Fischer–Tropsch plant [98]. The synthesis gas was produced by a commercial coal gasification plant, cleaned by a Rectisol® scrubber and conditioned by a water gas shift step. A second pilot plant was erected by Lurgi and EL Paso Natural Gas

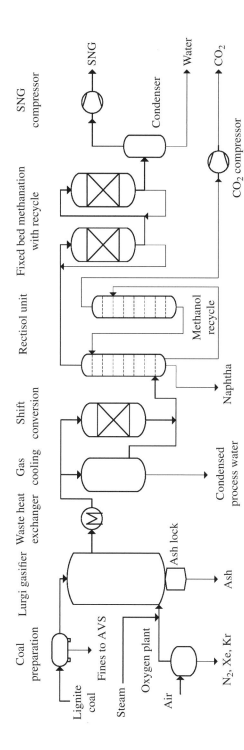

FIGURE 4.12 Block scheme of the Lurgi lignite to SNG process as built in Great Plains, North Dakota; from [48]. *Source*: Kopyscinski 2010 [48]. Reproduced with permission of Elsevier.

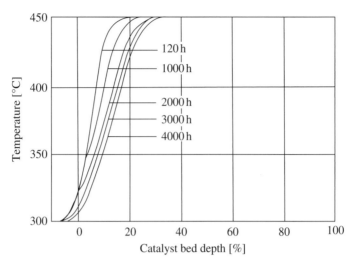

Catalyst bed depth [%]

FIGURE 4.13 Lurgi process with adiabatic fixed bed methanation reactor: temperature profile for the first fixed bed reactor (reprinted from [48], adapted from [99]). *Source*: Kopyscinski 2010 [48]. Reproduced with permission of Elsevier.

Corporation in Schwechat (Austria) converting synthesis gas from naphtha to methane. The pilot plants were operated for 1.5 years demonstrating the stability of the G-185 methanation catalyst developed by BASF for 4000 h. The adiabatic equilibrium temperature of 450 °C was reached after 20% of the catalyst bed for the fresh catalyst, and after 32% of the catalyst bed at around 4000 h time on stream; see Figure 4.13. This slight deactivation can be explained by the fact that the H_2 chemisorption decreased by approximately 50% and the nickel crystallite sized increased from 40 to 75 Å [96]. Based on these results, the first commercial large scale lignite to SNG plant in Great Plains, North Dakota, was erected. The plant is run since 30 years by the Dakota Gas Company and comprises 14 Mark IV gasifiers in parallel. Since 1999, the CO_2 separated in the Rectisol® scrubbing is fed to an oil field in Canada for enhanced oil recovery (EOR) [99–101].

The experimental condition and gas compositions for a typical run of the pilot plant are summarized in Table 4.6.

Recent publications indicate that Lurgi, now integrated into the Engineering and Construction division of Air Liquide, still or again is active in further developing the process, especially with regard to the new challenges when methanation is used for power to gas applications and when converting biomass [102].

4.2.1.2   TREMP (Haldor Topsoe A/S)   While most nickel based methanation catalyst suffer severely from temperatures above 600 °C, Haldor Topsoe A/S invested their broad knowhow in nickel based catalysts on developing a new methanation catalyst with significantly higher temperature stability (around 700 °C). This catalyst is at lower temperatures (suitable for methanation, e.g., reactor inlet 300 °C) way more active than nickel based steam reforming catalysts. Further, it keeps that

TABLE 4.6 **Operational parameters and gas composition at the pilot plant of the Lurgi process [96].**

	Feedgas	Fixed bed reactor R1		Fixed bed reactor R2	
		inlet	outlet	inlet	outlet
Temp., °C	270	300	450	260	315
Gas flow rate (wet), m^3_N/h	18.2	96.0	89.6	8.2	7.9
Dry gas composition, vol%					
H_2	60.1	21.3	7.7	7.7	0.7
CO	15.5	4.3	0.4	0.4	0.05
CO_2	13.0	19.3	21.5	21.5	21.3
CH_4	10.3	53.3	68.4	68.4	75.9
C_2+	0.2	0.1	0.05	0.05	0.05
N_2	0.9	1.7	2.0	2.0	2.0

Source: Moeller FW, Ros H, Britz B. Methanation of coal gas for SNG. Hydrocarbon Processing. 1974; 69–74.

activity at low temperatures even when the catalyst has experienced high temperatures up to 700 °C before [103] which is important when the temperature profile slowly moves downstream due to slow catalyst deactivation. The catalyst (MCR-2X) was described at this time as an alumina support with a stabilized micro-pore system to decrease nickel crystallite sintering, with a high nickel surface area and free of alkaline [103].

This temperature-stable catalyst allowed significantly decreasing the recycle gas cooling compared to other reactor concepts and therefore the necessary electricity consumption for the recycle gas compressor. Even more important, much higher steam parameters can be achieved in the first cooler, which leads to higher electricity yield in the energy integration; see Figure 4.14. This so-called TREMP® process ("Topsoes' recycle energy efficient methanation process") was originally developed for the so-called Adam and Eva project [94, 104]. In this project, long distance transport of high temperature heat from a nuclear reactor was considered by combining endothermic methane reforming at the nuclear reactor and exothermic methanation at the industrial site where the high temperature heat was needed or could be used for electricity production. This process concept was investigated by pilot and demonstration plants in Jülich and Wesseling (Germany) in the early 1980s and further developed for coal to SNG plants.

Haldor Topsoe A/S tested their catalyst in laboratory scale and pilot scale plants with respect to temperature stability and catalyst deactivation. Special attention was paid to secure that the predictions of tests in small adiabatic reactors are reliable and not biased [105]. Typical operation conditions of the reported pilot scale campaigns (duration several 1000 h) are H_2/CO ratio of 3, reactor inlet 300–330 °C (to safely avoid nickel tetra carbonyl formation), pressure of 30 bar$_g$ and space velocities of about 15 000 h^{-1} [106].

Figure 4.15a shows pilot scale experiments from 1980 [106]. Starting at 350 °C, a stoichiometric mixture of hydrogen and CO react that fast that equilibrium is reached

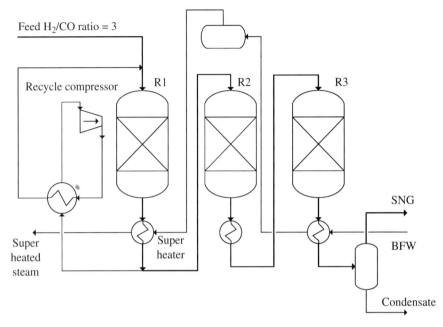

FIGURE 4.14 Scheme of the TREMP® process (*Topsoes's Recycle Energy* efficient *Methanation Process*); from [48]. *Source*: Kopyscinski 2010 [48]. Reproduced with permission of Elsevier.

at 700 °C within only 10 cm reactor length. Due to slow catalyst deactivation, after 2000 h, the temperature gradient is lower, still reaching the same hot spot temperature after around 20 cm. For these tests, thermal sintering is reported as the main cause for the slow catalyst deactivation which caused a loss of the support surface (measured by BET) from values of 52 to 30–35 m² g⁻¹. Further, the hydrogen chemisorption area decreased from values around 8 to 2–3 m² g⁻¹, while the nickel crystallite diameter (determined by XRD) increased from around 20 to 35–55 nm [106]. This leads to a loss of catalyst activity (measured at 250 °C) by nearly a factor of 20 within one year. As the nickel surface (determined by H₂ chemisorption) decreased in the same time by a factor of three to four, it means that the surface specific activity of the catalyst decreased by a factor of five to six which supports the conclusion that methanation is strongly structure sensitive [106].

Meanwhile, the catalyst was slightly modified to further improve the temperature stability and resistance against carbon deposition [107]. Figure 4.15b from [108] shows results from pilot experiments from 2010. The modified catalyst keeps its activity for more than 1000 h with only negligible deactivation and without showing signs of carbon deposition (amorphous carbon C$_\beta$ or "gum"). The characterization of the catalyst indicated that, within 2000 h, the BET area decreased from 45 to 22 m² g⁻¹ and nickel crystallite size (determined by XRD) increased from 15.5 to 22 nm. The normalized nickel surface (measured by sulfur adsorption) decreased to 29% of the fresh catalyst value; similarly the normalized activity was reduced by 69%.

It is concluded again that thermal sintering was the cause of catalyst deactivation, but that it proceeded so slowly that industrial operation would be possible for more than two years [108]. The reason is visible in Figure 4.15b: despite the significant loss of activity, the temperature profile hardly changes, which allows the conclusion that the reactor performance depends mainly on heat and mass transfer and not on catalyst activity under the chosen operating conditions.

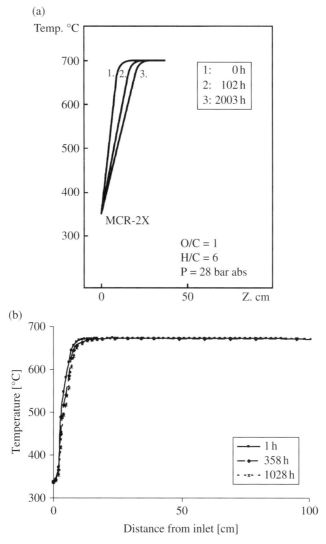

FIGURE 4.15 Temperature profiles from TREMP® pilot scale experiments with H$_2$/CO = 3 at 28/30 barg. (a) Data from 1980 [106]. (b) Data from 2010 [108]. *Source*: Nguyen 2013 [108]. Reproduced with permission of Elsevier.

With the emerging need for clean coal technologies in China, in the last few years, three coal (or coke oven gas) to SNG plants have been commissioned in Wuhai (Inner Mongolia), Quinghua (Xinjiang region) and Keshiketeng (Inner Mongolia); further projects are in the pipeline [109]. Of the mentioned plants, the first two are based on the TREMP process. The third one (Datang Kequi project) is erected by Davy Process Technology, a full subsidiary of Johnson Matthey plc, which developed the applied CRG methanation catalyst.

4.2.1.3 VESTA

4.2.1.3 VESTA While the Lurgi process decided for relatively high recirculation rates to control the temperature in the first reactor and Haldor Topsoe decreased the recirculation rates by means of a more temperature stable catalyst, Foster Wheeler and Clariant (catalyst supplier) developed for their VESTA process a different concept without recirculation [97]. They propose to separate the CO_2 only after methanation, therefore using it as ballast gas which dampens the temperature increase in all the adiabatic methanation steps. Additionally, steam is added as thermal ballast gas which also suppresses coke formation. The concept envisages a sulfur removal step followed by a high temperature water gas shift reactor and three reactors in series with intermittent cooling, but without recirculation cooling. For each of the methanation reactors, the steam addition is designed such that the outlet temperature is 550 °C, allowing for decent steam parameters, on the one hand, but simplified construction and therefore lower CAPEX on the other hand [97].

4.2.1.4 GoBiGas Although the GoBiGas project is presented within this book (see Chapter 6) as the biggest biomass to SNG project right now, the design of the methanation section will be discussed here, as it deviates from the standard TREMP scheme due to the special boundary conditions of biomass processes. It should be noted that the plant is still in commissioning (full operation was reported late in 2014 [110], but no details are known so far); therefore, here the design only, not real results, is discussed.

The main reason for changing the plant design from the original TREMP scheme is the fact that Rectisol® scrubbing and similar units are commercially not viable in biomass to SNG plants, as the latter are inherently small (10–50 MW inland, up to 200 MW at coasts) due to the logistic challenge in obtaining enough biomass feedstock. Therefore, thiophene and ethylene, which are present in the producer gas of the installed allothermal dual fluidized bed gasifier, cannot easily be removed without converting them. In the GoBiGas plant, a molybdenum sulfide based catalytic reactor (so-called olefin hydrogenation) followed by an amine scrubber to remove H_2S is installed. This way, most of the thiophene and parts of the ethylene are converted while half of the CO_2 and the complete H_2S are removed.

In the downstream shift reactor, the hydrogen to carbon monoxide ratio is adjusted increasing again the CO_2 content; refer to Figure 6.5 of Chapter 6 in this book, which shows a block flow diagram of the plant. The CO_2 helps to dampen the temperature increase in the following so-called pre-methanator where the major part of the CO is converted and the equilibrium is reached at a temperature of 455 °C [107]. While the CO_2 is completely removed in the following CO_2 scrubber, the remaining hydrogen, methane and carbon monoxide are sent to a series of three adiabatic fixed bed reactors with intermittent cooling. After condensation of the produced water, a last trim reactor converts the final remaining CO_2 to methane to secure the necessary high methane content to reach the gas grid specification [107]. Using the CO_2 as ballast gas in the pre-methanator helps to limit the reaction temperature in order to favor hydrogenation of the residual ethylene content before it causes carbon deposition at higher temperatures. This way, the usual high steam parameters can no longer be achieved in the intermittent cooling sections, which however might be less important in such a small plant with several scrubbing sections that need thermal regeneration.

4.2.2   Cooled Reactors

As discussed above, cooling of the catalytic reactors can help to overcome limitations by thermodynamic equilibrium (in the case of exothermic reactions such as methanation and water gas shift), to positively influence selectivity and to avoid or at least limit catalyst deactivation. Depending on the desired temperature level, different cooling media can be chosen. While steam production inside the reactor has a relatively broad range of potential operation temperatures (just coupled to the pressure level), the use of thermo-oils is limited to a range of maximum of 350–400 °C; the latter due to the thermal degradation of the thermo-oils. At higher temperatures, cooling media such as molten salts have to be considered. Cooling with gas is not applied in the very exothermic methanation reactors because of the necessary high volumetric flow rates and heat exchanger areas caused by the low gas volumetric heat capacity and the low gas-side heat transfer coefficient, respectively. Regarding the reactor types, cooled fixed bed reactors and multiphase reactors with moving catalyst particles have to be distinguished.

4.2.2.1   *Cooled Fixed Bed Methanation Reactors*   Cooling of fixed bed methanation reactors is accomplished: (i) by inserting cooling coils or tubes into the random bed of catalyst particles, (ii) keeping the catalyst inside tubes and the cooling medium on the shell side, or (iii) by coating the catalyst to surfaces which are cooled from the other side.

Examples for reactors with catalyst on the shell side are, for example, Linde's isothermal reactor concept [111] (see Figure 4.16) or the very recent concept of etogas GmbH, Stuttgart [112]. In their methanation reactors proposed for power to gas applications, hollow structured plates with cooling medium flow are inserted into the pressurized vessels, which leads to lower capital costs. The challenge in this reactor type is to ensure equal distribution of the gas flow over the reactor cross-section such that no preferential pathways develop and all heat transfer area is used efficiently. Optimal packing of the catalyst particles has therefore to be checked during each filling of the reactor with catalyst particles. Since even a few centimeters distance between the heat transfer areas can lead to significant hot spots in the catalyst bed between the two areas, the space between the heat transfer coils/plates is relatively narrow which does not facilitate careful packing of the particles.

Steam   Syngas

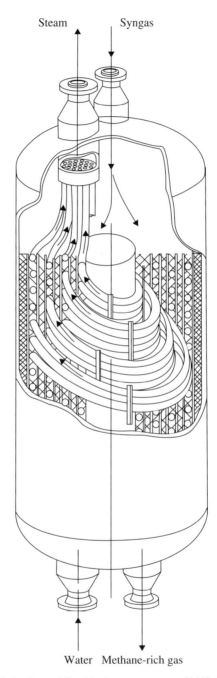

Water   Methane-rich gas

FIGURE 4.16   Linde's isothermal fixed bed reactor concept [111]. *Source*: Lohmüller 1977 [111]. Reproduced with permission of Elsevier.

Catalyst Inside Cooled Tubes   By far the most common type of cooled fixed bed reactors are multi-tubular reactors, where the catalyst is filled into the tubes while the cooling medium is on the shell side. Such reactors are used in partial oxidation reactions (e.g., maleic or phthalic anhydride) or Fischer–Tropsch processes (e.g., Shell's middle distillate synthesis) and can comprise several tens of thousands of parallel tubes filled with catalyst. Although catalyst change is not simple in such reactors, as an equal pressure drop in all tubes has to be ensured to avoid preferential pathways, the advantages such as scalability, large heat transfer areas, and relatively short transport distances within the bed to the next heat transfer area prevail. In consequence, a wealth of scientific literature exists that covers heat transfer correlations for all common catalyst shapes, modeling, optimization of the pressure drop, and the replacement of random catalyst beds by structured catalyst supports [115–123].

In the methanation for SNG, this reactor concept has so far been applied by ZSW/etogas in the Werlte power to gas pilot plant (see Chapter 7 in this book) and by polytropic fixed bed methanation (see [88] and Chapter 11 in this book). Recent modeling studies [91, 122] and experimental results [123] show that, in fixed bed methanation reactors, thermal run-away is difficult to avoid due to the exothermicity, even if cooling is applied. This is shown in Figure 4.17 where the maximum temperature predicted by a one-dimensional pseudo-homogeneous model for a cooled fixed bed methanation reactor is plotted as a function of cooling temperature and tube diameter. One can see that stable operation is possible only below 300–350 °C (where yield is unsufficient) and above around 700 °C (where is thermodynamic equilibrium is reached).

In case of polytropic fixed bed methanation, this is used to create a high temperature spot at the reactor inlet that helps to convert hydrocarbons. There, the temperature increase is limited to around 500 °C by the ballast gases (steam, methane, carbon dioxide) which are part of the fed producer gas from allothermal biomass gasification.

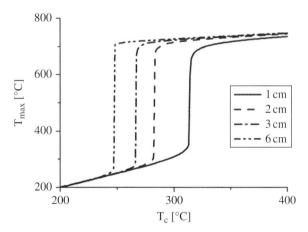

FIGURE 4.17   Maximum temperature predicted by a one-dimensional pseudo-homogeneous model for a cooled fixed bed methanation. *Source*: Schlereth 2014 [122]. Reproduced with permission of Elsevier.

This way, catalyst deactivation by thermal sintering is limited. Rather, to simplify the gas cleaning sections, slow catalyst deactivation is accepted which is mainly a function of the content of olefins, tars, and sulfur species.

Multiple Feed Injection   In the CO_2 methanation reactors used for power to gas applications (see Chapter 7 in this book), besides staged cooling, also multiple feed injection is applied to limit the temperature rise in the hot spot. Still, [91] showed in their modeling study that the feed distribution has to be optimized carefully. They simulated a reactor for stoichiometric CO_2 methanation where the hydrogen is fed at the inlet and the CO_2 is distributed over several injection points. Too high CO_2 addition at too low CH_4 content caused temperature run-away. Therefore the addition of the CO_2 has to be realized such that the CH_4 content produced already upstream is high enough to serve as an efficient ballast gas and to dampen the temperature increase. Reference [122] modeled a membrane reactor with continuous addition of CO_2 to the H_2 flow and found that, by this, nearly isothermal conditions can be achieved if the operation conditions are chosen such that accumulation of CO_2 (and thus the exothermal potential) at the reactor beginning is controlled.

Applying Structured Catalyst Supports   Besides multiple feed injections, also improvement of radial heat transport in methanation reactors by applying structured catalyst supports is investigated. Reference [91] studied in a pseudo-homogeneous one-dimensional model the application of so-called closed channel cross-flow structures (see Figure 4.18) as catalyst supports which direct the fluid flow against the cooling wall and minimize the thickness of the laminar film which is the main heat transfer resistance in the system. These structures show a 30–55% higher heat transfer coefficient than beds of catalyst particles in single phase flow [92] and even 60–100% in gas–liquid flow [93] and therefore higher space–time yield in other exothermic reaction systems (e.g., Fischer–Tropsch reactor simulation [124]). However, they do not allow avoiding a temperature run-away when stoichiometric methanation feed mixtures are applied. Still, compared to a fixed bed of catalyst particles, an increased flexibility with respect to cooling temperature is observed. The lower

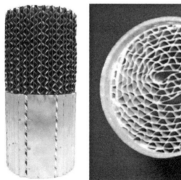

FIGURE 4.18   Closed cross-flow structures (left), rolled metal monoliths (right).

catalyst hold-up (voidage of about 85%) and the better heat transfer coefficient of these structures allow at similar maximum temperature a higher overall temperature level (inlet and cooling temperature) and therefore a higher productivity per catalyst mass. The high voidage causes a significantly lower pressure drop which is well defined by the type and number of structures applied. Compared to fixed beds of particles where equal pressure drop over each tube has to be secured individually during catalyst change, the relatively simple exchange of a certain number of structures may shorten shut-down time.

Highly Conducting Metal Monoliths as Catalyst Supports   While the application of cross-flow structures focuses on directing the convective flow against the wall, the use of metal monoliths as catalyst supports aims at exploiting the high thermal conductivity of either aluminium or copper. The concepts has been developed and experimentally proven by E. Tronconi's group [119, 125–127]. The challenges are the necessary relative high metal content in the reactor, a tight contact between metal monolith catalyst support and the reactor tube inner wall and the therefore needed high catalyst stability. The same group simulated the use of such metal monoliths as catalyst supports for both CO and CO_2 methanation [128]. They found that, with a metal (aluminium) volume fraction of 0.25 resulting in a radial heat transfer coefficient of $200\,W\,m^{-1}K$, it is possible to limit the temperature increase in a ten-inch (25-cm) tube to below $200\,K$.

While the simulation study [128] applied metal monoliths with square channels, the Engler-Bunte-Institut at Karlsruhe Institute of Technology aim at rolled metal monoliths (known from exhaust gas cleaning) due to the easier large-scale production. Based on simulations and their own experiments, they state [129] they have achieved similar hot spots in a cooled fixed bed reactor and a metal monolith reactor of around four times higher diameter. As they aim at de-centralized application for power to gas applications (both biogas and gasification-derived producer gas), this would reduce the number of necessary tubes, simplify reactor construction, and reduce capital costs.

Plate Reactor Concepts   The maximum possible heat transfer can be achieved by coating the catalyst on a plate which is cooled from the back. This way, the only heat transfer mechanism is conduction through the relatively thin walls, and an overall heat transfer coefficients of several $1000\,W\,m^{-2}K$ can be achieved [130]. Further also in axial and/or lateral direction, the thermal conductivity of the support materials helps to even the temperature profiles. The concept is well known and has been applied successfully for autothermal reforming and Fischer–Tropsch, for example, by Velocys Ltd. [131] where it helps to keep the system isothermal and to keep C5+ selectivity high. Only once so far has it been applied for methanation, and only on a microreactor scale, such that no conclusions for a technical application can be drawn [132]. Therefore, it stays unclear which advantages the concept may have over other isothermal reactor concepts (e.g., fluidized beds), while a stable catalyst coating on expensive microchannel structures and avoiding catalyst poisoning is an important challenge.

For controling catalyst deactivation, the cooled reactor concepts have the advantage of avoiding too high temperature peaks (hot spots) which decreases the rates of thermal sintering. The most important threat to nickel catalyst, that is, poisoning by sulfur species, is more or less independent of the temperature in the temperature range favorable for methanation; for any methanation process with nickel catalyst, the sulfur species concentration has to be minimized by upstream gas cleaning. An important drawback of all fixed bed methanation reactors, and therefore also of the cooled fixed bed reactors, is vulnerability of the catalyst to carbon and coke formation by unsaturated species, especially olefins. Recently, [59] showed that even very low amounts of ethyne (acetylene, C_2H_2) lead to severe carbon formation and catalyst deactivation already at 300 °C. Keeping in mind that the dry gas concentration of ethyne can have values up to 0.5 vol% in the producer gas of allothermal dual fluidized bed gasification, gas cleaning upstream of cooled fixed bed reactors has to either convert or remove these unsaturated species, or the process concept has to accept a certain level of continuous catalyst activity loss.

4.2.2.2 Multiphase Reactors with Moving Particles

As mentioned in the introduction to this chapter part, dispersing the heat over a larger reactor volume, such that a significantly increased heat transfer area can be used, is another option to reach sufficient heat removal and to allow for nearly isothermal operation. As the heat of the exothermic reaction is released on the catalyst surface, the best option to move the heat source is to move the catalyst particles. Depending on the movement of the particles, this concept can outperform other heat transport mechanisms, such as conduction. Moreover, the movement of particles in a fluid also increases the heat transfer to cooling surfaces, because laminar films at these surfaces are efficiently disturbed by the moving particles and fluid. Technically, two main forms have been developed: (i) bubbling fluidized bed methanation reactors where the catalyst particles are suspended in the upwards flowing reacting gas, and (ii) bubble columns where an inert liquid, in which both particles and gas are suspended, is used to increase the thermal inertia of the system and to improve the heat transport.

Fluidized Bed Methanation Reactor Developed by the Bureau of Mines   The first development of a cooled fluidized bed methanation reactor was started as early as 1952 [133, 134]. It was a relative slender reactor of low diameter (one inch; 2.5 cm), with > 1 m bed height, which was cooled by thermo-oil in a jacket and filled with Geldart A type catalyst particles. The feed consisted of hydrogen/carbon monoxide mixtures; the operation conditions were 20 bar and around 400 °C. When fed only from the bottom, relative strong axial temperature profiles (up to 100 K) were observed. To achieve more isothermal conditions, in a second version, three feed injections were used.

Considering the very small diameter, one can speculate that axial particle transport was quite limited, because the necessary development of a (usually more central) particle rising zone in the wake of rising bubbles and a down-coming zone is most probably hard to imagine. Rather, slugging has to be expected in such a thin reactor tube at the chosen operation conditions; therefore sufficient axial transport of heat throughout the bed was not possible.

Bi-Gas Project   The next fluidized bed methanation reactor development was conducted within the Bi-Gas project that was initiated in 1963 by Bituminous Coal Research Inc. (BCR, USA). A 150 mm diameter reactor with about 2 m bed height and two bundles of immersed heat exchanger tubes (about $3\,m^2$ heat exchanger surface) was operated for several 1000 h [135–140]. It had to feed injection points, one at the bottom and the second between the first and the second heat exchanger tube bundle. Typical operation conditions were temperatures from 430 to 530 °C, pressures from 69 to 87 bar and a catalyst charge of 23 to 27 kg. The fluidization numbers (i.e. the ration between actual volumetric flow and the minimum one to lift the particles to enable the fluidization mode) were chosen from 8 to 18, which can be considered as a strongly bubbling fluidized bed. While a high CO conversion (up to 99.3%) was reached, the catalyst produced (besides the main product CH_4) considerable amounts of CO_2 and ethane, that is, it was also active for water gas shift reaction and ethane formation.

Comflux Process   The biggest fluidized bed methanation reactors so far have been erected and operated within the Comflux project in Germany from 1975 to 1986 [141–147]. Thyssengas GmbH aimed at a process from coal to SNG that should deliver SNG at 10% lower costs than fixed bed methanation concepts. The lower costs should be achieved by applying only one reactor for combined water gas shift and methanation, therefore allowing for significantly lower equipment and capital costs (–30%). They chose a bubbling fluidized bed with recycle cooling and immersed vertical heat exchanger tubes, in which steam was raised (see Figure 4.19).

The technology development was based on a broad research program at the Engler-Bunte-Institut of the University of Karlsruhe, which conducted laboratory scale research (see, e.g., the isothermal reactor in Figure 4.20) on catalyst deactivation, kinetics, attrition resistance of catalysts and methanation in the presence of sulfur species [144–147].

The attrition test (both cold with air and hot with hydrogen) showed that attrition increases with increasing flow rates and is a function of the gas distributor plate at the bottom of the reactor, which should avoid jet formation. Further, it was found that commercial nickel catalysts are available with attrition of not more than 0.04 wt% per day, far below the limit of 1 wt% per day, which was considered as an economic limit. The work on methanation in the presence of up to 140 ppm of H_2S with nickel molybdenum (NiMo) and nickel tungsten (NiW) based catalysts, both in metallic and sulfidic form, showed that these generally allow methanation in the presence of sulfur species, but that high pressures and very high temperatures (600–750 °C) are necessary to reach equilibrium conversion.

Based on the promising results in the laboratory scale reactor, see Figure 4.20, a pilot plant (diameter 40 cm) was engineered by Didier Engineering (Essen, Germany) and operated from 1977 to 1981. The results obtained there lead to the engineering (again by Didier Engineering), erection and successful operation of a 20 MW$_{SNG}$

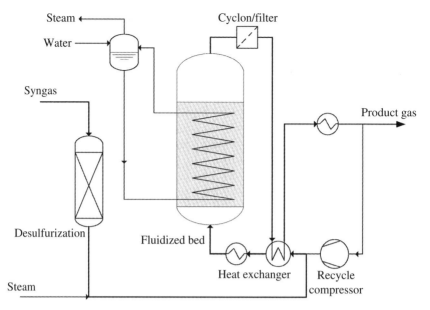

FIGURE 4.19 Thyssengas process flow diagram, adapted from [143]. *Source*: Lommerzheim 1978 [143]. Reproduced with permission of Elsevier.

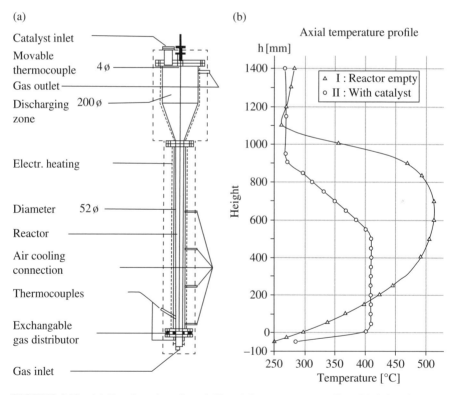

FIGURE 4.20 (a) Bench scale unit and (b) axial temperature profile with (triangles) and without (circles) catalyst, adapted from [146].

demonstration plant (inner diameter 1 m) on the site of Ruhrchemie in Oberhausen, Germany, which was financially supported by the German Ministry of Research and Technology. A commercial plant was expected to have about 3 m diameter. The scale-up was supported by extended cold flow tests in the University of Erlangen, Germany. After several tests, including long duration tests up to 2125 h, the work was discontinued when the price of oil dropped in the mid-1980s. Still, it could be shown that the reactor concept allows converting even under-stoichiometric mixtures of hydrogen carbon monoxide by steam addition in one practically isothermal step to methane rich gases with a very broad load range (30–100%) and without important catalyst deactivation or attrition issues while producing valuable high pressure steam. Table 4.7 gives the dimensions and operation conditions for the systematic experiments which were conducted in the pilot and in the demonstration unit, respectively.

Paul Scherrer Institut   Based on the technically successful development in the Comflux project, Paul Scherrer Institut (PSI) in Switzerland chose the bubbling fluidized bed reactor type as methanation step in the process chain from wood gasification to SNG. During the research, it was demonstrated that, besides its simplicity, this reactor type has advantages for converting the specific gas composition of gasification-derived producer gas.

TABLE 4.7   Experimental Conditions for the Thyssengas Pilot Plant and Demonstration Plant. Data from References [141, 143].

	Pilot Scale Plant	Demonstration Plant
Engineering	Didier	Didier
Operation years	1977–1981	1982–1983
Diameter [m]	0.4	1.0
Bed height [m]	2–4	2–4
Amount of catalyst [kg]	200	1000–1800
Particle diameter [μm]	50–250	10–400
Temperature [°C]	300–500	420–550
Pressure [bar]	20–60	20, 30, 60
Gas mixture	CO, H_2, (H_2O)	CO, H_2, (H_2O)
CO feed flow [m³/h]		1000, 1750, 2500
H_2/CO ratio	2–3	2.0, 2.5, 3.0
Maximum flow of SNG [m³/h]	400	3000
Recycle/feed ratio	0–2	0, 0.1, 0.3
Steam pressure [bar]		70, 90, 120
Maximum steam temperature [°C]		475 °C
Duration of tests [h]	>1000	Up to 2125
Special aspects		Addition and removal of catalyst during pressurized operation

In the process chain from dry biomass to SNG, the application of low temperature gasification (around 850 °C) is advantageous, as the low gasification temperatures favor the production of methane in the gasification. Producer gas of typical allothermal low temperature gasification contains around 10% (dry basis) methane. Methane not converted by endothermic steam reforming to carbon oxides and hydrogen in the gasification means that less feedstock has to be burned to heat the gasification step. Further, less carbon oxide has to be converted by exothermic methanation reaction in the methanation step. This in turn decreases the loss of chemical energy content and therefore the overall chemical or cold gas efficiency, that is, the ratio between the lower heating value (LHV) of the produced SNG mass flow to the LHV of the gasified wood mass flow. Compared to process concepts where high temperature gasification with little methane formation is applied, the cold gas efficiency is significantly higher and can reach 60–70%, depending on the water content of the wood and the heat integration.

Unfortunately, the high methane content in gasification producer gas is usually connected to a significant concentration (a small percentage) of ethane, ethene and ethyne and traces of aromatic species which can cause severe carbon deposition on nickel based methanation catalysts. As discussed in more detail in Section 4.1.3.2 and Chapter 8 in this book, the fluidized bed reactor not only provides the simplicity of having one reactor for water gas shift and methanation, but has the nearly unique capability to convert the unsaturated species to saturated hydrocarbons, that is, methane (CH_4) and ethane (C_2H_6), and to limit carbon deposition. This in turn allows significantly simplifying the gas cleaning, because it is not necessary, unlike in the case of fixed bed methanation reactors, to remove or convert the unsaturated hydrocarbons upstream of the methanation.

Deviating from the Comflux concept, the recycle compressor is completely omitted to further simplify the plant. This is possible because gasification producer gas usually is not a stoichiometric H_2/CO mixture, but contains significant amounts of CO_2 and CH_4 which dampen the temperature increase. Further, the under-stoichiometric H_2 to CO ratio asks for the addition of steam, which again works as ballast gas. The operation conditions had to be adapted to the specific challenges of converting the producer gas with the high fraction of C_2 species; see Section 4.1.3.2.

The concept was tested extensively in a laboratory scale reactor of the same dimension as the one shown in Figure 4.20. Under laboratory conditions, the operating conditions were systematically varied while an axially movable sampling tube allowed collecting axial concentration profiles to better understand the interaction of chemical reactions and hydrodynamics and to validate computer models [148–150]. Further, a fully automated set-up of the same dimensions, but a higher pressure range, was used in field tests connected to a slip stream of the commercial wood gasifier in Güssing, Austria. While the gas cleaning, especially removing the thiophenic sulfur species was a challenge, the methanation proved to be very stable and robust, but also flexible with respect to load changes [51]. Finally, after several improvements of the desulfurization, a long duration test over 1000 h was conducted successfully [48].

In a next step, within the European Union funded project "Bio-SNG" (2006–2009) and with substantial support of the Swiss electricity producers, CTU AG

(Winterthur, Switzerland) engineered and constructed with the support of repotec (Vienna, Austria) a 1 MW$_{SNG}$ Process development unit including complete gas cleaning, methanation step, and gas up-grading to pipeline-ready SNG. This plant was erected next to the gasifier in Güssing, Austria and commissioned with support, especially for analytics and process diagnostics, by PSI and TU Vienna. The plant was operated at a load range from 30 to 100% and converted up to 20% of its producer gas to SNG (H$_2$gas quality, Wobbe index = 14.0, HHV = 10.67 kWh m^{-3}N) that was successfully used for field tests with CNG cars [48]. The most important findings of the pilot scale test runs are the flexibility with respect to load range and hydrogen recirculation, the nearly isothermal operation, the fast start-up times (less than 30 min) and the possibility to convert major parts of the ethylene in the producer gas feed to valuable ethane which increases the LHV of the SNG [151].

Ongoing research continues to both deepen and broaden the knowhow on this process in order to consolidate experience and further minimise risks during scale-up. The work focuses on the hydrodynamics of bubbling fluidized beds with vertical internals, catalyst deactivation, kinetics of side reactions, gas cleaning and modeling/simulation. Further, the ability of the reactor concept to convert hydrogen with CO$_2$ to methane within power to gas applications has been shown by modeling and in laboratory experiments and will be tested in (dynamic) pilot scale tests.

Meanwhile, the French project GAYA (aiming at SNG production from lignocellulosic biomass in the 20–80 MW$_{SNG}$ scale) considers allothermal dual fluidized bed gasification and fluidized bed methanation as a promising technology combination [152, 153].

Recent Developments in China As mentioned in Section 4.2.1.2, China is by far the biggest and fastest growing market for SNG production due to the need for clean coal technologies. Besides the construction of coal to SNG plants based on state of the art adiabatic fixed bed methanation (see Section 4.2.1), also domestic research is triggered on both fixed bed and fluidized bed methanation, including pilot and demonstration scale plants [109]. There are several recent publications on fluidized bed methanation investigations at the laboratory scale [154, 155], but also the overview given in [109] mentions the advantages of fluidized bed technology, especially with respect to carbon and coke deposition.

Bubble Columns Introducing a liquid phase into a fluidized bed methanation reactor has a couple of advantages. Due to the high density, the liquid phase acts as a strong thermal ballast medium which efficiently dampens any local hot spot formation and may serve as a heat reservoir to keep the reactor above "ignition" temperature when frequent start-up and shut-down are necessary, for example, in power to gas applications. Moreover, the gas bubbles rising in the liquid induce strong mixing patterns, which increase the high heat dispersion throughout the reactor and the heat transfer to immersed heat exchanger surfaces. Further, as the catalyst particles have to be small to be suspended in the gas–liquid mixture, intra-particle diffusion limitation may not be significant.

On the other hand, this reactor type is usually connected to a few challenges, too. While the catalyst hold-up is about 50–60 vol% in a fixed bed reactor and easily can reach 40 vol% in a bubbling fluidized gas–solid reactor, in bubble columns often relative low values are found, although 40 vol% would be technically possible. At very low solid hold-ups, the available catalyst mass and its intrinsic reaction rate are limiting; however at higher solid hold-ups, the gas–liquid mass transfer becomes limiting (see Figure 4.21), because the catalyst particles are surrounded by liquid and the gas first has to dissolve in the liquid and then diffuse to the catalyst surface. Further, high solid contents may increase bubble coalescence. To increase gas solubility, usually high pressures are applied.

Another challenge is the choice of a proper liquid phase which should have a very low vapor pressure, a high thermal stability and low chemical degradation rates, high solubility for the reaction gases and should not react with the involved species nor deactivate the catalyst. Liquids with high density, low viscosities and surface tension are favorable as they allow for smaller bubbles and therefore higher mass transfer rates [156]. Further, for typical gas–liquid–solid systems such as Fischer–Tropsch reactors, internal recirculation patterns of reactants and products, and in consequence, lower selectivities are reported.

Chem System Inc. (USA) developed in the late 1970s a catalytic bubble column reactor (so-called liquid phase methanation reactor; LPM) to convert syngas produced in a coal gasifier to SNG [158, 159]. The produced gas, mainly methane and carbon dioxide with some unconverted hydrogen and carbon monoxide, was separated in a liquid phase separator and analyzed; as liquid, mineral oil was used. Until 1981, laboratory and pilot scale experiments were conducted, see Table 4.8. Recently, applying catalytic bubble columns was considered for CO_2 methanation in power to gas applications, mainly because of the options for efficient heat management, which is of advantage at dynamically changing and part-load operation [157].

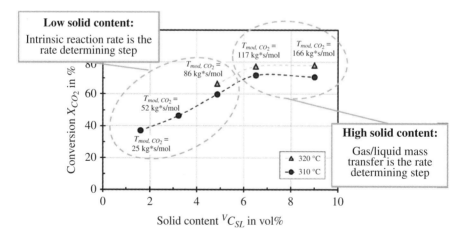

FIGURE 4.21 Influence of catalyst concentration on conversion in a gas–liquid–solid reactor for CO_2 methanation (20 bar, 310–320 °C, particle diameter ≤ 100 µm) for power to gas application [157]. *Source*: Götz M. Recent Developments in Three Phase Methanation. Presentation at the 2nd Nuremberg Workshop Methanation and Second Generation Fuels. June 2014, Nuremberg/Germany.

TABLE 4.8   Liquid phase methanation reactors and operational conditions [158–161].

	Bench scale unit	PDU	Pilot plant
Reactor diameter, cm	2.0	9.2	61.0
Reactor height, m	1.2	2.1	4.5
Gas flow, m_3N/h	0.85	42.5	425–1534
Catalyst bed height, m	0.3–0.9	0.61–1.8	
Catalyst mass, kg			390–1000
Pressure, bar	20.7–69.0		34–52
Temperature, °C	260–380		315–360
H_2/CO	1–10		2.2–9.5
Catalyst size, mm	0.79–4.76		

So far, laboratory scale experiments with reactors of 25 and 55 mm at pressures of 5–20 bar and temperatures from 230 to 320 °C have been conducted [156, 157].

4.2.3   Comparison of Methanation Reactor Concepts

In Table 4.9, the different methanation reactor concepts are compared with respect to a number of aspects which reflect chemical reaction engineering, heat integration/recuperation and costs. The comparison does not claim to identify the best reactor concept; this will always depend on the circumstances like feedstock, applied gasification and gas cleaning technologies, availability and price of utilities and commodities, experience of suppliers and the weighting of the different aspects shown in the table. Therefore, thorough engineering will always be necessary for which the comparison may serve as starting point.

The *simplicity of the reactor* refers to not only the complexity in manufacturing the reactor (i.e., necessary capital costs), but also to the challenges in up-scaling the reactor.

The *number of necessary units* (and therefore capital costs for reactors, compressors and heat exchangers) will be low in cases where full conversion can be reached with one reactor, and high when a series of reactors with intermittent and recycling cooling is applied.

The *temperature level of the cooling* is an important parameter to estimate the quality of the reaction heat to be recovered from the methanation reactor. It decides about the steam parameters and therefore the electricity production that could be reached within the plant and therefore on the operation costs and revenues.

The *flexibility of the reactor concepts* refers to several aspects such as part-load flexibility, dynamic start-up/shut-down and flexible and dynamic addition of hydrogen. All these aspects are not important for stationary coal to SNG plants with high capacity, but very important for methanation units in power to gas applications or in polygeneration concepts, where gasifiers run stationary, but their producer gas can be flexibly used for either electricity (and heat) generation or SNG production.

Possibility to add olefins is then favorable when the producer gas from the gasification contains olefins (ethylene, acetylene) and the capacity of the plant and the

TABLE 4.9　**Comparison of Different Methanation Reactor Concepts.**

Name of Methanation Process or Supplier/ Inventor	Lurgi	TREMP (HT A/S), DPT	Vestas (Foster Wheeler/Clariant)
Type of reactor	Series of adiabatic fixed beds with intermittent and recirculation cooling	Series of adiabatic fixed beds with intermittent and recirculation cooling	Series of adiabatic fixed beds with intermittent and recirculation cooling
Specialities		Higher catalyst stability, lower recirculation rate	Using CO_2 as ballast gas
Simplicity of reactor	+	+	+
Low number of units	–	–	–
High temperature of cooling	+	++	+
Flexibility	o	o	o
Addition of olefins possible	–	–	–
Sufficient mass transfer	+	+	+
Good heat transfer	n.a.	n.a.	n.a.
Low challenges for catalyst	o	-	o
Maturity (technical readiness)	9	9	7–8
Commercial plants (TR level 9), Pilot and demonstration units (TR level 7–8)	Great Plains, USA, 1984 (>1 GW_{SNG})	Several in China, 2013/2014 (>1 GW_{SNG}), GoBiGas, Sweden, 2014 (20 MW_{SNG})	Nanjing, China, 2013/2014

++, Very much given; +, given; o, less given - not given; –, not given at all; n.a., not applicable.

economics do not allow for low temperature scrubbers at high pressures. In such cases, gas cleaning and conditioning steps are significantly simplified when the methanation unit can handle unsaturated hydrocarbons.

Sufficient mass transfer and *good heat transfer* are measures for the size of the reactor, defined by either the necessary mass transfer or the heat transfer area while the catalyst activity usually is not limiting.

The *maturity of reactor concepts* is expressed as Technical Readiness Level (TRL). This concept, developed by NASA, is meanwhile applied by many funding

Comflux, PSI	Etogas/ZSW	EBI	Agnion	EBI
Isothermal bubbling fluidized bed reactor	Polytropic fixed bed with several injection points and cooling zones	Isothermal bubble column reactor	Polytropic fixed bed with partial cooling	Polytropic fixed bed with conductive catalyst support
Attrition resistant catalyst necessary		Liquid phase with thermal/chemical stability and low vapour pressure necessary	Slow catalyst deactivation by olefins, sulfur and tars accepted	Metal monolith coated with catalyst in cooled tube
−	−	−	o	o
+	+	o	++	+
−	o	−	o	−
++	+	++	o	+
++	−	?	+	−
+	+	−	+	+
++	o	++	o	+
−	o	+	−	o
7, 8	8	4	5	4
Oberhausen, Germany, 1982 (20 MW_{SNG}) Güssing, Austria, 2009 (1 MW_{SNG})	Werlte, Germany, 2013/2014 (3.5 MW_{SNG})			

organizations such as the U.S. Department of Energy. While TRL 1 signifies an idea, TRL 9 means that a commercial scale plant is built. Starting with research (TRL 1 to TRL 4, laboratory experiments with model substances), the development continues with long duration tests with real feedstocks in laboratory scale (TRL 5), pilot plants with either model compounds or real feedstock (TRL 6 or TRL 7, respectively) and demonstration plants (TRL 8). At the bottom of the table, examples for commercial plants or pilot and demonstration units and the year of commissioning are given.

4.3   MODELING AND SIMULATION OF METHANATION REACTORS

In the development of chemical reactors for certain process chains, modeling and simulation of chemical reactor can be considered as an important and useful second pillar that is complementary to the first pillar, the experimental work. Although only reality, that is, (long duration) experimental campaigns at pilot scale can prove and demonstrate the performance of a reactor as well as the technical and economic feasibility of a process, the synergetic use of modeling along with the experimental work will improve and accelerate the reactor and process development and limit the risks during scale-up. The level of detail that is implemented in the model depends thereby on the statements or predictions that can be supported by means of the modeling results.

Already based on early laboratory results or, if necessary, even with a thermody-namic equilibrium assumption, a basic model of the reactor can be developed which uses standard or simplified assumptions for hydrodynamics, mass and heat transfer. With such basic models, sensitivity studies can be conducted that allow under-standing where deviations from the used assumptions significantly change the reactor performance. From these results, the further experimental needs can be derived and the experimental work can be focused on the most sensitive aspects (e.g., kinetics or heat/mass transfer) and on the operation parameter regions with the strongest deviations. Modeling/simulation is also useful to interpret experimental results, when several different physico-chemical phenomena could explain a finding. In such cases, a simulation, in which all important phenomena have to be described quantita-tively, may help to understand which sub-process is limiting.

With the results and knowledge gained by these experiments, an improved so-called rate based model should be developed that properly describes the rates of all important sub-processes in the reactor (e.g., chemical reaction kinetics, heat/mass transfer, hydrodynamics, formation of phases, etc.). The rate based model will be helpful to optimise operation conditions, to design the necessary pilot scale facilities and to confine the range of operation parameters to be tested during the pilot scale tests. With the model, if then validated at pilot scale, it should be possible to extrap-olate beyond the pilot scale with decreased risk, because a rate based model can predict the limiting or dominating sub-process changes during up-scaling.

A validated rate based model can further be applied within the thermo-economic optimization of complete process chains where the interdependencies of the different process units are represented to enable the simulation of different process configura-tions, including varied operating conditions [162]. With such tools, efficiencies, mass balances and costs can be calculated and so-called pareto curves of, for example, costs versus efficiency are obtained also for SNG production [163, 164]. As recently shown by [165], integrating a rate based model of the methanation reactor, if necessary as surrogate models to minimize computation times, can significantly change the outcome of thermo-economic analyses and optimization efforts.

In the following section, some options to properly determine reaction kinetics will be presented, which is a challenge for highly exothermic and potentially carbon depositing reactions such as methanation. Further, the status of modeling the two main methanation reactor types (fixed bed and fluidized bed) will be discussed, including aspects of determining hydrodynamics and consideration with respect to conducting experiments for model validation.

4.3.1 How to Measure (Intrinsic) Kinetics?

The most important aspect of experimentally determining chemical kinetics is the absence of physical limitations such as mass transfer and heat transfer. Only then, the rate based model will be able to properly predict if the dominating sub-process in the reactor changes during scale-up. More specifically, the experimenter must ensure that the measured temperature is representative for the temperature of the catalyst bed and that the reaction rate is not limited by intra-particle diffusion limitation, film diffusion limitation or some residence time effects. While potential mass transfer limitations can be excluded by calculating, for example, the Thiele modulus or the second Damköhler number, Da_{II}, keeping the catalyst more or less isothermal is not effortless in the case of methanation which is strongly exothermic. Further, certain reactor types are more prone to contribute to carbon deposition and loss of activity (e.g., due to the presence of unsaturated and aromatic species) than others. Therefore, a couple of typical or potential reactors for kinetics determination are discussed in more detail.

Tubular Fixed Beds of Catalyst Particles As discussed in the second part of this chapter, using adiabatic fixed bed reactors always carries the danger that axial temperature profiles develop which practically can impede proper data evaluation. The development of significant temperature profiles is inherently connected to strong axial concentration profiles, unless very small tube diameter and catalyst particles as well as catalyst dilution are applied [166]. In consequence, mass transfer limitations and deviation from plug flow may occur locally, but cannot be detected by calculations of the relevant criteria (Thiele modulus, Damköhler number, Peclet number) using the overall conversion over the catalyst bed. Further, it has to be excluded that thermodynamic equilibrium is reached within the catalyst bed. Moreover, if fixed bed reactors are cooled, besides the axial also radial temperature and concentration profiles may develop. Therefore, experiments in fixed bed laboratory reactors with significant temperature profiles may be used for screening of catalysts only, but cannot be recommended for the determination of kinetic data. In the case of catalyst screening, to enable at least fair comparability, an adiabatic reactor should be used; further, the inlet temperature, pressure, catalyst mass, and molar flows (i.e., the weight hourly space velocity; WHSV) should always be kept constant. The exit temperature should be recorded by a second thermocouple downstream as it is, under adiabatic conditions, a measure of the catalyst activity in exothermic reactions.

Berty Type of Reactors A perfect way of avoiding concentration and temperature gradients is the use of a Berty type reactor (with and without gas recycle) that contains a basket filled with catalyst and an impeller to fully mix the gas phase. Alternatively, high recirculation through a fixed catalyst bed may be used [102].

Reference [167] studied systematically the operation parameters of a commercial Berty reactor using methanation as the model reaction. Reference [39] cites an important number of kinetic studies on nickel based methanation catalysts using Berty type reactors that therefore could be considered a standard tool. Still, the system has some limitations. When the system is used as differential reactor, relatively low conversion (often below 10%) is required. This and the need to avoid too strong temperature gradients over the bed motivate the dilution of catalyst beds and/or the reactant gases. Further, very sensitive analytics is necessary which may suffer from significant relative errors.

Channel Reactor with Catalytically Coated Plate One approach to overcome some of the limitations mentioned for the other reactor types is to combine differential and integral reactor in one apparatus. This is accomplished by appropriately taking the axial concentration and temperature profiles while controling the radial gradients. This can be accomplished by fixing the catalyst on a monolith or foam and axially moving a thermocouple and a gas sampling tube through a drilled channel in the monolith or foam [168–172]. A channel reactor, where the catalyst is coated on the wall, is another option to fulfil these criteria. Moreover, the catalyst coated on the wall can be heated and cooled from the outside, aiming at close to isothermal conditions, which makes this approach especially suitable for the determination of kinetic data. In a tubular geometry, the developing temperature profile can be measured by one slidable or several fixed thermocouples in the centerline of the reactor, for example, see [173]. With a thin catalyst coating, the intra-particle concentration gradients and therefore the mass transfer limitation can be minimized. The external mass transfer, that is, the diffusion through the laminar layers inside the reactor tube has to be carefully checked (i.e., the tube inner diameter should be in the range of only a few millimeters), but is easily calculable due to the very simple geometry. The concentration profile has to be measured by a slidable sampling tube in the centerline.

An even better control over the reaction conditions can be achieved by an optical access that allows measuring the catalyst surface temperature by IR thermography. This flat channel concept had been developed by [174, 175] and was applied to determine the rates of methanation and water gas shift reaction by [37], see Figure 4.22. Figure 4.4 shows an example of the concentration profiles that can be measured with this reactor type. By minimizing the temperature gradient and with careful 1-D and 2-D simulation, the influence of mass transfer limitations and axial dispersion on the determined kinetics could be excluded for this set-up [37]. It was possible to calculate for each data point (i.e., the conversion from one axial measurement position to the next) the Weisz modulus and the Carberry number (second Damköhler number Da_{II} times the catalyst effectiveness factor) to control intra-particle and external diffusion limitation, respectively. By this and the temperature measurement by means of the IR camera, the (nearly) isothermal data points without mass transfer limitation could be selected to be used for the determination of the kinetic parameters [37]. Although the thickness profile of the catalyst coating layer had to be determined after the experiment to account for the slightly varying catalyst mass per length ratio, this

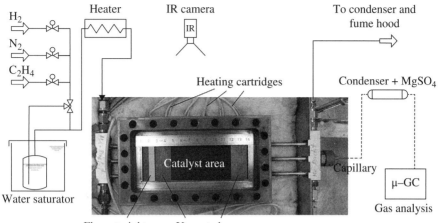

FIGURE 4.22 Experimental set-up for measurement of methanation kinetics: catalytically coated plate with temperature measurement by IR camera through quartz window and axially movable sampling tube to obtain concentration profiles; adapted from [52]. *Source*: Kopyscinski 2013 [52]. Reproduced with permission of Elsevier.

reactor concept was able to combine the advantages of differential and integral reactors for an exothermic reaction like methanation.

Fluidized Bed Slightly Above Minimum Fluidization As mentioned in [52] and shown in Figure 4.4, the channel reactor with a catalytically coated plate was used to investigate the reaction network in the hydrogenation of ethylene. However, it could not be used for the determination of kinetics as the catalyst suffered from slow catalyst deactivation, indicated by an axial movement of the reaction front (temperature peak of 3–4 K) by 20 mm in 5 h. From the discussion in the first part of this chapter, the slow formation of nickel carbide and/or amorphous carbon can be expected as the reason for the slow catalyst deactivation. Therefore, a reactor suited for kinetic experiments with ethene should not be a fixed bed to enable sufficient catalyst stability. Generally, fluidized beds have proven to limit catalyst deactivation by carbon deposition (see the first parts of this chapter) and allow for excellent temperature control. However, the mass transfer has to be considered carefully because the formation of voids in the fluidized bed (often referred to as "bubbles") may lead to mass transport limitations, especially in case of Geldart B particles.

Previous authors [176] combined a bed of larger catalyst particles with fluidized fine particles of inert material to benefit from the high heat transfer of the fluidized inert fines. A very recent study [177] investigated systematically the potential of micro-fluidized bed reactors with a means to measure axial temperature and concentration profiles for kinetic experiments. In bubbling fluidized beds, a compromise has to be found between the need for high heat transfer and catalyst particle mixing (both of which are improved by a more intense movement of bubbles) on the one hand, and the need, on the other hand, to avoid reactant breakthrough and therefore limited mass transfer by keeping the bubble diameter small. Reference [177] varied the fluidization number (ratio of applied linear gas velocity to the minimum one necessary for fluidization) at a constant space velocity to find an absence of hot spot formation and mass transfer limitation for very low fluidization numbers. This means that the size of the developed bubbles in the micro-fluidized bed reactor (bed height several centimeters) are too small to impede mass transfer and to influence the reaction rate. Therefore, if experiments are conducted carefully, a micro-fluidized bed slightly above minimum fluidization can be considered as a useful concept for kinetics determination of exothermic reactions potentially suffering from carbon deposition.

4.3.2 Modeling of Fixed Bed Reactors

When the reaction kinetics are known or have been determined, the next step is to decide on the assumptions that allow building of a rate-based reactor model that properly represents the hydrodynamic aspects, such as heat and mass transfer. As mentioned in the introduction to this third part of the chapter, the level of detail in the model depends on what quality of the prediction should be accomplished and which purpose the modeling/simulation is conducted for.

Especially for a relative comparison between different design options and operation conditions, more simplification may be appropriate. When absolute values become important [e.g., for a thorough determination of reactor sizes (for cost estimation) and for model validation with experimental data], more details and therefore more complex models will be necessary. More complex models however usually contain more parameters (e.g., mass transfer coefficients) which might limit the use of the more complex models, if the data base for these parameters is not of adequate quality. Therefore, it is mandatory to explore the reliability of model predictions by sensitivity analyses for the applied parameter values and by calculation of the reactor engineering design criteria such as Thiele modulus, and so on.

In the case of fixed bed methanation reactors, several simplifying assumptions will be possible and reasonable; one may refer to textbooks such as Levenspiel [178], Froment-Bischoff [179] or Baerns-Hoffmann-Renken [180]. If the reactor inner diameter is at least 10 times the particle diameter and the particle bed is at least 10 times the reactor inner diameter, maldistribution should be negligible and plug flow behavior can be assumed.

In the case of adiabatic fixed beds, like the state of art methanation reactors in coal to SNG processes, a one-dimensional model is fully sufficient, because no radial gradients of concentrations or temperature have to be expected due to the absence of wall cooling. For power to gas applications, where cooled fixed bed reactors have been realized (refer to Chapter 7 in this book), a choice has to be made between simpler one-dimensional models which assume all heat transfer resistance to lie at the inner tube wall and more complex two-dimensional models which then should describe both the apparent radial bed conductivity and the wall heat transfer coefficient. Further, in all cases, an appropriate representation or (justified) neglecting of external mass transfer (leading to particle effectiveness factors lower than one) and of intra-particle gradients has to be considered.

Reference [181] used a one-dimensional model to simulate an adiabatic fixed bed methanation reactor for the conversion of biomass gasification derived producer gas to SNG based on the kinetics by [182], but adjusted to the performance of the commercial catalysts used for their experiments. Their model correctly predicts the temperature level to be reached when the thermodynamic equilibrium is reached, but slightly overestimates the temperature gradients during run-away of the exothermic methanation, as observed in TREMP reactors; see Figure 4.15. The authors found some deviation between the heterogeneous and pseudo-homogenous models. In their heterogeneous model, the exothermic methanation reaction progresses faster than the endothermic reverse water gas shift, leading to a local hot spot of about 150 K for 1 cm length. It is however debatable how important this deviation is as the model did not consider axial dispersion of heat (e.g., by radiation and conduction through the solids) or mass and the temperature profiles of such reactors (Figure 4.15) do not show any hot spots. In consequence, the authors used the pseudo-homogenous model for their further sensitivity analyses with respect to catalyst effectiveness factors and CO_2/CO ratio at the reactor inlet. They found that, according to the model results, the addition of CO_2 is beneficial as it dampens the temperature increase and decreases the amount of CO converted with the steam from the methanation to CO_2 by water gas shift reaction.

Similarly, [91] re-established the same model for sensitivity analysis of cooled multi-tubular fixed bed reactors with relative small tube diameter (19 mm) used for CO_2 methanation in power to gas applications. In this study, they investigated the influence of applying structured catalyst supports with high radial heat transport properties [93], multiple feed injections or converting biogas to control the temperature increase in the reactor. The model was expanded by a heat transfer term leading to the following mass and energy balances for the fixed bed reactor:

$$\frac{d\dot{n}_i}{dz} = v_{ij} r_j \cdot A_{cross} \cdot \rho_{cat} \tag{4.12}$$

$$G \cdot c_p \cdot \frac{dT}{dz} = \sum_j r_j \cdot \Delta H_{R,j} \cdot \rho_{cat} + \frac{4 \cdot U}{d_i} \cdot \left(T_{wall} - T \right) \tag{4.13}$$

Where A_{cross} is the cross-sectional area of the reactor, ρ_{cat} is the catalyst mass per volume, G is the mass flow per cross sectional area, c_p is the heat capacity, U is the overall heat transfer coefficient and T_{wall} is the temperature of the cooling wall. The initial conditions applied are: $X_{(z=0)} = 0.0$, $T_{(z=0)} = T_0$. At each step, the overall heat transfer coefficient U is calculated based on the actual gas composition according to the correlation determined for structured catalyst supports [92] and the particle bed [183].

Figure 4.23 shows the temperature and concentration profiles to be expected in such reactors based on the simplified pseudo-homogeneous one-dimensional model. It can be seen that, in such reactors, nearly full CO_2 conversion and negligible CO content can be achieved if the reactor is long enough to cool down. This shows that

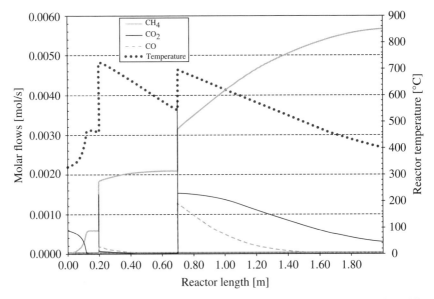

FIGURE 4.23 Molar flows (full and dashed lines) and reactor temperature (dotted line) for on tube (diameter 25 mm) with slightly over-stoichiometric (4.1 to 1) mixture of H_2 and CO_2 at 10 bar with inlet temperature 280 °C and cooling temperature 340 °C: Results for adding 10% of the CO_2 at the inlet, next 25% at 20 cm and the remaining 65% at 70 cm; from [91].

the reactor size is completely dependent on the heat transfer as the reaction is following the thermodynamic equilibrium and therefore the temperature level. Further the authors concluded from their sensitivity analysis that it is challenging to limit the temperature level without adding ballast gas such as CH_4 (e.g., by conversion of biogas) or applying several cooling zones because local heat production by exothermic methanation easily exceeds the (locally restricted) cooling performance of fixed bed reactors.

Reference [122] conducted a valuable and thorough investigation on modeling cooled multi-tubular fixed bed reactors for CO_2 methanation comparing one-dimensional and two-dimensional models as well as pseudo-homogeneous and heterogeneous models. Based on the original kinetics from [182], they especially investigated for different reactor tube diameters the cooling temperature at which the cooling is such insufficient that the thermal run-away sets in; see also Figure 4.17. A two-dimensional model shifts this predicted "ignition" temperature by about 5–10 K to lower cooling temperatures, which is caused by the earlier development of a hot spot due to radial temperature gradients. A heterogeneous model, assuming 5 mm pellets and a pore diameter of 20 nm, predicted up to 20 K temperature difference between the gas phase and the pellet, but also a significantly shifted "ignition" temperature (increased by 20 K) due to the limited mass transfer and therefore low catalyst effectiveness factors in the thermal run-away. When the pore diameter in the heterogeneous model was assumed to be 50 nm, the model predicted however similar results as the pseudo-homogeneous model. In summary, they confirmed that the use of simplified pseudo-homogenous one-dimensional models is justified for sensitivity analyses. Further, it can be concluded that for proper sizing and exact knowledge of the onset of temperature run-away and the resulting maximum temperature, most probably a heterogeneous two-dimensional model is necessary, but needs to be validated with real data because of the uncertainty of mass and heat transfer parameters in real systems.

Model Validation From the studies presented in the literature, it can be concluded that, in fixed bed methanation reactors, three zones exist:

1. An induction zone where conversion and heat release slowly progress until the temperature is high enough to initiate thermal run-away.
2. In the second zone, where the thermal run-away takes place and temperature rises until thermodynamic equilibrium is reached in a short length of the reactor, mass transfer is the limiting factor while cooling (if applied) has hardly any impact due to the low local cooling surface.
3. In the third zone downstream of the temperature peak, the reaction is limited by the equilibrium, which in turn depends on the temperature level.

In consequence, the third zone of the reactor and therefore the length of such cooled fixed bed reactors are fully limited by the heat transfer.

Based on these considerations, a detailed rate based model used to predict reactor performance, maximum temperature (as boundary condition for catalyst stability) and necessary reactor design should be able to simulate correctly the onset of thermal run-away, the axial position and level of the temperature peak and the exit temperature and conversion. As in the second zone, only mass transfer is dominant and in the third zone only heat transfer (while the reaction is equilibrium limited); it is therefore sufficient to measure in a pilot scale plant the axial temperature profile in the reactor and the outlet concentrations. Due to the inherent connection between conversion and heat production under these circumstances, these data allow for validation of such models.

It should be noted that, for the sizing of (adiabatic) fixed bed reactors, the necessary catalyst reserve to compensate for slow catalyst deactivation dominates rather than the amount of catalyst needed to reach equilibrium conversion; refer to Figure 4.15. Therefore a model only considering the actual conversion reached in a reactor with a fully active catalyst and neglecting deactivation phenomena will underestimate the necessary reactor size.

4.3.3 Modeling of Isothermal Fluidized Bed Reactors

Contrary to fixed bed reactors, the temperature profile in fluidized bed reactors is of less importance, because, compared to a fixed bed, only relatively weak temperature gradients are observed, if the aspect ratio of the bed, the immersed heat transfer area and the degree of fluidization are chosen appropriately; see also Figure 4.20

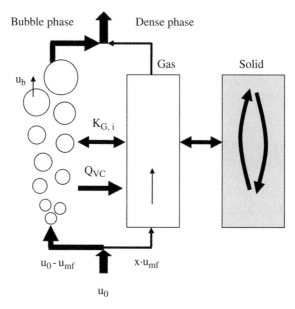

FIGURE 4.24   Two-phase model for the simulation of fluidized bed methanation reactors.

(adapted from [146], second part of this chapter). Bench-scale experiments with 100–200 g of very active catalyst [149] seemed to show hot spots of up to 100 K close to the distributor plate, but careful re-evaluation of these experiments and control experiments rather suggests that the observed hot spots are an artefact caused by the used moveable sampling tube that immobilized catalyst particles and thereby formed locally a fixed bed [165].

The main challenge in modeling fluidized beds is the formation of voids (also referred to as "bubbles") caused by gas flow exceeding the minimum gas flow needed to suspend the particles. These voids rise independently of the gas flowing around the catalyst particles which forms together with particles the so-called dense phase. As the voids contain no or only a small amount of catalyst particles, the gas molecules contained in the voids first have to be transferred to the dense phase before they can react on the surface of the catalyst particles. This mass transfer is an important resistance which has to be properly described to avoid over-estimation of mass transfer or unexpected reactant breakthrough in up-scaled units. This leads to the so-called two-phase model developed by [184, 185]. This approach was applied to fluidized bed methanation reactors by [186, 187] and recently in modified form by [150] using the kinetics from [37].

The model (see Figure 4.24) relies on a number of assumptions:

- The bubbles are solid free (no reaction).
- Bubble and dense phase are in plug flow.
- The bubble diameters and other hydrodynamic correlations (hold-up, bubble rise velocity and mass transfer $K_{G,i}$) are derived from literature data.

- The dense phase is at minimum fluidization u_{mf} and comprises a constant volume flow.
- An additional mass transfer Q_{VC} from the bubble phase to the dense phase is included to account for the volume contraction during the methanation reaction.
- The kinetics of methanation and water gas shift were taken from independent experiments.

This leads to the following balance equations [150]:

$$0 = -\frac{d\dot{n}_{b,i}}{dh} - K_{G,i} \cdot a \cdot A \cdot \left(c_{b,i} - c_{e,i}\right) - \dot{N}_{vc} \cdot x_{b,i} \tag{4.14}$$

$$0 = -\frac{d\dot{n}_{e,i}}{dh} + K_{G,i} \cdot a \cdot A \cdot \left(c_{b,i} - c_{e,i}\right) + \dot{N}_{vc} \cdot x_{b,i} + \left(1 - \varepsilon_b\right) \cdot \left(1 - \varepsilon_{mf}\right) \cdot \rho_P \cdot A \cdot R_i \tag{4.15}$$

$$\underbrace{\frac{mol}{m \cdot s}}_{\text{convection}} + \underbrace{\frac{m}{s} \cdot \frac{1}{m} \cdot m^2 \cdot \frac{mol}{m_{bed}^3}}_{\text{mass transfer}} + \underbrace{\frac{mol}{m \cdot s}}_{\text{bulk flow}} + \underbrace{\frac{m_{dense}^3}{m_{bed}^3} \cdot \frac{m_{solid}^3}{m_{dense}^3} \cdot \frac{kg_{cat}}{m_{solid}^3} \cdot m^2 \cdot \frac{mol}{s \cdot kg_{cat}}}_{\text{reaction}}$$

Here, a is the is specific mass transfer area, A is the cross sectional area of the reactor, $K_{G,i}$ is the mass transfer coefficient, $x_{b,i}$ is the molar fraction in the bubble phase, ρ_P is the particle density, $(1 - \varepsilon_b)$ is the volume fraction of the dense phase and $(1 - \varepsilon_{mf})$ is the volume fraction of the particles. The total bulk flow from the bubble to the dense phase $\dot{N}_{vc}$ (that compensates for the volume contraction of the reactions) is described as the sum of the molar losses due to the reaction and mass transfer in the dense phase [150].

$$\dot{N}_{vc} = \frac{\dot{n}_{vc}}{dh} = \sum_i K_{G,i} \cdot a \cdot A \cdot \left(c_{b,i} - c_{e,i}\right) + \left(1 - \varepsilon_b\right) \cdot \left(1 - \varepsilon_{mf}\right) \cdot \rho_P \cdot A \cdot \sum_i R_i \tag{4.16}$$

The boundary conditions at $h = 0$ (inlet) are

$$\dot{n}_{b,i}\big|_{h=0} = \dot{n}_{b,i,feed} \tag{4.17}$$

$$\dot{n}_{e,i}\big|_{h=0} = \dot{n}_{e,i,feed} \tag{4.18}$$

To properly describe the mass transfer (which generally is linear to the concentration difference between bubble and dense phase), hold-up, size and rise velocity of the bubbles have to be known. For all these parameters and for the rules to be followed during scale-up of fluidized beds, a wealth of hydrodynamic correlations exists in the literature, especially in the work of Grace, Werther, Horio, Darton, Rowe, Mori and their coworkers, to mention only a few. To manage complexity, common assumptions for these correlations, which were derived from experimental data, are: (i) that all bubbles in a specific height of a fluidized bed have the same diameter, (ii) that bubbles grow by radial coalescence, (iii) that larger bubbles rise faster and (iv) that all bubbles have a spherical shape.

Alternatively, fluidized bed methanation reactors could be modeled applying CFD techniques. Reference [188] simulated successfully an isothermal fluidized bed reactor (52 mm inner diameter) as a Eulerian–Eulerian two-fluid model using the methanation and water gas shift kinetics from [37] and a grid with 2 mm cells. With this model, the authors succeeded in predicting the bed height and the exit concentrations measured by [149] in a reactor of the same diameter. Predicting the inlet region of the reactor in these experiments is anyway not possible because the axial gas phase concentration measurements by means of a gas sampling tube were, as mentioned above, biased by the interaction of the gas sampling tube at low positions and the distributor plate.

While the predictions of CFD models with respect to flow patterns have to be validated with either tomography measurements and/or hydrodynamic correlations, the macroscopic two-phase model depend on the quality of the underlying assumptions and hydrodynamic correlations.

Unfortunately, several aspects limit the application of these correlations for modeling of fluidized bed methanation reactors. While most of the correlations have been determined for freely bubbling fluidized beds, isothermal methanation reactors comprise a dense packing of vertical heat exchanger tubes. Pressure fluctuation measurements in a cold flow model indicated [189] that, by these vertical tubes, the bubbles growth is decreased and a smoother fluidization can be observed. Optical probe measurements showed that indeed, the *average* pierced chord length of the bubbles and rise velocity increase with increasing bed height, but that *distributions* exist of both bubble diameter and bubble rise velocity; see Figure 4.25. It was further observed that both depend on the geometry of the cooling tubes packing, that is, tube diameter and tube distance, but hardly on the arrangement (quadratic versus triangular) of the tubes [189]. It has to be noted, that for proper prediction of reactant breakthrough (and therefore of the reactor performance), the hydrodynamic correlations for the average values may not be sufficient; therefore, the distributions have to be considered, as the fastest and/or largest bubbles (rather than the average bubbles) have the highest probability of reactant breakthrough.

Very recent studies in cold flow models applying X-ray tomography and subsequent image reconstruction elucidate that the bubbles between the vertical tubes are not spherical but significantly elongated and seem to grow by vertical coalescence as the radial movement is limited [190]; see Figure 4.26. Further, it was observed that fast, small bubbles and slow, big bubbles coexist at all heights, although the *number based averages* follow the classic assumption that the bubbles grow and rise faster with increasing height. This emphasizes the need to measure these hydrodynamic parameters in pilot scale plants with vertical internals in the necessary pressure range and to derive representative correlations from these data that can be introduced to the two-phase model. To understand the influence of the volume contraction of the methanation, it would be interesting to repeat such hydrodynamic measurements also during ongoing reaction which in turn asks for measurement devices that can withstand the conditions of reactive operation.

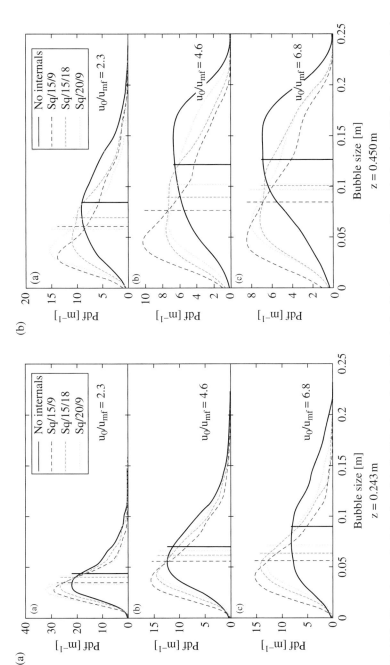

FIGURE 4.25 Kernel smoothed density of the bubble pierced chord length distribution measured by optical probes. Tube bank configurations are Sq/15/9 (reference: 15 mm tube diameter, 9 mm tube to tube distance), Sq/15/18 (wider tube to tube spacing), and Sq/20/9 (larger tube diameter), as well as no internals. Optical probe heights are: (a) 0.243 m and (b) 0.450 m. Reprinted from [189]. Copyright (2012) American Chemical Society. *Source:* Schildhauer 2012 [189]. Reproduced with permission of American Chemical Society.

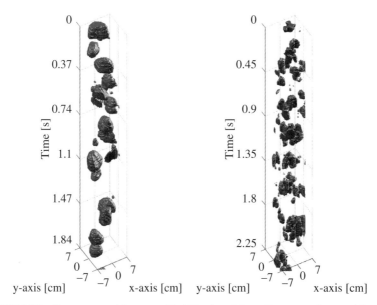

FIGURE 4.26   Reconstructed images of bubbles in a 15 cm diameter column without (left) and with (right) vertical internals at $u/u_{MF} = 3$ [190].

Sectoral Scaling   An interesting feature of bubbling fluidized beds with vertical internal is that they can be represented by one-dimensional models, if the hydrodynamic correlations have been determined on a pilot scale. The reason is that, different from freely bubbling fluidized beds without internals, the vertical internal form within a few tube banks a local environment that defines the hydrodynamic situation which is independent from the overall reactor diameter. This allows using, for example, the classical scale-up criteria of Glicksmann [191] (Reynolds number, Froude number, gas-solid density ratio, bed geometry ratio, ratio between particle and bed diameter, sphericity of the particles and particle size distribution) whereby the reactor diameter is replaced by the hydraulic diameter.

$$\frac{u_0 \rho_g D}{\eta}, \frac{u_0^2}{g\, D}, \frac{\rho_g}{\rho_p}, \frac{D}{H}, \frac{d_p}{D}, \phi, psd \tag{4.19}$$

This was shown by comparison of pressure fluctuation data from the pilot scale plant in Güssing with cold flow experiments [192]. Further, optical probe measurements in fluidized beds of different sizes and different number of vertical internals (but identical tube arrangement) showed that at least two rows of tubes should surround the measurement position to allow neglecting outer wall effects and reaching similar hydrodynamic properties [193].

Model Validation   The excellent heat transfer is a major advantage of fluidized bed reactors and leads to nearly isothermal conditions. In consequence, the temperature profile cannot be used for model validation. Still, the temperature profile should be measured at several axial positions. If the vertical cooling tubes are equally distributed, such that no preferential paths develop, the absence of radial gradients is a reasonable assumption. Slight deviations from isothermicity in axial direction however can be expected due to the fact that cooled thermo-oil is pumped downwards in one branch of a cooling tube and upwards in the other branch. This means that, close to the catalyst particles, cooling media with not necessarily negligible axial temperature gradients (in the order of very few tens of degrees over the length of a few meters) exist. Such conditions make an exact prediction of the bed temperature very challenging, and computational fluid dynamics methods would have to be used to take local turbulence into account. Therefore, slight axial bed temperature gradients could either be used as model input, or the average temperature could be used as an assumption.

Concentrations and flows are therefore the main information to be used for model validation. It would be desirable to have proper axial profiles of the concentrations. However, due to the formation of the second phase, the bubbles, some uncertainty has to be taken into account. While in the freeboard above the catalyst bed, the gas flow contains gas from both bubble and dense phase, inside the bed it is not known when bubbles are sampled and when gas from the dense phase. The impact of this effect can be estimated by two considerations: first, the model should give an indication, up to what axial position do the concentrations in the bubble and dense phase differ significantly; second, the bubble hold-up gives a certain indication of the probability to sample a bubble and not the dense phase. Still, it should be kept in mind that, due to the tendency of the bubbles to rise in the middle of the bed, the resulting radial distribution of the local bubble hold-up weakens the second approach to estimate the ratio between bubble and dense phase in the sampled gas.

For reliable model validation, therefore the free board concentrations upstream of the filter are the important experimental findings, which a model should predict properly. They should be measured carefully, for example, by using an inertial filter. Within the bed, the sampled gas concentrations should lie between the predicted values of bubble and dense phase.

Using a sampling tube to measure at different axial positions has additionally the advantage to allow early detection of slow catalyst deactivation. Usually, the catalyst hold-up is so big that thermodynamic equilibrium is reached even before the reactor outlet. A measurement position in the bed will therefore see the slow movement of the reaction front before it is visible in the freeboard. Still, one should be careful to avoid immobilization of catalyst particles either at the tip of the sampling tube (e.g., by too high, not iso-kinetic suction velocity) or between the sampling tube and the distributor plate when measuring close to it. In both cases, a small fixed bed is created leading to local hot spots and non-representative concentration measurements. A good practice is to backflush the sampling line each time before starting the next measurement.

A further advantage of fluidized beds is the possibility to take catalyst samples during operation, even without them contacting the air [77]. This allows the validation of models that predict the hold-up of carbon depositions or the degree of sulfur poisoning.

4.4   CONCLUSIONS AND OPEN RESEARCH QUESTIONS

Since applying the methanation of carbon monoxide as a technical process was considered for the first time in the 1930s [194], then to detoxify town gas, the conversion of carbon oxides to synthetic natural gas (SNG) went through several phases. Prior to the introduction of natural gas in Europe during the 1960s, the detoxification of town gas was an important topic. The aim was to convert the high content of CO, which was responsible for fatal accidents and facilitated suicides. However, instead of methanation, the problem was solved by applying the water gas shift reaction on a large scale, until the introduction of natural gas.

Later, the potential shortage of natural gas and the efficient use of domestic coal reserves caused concern, especially in the 1970s and then reinforced by the oil crisis. Important research efforts were started in the United States, the United Kingdom, and Germany to develop an efficient process from coal to SNG. At this time, very broad knowledge concerning methanation was acquired on which research today is based. Besides catalysts, kinetics, reaction mechanisms and deactivation phenomena, nearly all possible reactor types were investigated to find the optimal compromise between costs, heat management, catalyst stability, and energy efficiency: adiabatic fixed beds, isothermal bubble columns, and fluidized beds. The large number of concepts developed up to pilot and even demonstration scale is a measure for the initial uncertainty of the best solution.

The possibility to raise high pressure steam (once high temperature stable catalysts had been developed) and to generate thereby important economic synergies, favored the application of adiabatic fixed bed reactors for large scale coal to SNG plants. With their capacity of more than 1 GW, even expensive gas cleaning technology such as Rectisol® could be applied cost effectively in such plants. Series of adiabatic fixed bed methanation reactors were applied in the first large scale coal to SNG plant (Great Plains, North Dakota, in operation since 1984) and intended for a number of coal to SNG projects in the United States in the first decade of the 21st century (just before the emergence of shale gas exploitation). This reactor concept can therefore be considered as state of the art and fully commercially deployed technology, and is also applied in several recent large plants in China, where the need for clean use of domestic coal reserves favors the conversion to cleaner energy carriers and chemicals via gasification.

In the last 15 years, the efficient use of biomass has been increasingly investigated in Europe, as biomass is considered as a domestic and CO_2 neutral resource. This has also triggered research in converting ligno-cellulosic biomass to SNG via gasification and subsequent methanation. As such wood to SNG plants are significantly smaller than coal to SNG plants due to biomass logistics limitations, state of the art gas cleaning technologies are no longer cost effective. Further, the importance to generate steam at the highest pressure level is reduced due to completely different energy integration options. In consequence, new combinations of gas cleaning and methanation steps are being developed (and presented in further chapters of this book). Some teams focus on using the relatively simple state of the art fixed bed methanation reactors and on developing specific, although relatively complex gas cleaning technologies which convert the unsaturated hydrocarbons. Others aim at simplifying the gas cleaning steps to lower costs and increase efficiency. They test therefore novel methanation reactor types that they hope can offer more robustness and flexibility.

Flexibility, now in terms of dynamic start-up/shut-down and part-load operation, is also important for the newest potential application of methanation in so-called power to gas applications. Depending on the hydrogen generation rate and the storage concepts, significant new challenges for the methanation reactors evolve.

While there are only very few open research questions, but a wealth of experience within the companies in the field of coal to SNG plants, in the case of biomass to SNG processes and power to gas applications, important questions are not yet completely answered and should trigger specific research, by reactor simulation/modeling and pilot/demo scale experimental campaigns.

So far, there is only limited experience with the performance of gas cleaning technology combinations for biomass to SNG with a series of fixed bed methanation reactors. Is it possible to convert efficiently (in terms of costs and energy loss) the unsaturated hydrocarbons in the producer gas from biomass gasification to guarantee sufficiently long lifetime of the vulnerable methanation catalyst?

Similarly, real long term experience (several thousand hours) with the chosen combinations of gas cleaning technologies in wood to SNG processes is missing with respect to removing organic sulfur species effectively. Which of the technologies is the best option, cold scrubbers or hot gas cleaning with catalytic conversion steps or even a combination? Is it a better option to develop methanation reactor concepts and catalysts that are more robust with respect to sulfur poisoning and carbon deposition and therefore allow significant simplification of gas cleaning?

While scale-up and simulation of adiabatic fixed bed reactors with bed of particles is generally solved, appropriate modeling of the heat transport in cooled fixed bed reactors and in methanation reactors with conductive catalyst supports, but also model validation with (pilot scale) experimental data is missing.

This holds in similar way also for bubble columns and fluidized bed reactors. Although pilot scale and even demo scale plants showed in the past that scale-up of these reactor types is possible, detailed rate based modeling and validation of the model with pilot scale data still have to be shown. Can the challenges for the proper prediction of hydrodynamics be solved such that reliable simulation tools to de-risk scale-up can be developed?

For the power to gas applications, frequent start-up and shut-down cycles as well as efficient part-load operation may be necessary. Further, a variety of carbon oxide sources is discussed to convert hydrogen to storable methane: conversion of purified CO_2 from atmosphere and industrial flue gases, direct methanation of biogas from fermentation, flexible hydrogen addition to producer gas from wood gasification.

Which of the reactor concepts allows the best heat management under stationary conditions and has the best performance with respect to avoiding catalyst deactivation by carbon deposition or by sintering due to frequent temperature changes? Is there a synergy to be realized between power to gas and biomass to SNG processes and do some reactor concepts have a specific advantage for that?

Based on the broad variation of requirements for methanation reactors, it may be expected that, after solving these research questions, not one reactor concept will outperform the others for all applications. Rather, there will be number of application-specific technology solutions, including at least adiabatic and cooled fixed beds as well as fluidized bed methanation reactors.

4.5   SYMBOL LIST

A	m^2	cross-sectional area of the reactor
a	$m^2 \cdot m^{-3}$	specific mass transfer area
c_i	$mol \cdot m^{-3}$	concentration of species i
$c_{p,i}$	$kJ \cdot K^{-1} \cdot mol^{-1}$	specific heat
D	m	reactor diameter
d_i	m	inner reactor tube diameter
d_p	m	particle diameter
G	$kg \cdot m^{-2} \cdot s^{-1}$	mass flux $= u_0 \rho_G$
g	$m \cdot s^{-2}$	acceleration due to gravity (9.80665)
h, H	m	height or distance from the gas distribution
$K_{G,i}$	$m \cdot s^{-1}$	mass transfer coefficient of species i
$\dot{N}_{vc}$	$mol \cdot s^{-1} \cdot m^{-1}$	total bulk flow from the bubble to the dense phase
$\dot{n}_i$	$mol \cdot s^{-1}$	molar flow of species i
p_i	bar or Pa	partial pressure of species i
psd	–	particle size distribution
$\Re$	$J \cdot mol^{-1} \cdot K^{-1}$	gas constant $= 8.314$
r_j	$mol \cdot kg_{cat}^{-1} \cdot s^{-1}$	rate of reaction 1, 2 (methanation and water gas shift)
R_i	$mol \cdot kg_{cat}^{-1} \cdot s^{-1}$	rate of disappearance or formation of species i
T	K or °C	temperature
u_0	$m \cdot s^{-1}$	superficial gas velocity based on empty tube diameter
u_{mf}	$m \cdot s^{-1}$	minimum fluidization gas velocity based on empty tube diameter
U	$W \cdot m^{-2} \cdot K^{-1}$	overall heat transfer coefficient
x_i	–	molar fraction of species i
X_i	–	conversion of species i
z	m	reactor tube length

Greek symbols

ΔH_R	$kJ \cdot mol^{-1}$	heat of reaction
ε	–	void fraction or porosity
η	$Pa \cdot s$	gas viscosity
ν_{ij}	–	stoichiometric factor of species i in reaction j
ρ	$kg \cdot m^{-3}$	density
Φ	–	sphericity of the catalyst particle

Subscripts and superscripts

b	bubble phase
cat	catalyst
e	emulsion or dense phase
G	gas
i	species i
j	reaction j
mf	minimum fluidization condition
p	particle
ref	reference
tot	total

REFERENCES

[1] Deutscher Verein des Gas- und Wasserfachs (DVGW). Arbeitsblatt G260 *Gasbeschaffenheit*, Bonn, Germany, 2013.

[2] Schweizerischer Verein des Gas- und Wasserfachs (SVGW). Arbeitsblatt G13d, *Richtlinien für die Einspeisung von Biogas*, Zürich, Switzerland, 2008.

[3] Dufour A, Masson E, Girods P, Rogaume Y, Zoulalian A. Evolution of Aromatic Tar Composition in Relation to Methane and Ethylene from Biomass Pyrolysis-Gasification. *Energy Fuels* **25**:4182–4189; 2011.

[4] Sabatier P, Senderens JB. New methane synthesis. *Academy of Sciences* **314**:514–516; 1902.

[5] Rönkkönen EH. *Catalytic clean-up of biomass derived gasification gas with zirconia based catalysts*. Dissertation, Aalto University, Finland, 2014.

[6] Rabou LPLM, Bos L. High efficiency production of substitute natural gas from biomass. *Applied Catalysis B: Environmental* **111/112**:456–460; 2012.

[7] Specht M, Baumgart F, Feigl B, Frick V, Stürmer B, Zuberbühler U, Sterner M, Waldstein G. Storing bioenergy and renewable electricity in the natural gas grid. *FVEE AEE Topics* 2009:12–19; 2009.

[8] Anon. *DIPPR Project 801 database*. Design Institute for Physical Properties, 2012.

[9] Outotec. *HSC© 7.0*, Outotec OyJ, Espoo, Finland, 2010.

[10] Eckle S, Denkwitz Y, Behm RJ. Activity, selectivity, and adsorbed reaction intermediates/reaction side products in the selective methanation of {CO} in reformate gases on supported Ru catalysts. *Journal of Catalysis* **269**(2):255–268; 2010.

[11] Ross JRH. *Metal catalysed methanation and steam reforming*. In: Bond GC, Webb G (eds) Catalysis, vol. **7**. Royal Society of Chemistry. pp. 1–45; 1985.

[12] Kopyscinski J. *Production of synthetic natural gas in a fluidized bed reactor – Understanding the hydrodynamic, mass transfer, and kinetic effects*. Dissertation, ETH Zürich, Nr. 18800, 2009.

[13] Klose J. *Reaktionskinetische Untersuchungen zur Methanisierung von Kohlenmonoxid*. Dissertation, Ruhr-Universität Bochum, 1982.

[14] van Meerten RZC, Vollenbroek JG, de Croon MHJM, van Nisselrooy PFMT, Coenen JWE. The kinetics and mechanism of the methanation of carbon monoxide on a nickel-silica catalyst. *Applied Catalysis* **3**(1):29–56; 1982.

[15] Schoubye P. Methanation of CO on some Ni catalysts. *Journal of Catalysis* **14**(3): 238–246; 1969.

[16] Schoubye P. Methanation of CO on a Ni catalyst. *Journal of Catalysis* **18**(2):118–119; 1970.

[17] William W Akers, Robert R White. Kinetics of Methane Synthesis. *Chemical Engineering Progress* **44**(7):553–566; 1948.

[18] Van Herwijnen T, Van Doesburg H, De Jong WA. Kinetics of the methanation of CO and CO_2 on a nickel catalyst. *Journal of Catalysis* **28**(3):391–402; 1973.

[19] Ho SV, Harriott P. The kinetics of methanation on nickel catalysts. *Journal of Catalysis* **64**(2):272–283; 1980.

[20] Inoue H, Funakoshi M. Kinetics of Methanation of Carbon Monoxide and Carbon Dioxide. *Journal of Chemical Engineering Japan* **17**(3):602–610; 1984.

[21] Ibraeva ZA, Nekrasov NV, Yakerson VI, Gudkov BS, Golosman EZ, et al. Kinetics of methanation of carbon monoxide on a nickel catalyst. *Kinetics of Catalysis* **28**(2): 386; 1987.

[22] Sughrue EL, Bartholomew CH. Kinetics of carbon monoxide methanation on nickel monolithic catalysts. *Applied Catalysis* **2**(4/5):239–256; 1982.

[23] Kai T, Furusaki S. Effect of volume change on conversions in fluidized catalyst beds. *Chemical Engineering Science* **39**(7/8):1317–1319; 1984.

[24] Hayes RE, Thomas WJ, Hayes KE. A study of the nickel-catalyzed methanation reaction. *Journal of Catalysis* **92**(2):312–326; 1985.

[25] Araki M, Ponec V. Methanation of carbon monoxide on nickel and nickel-copper alloys. *Journal of Catalysis* **44**(3):439–448; 1976.

[26] Ponec V. Some Aspects of the Mechanism of Methanation and Fischer-Tropsch Synthesis. *Catal Rev -Sci Eng.* **18**(1):151–171; 1978.

[27] Galuszka J, Chang JR, Amenomiya Y. Disproportionation of carbon monoxide on supported nickel catalysts. *Journal of Catalysis* **68**(1):172–181; 1981.

[28] Yadav R, Rinker RG. Step-response kinetics of methanation over a nickel/alumina catalyst. *Ind Eng Chem Res.* **31**(2):502–508; 1992.

[29] Marquez-Alvarez C, Martin GA, Mirodatos A. *Mechanistic insights in the CO hydrogenation reaction over Ni/SiO₂.* In: Parmaliana A, Sanfilippo D, Frusteri F, Vaccari A, Arena F (eds.), Studies in Surface Science and Catalysis, Elsevier, pp. 155–160; 1998.

[30] Otarod M, Ozawa S, Yin F, Chew M, Cheh HY, Happel J. Multiple isotope tracing of methanation over nickel catalyst: III. Completion of 13C and D tracing. *Journal of Catalysis* **84**(1):156–169; 1983.

[31] Wentrcek PR, Wood BJ, Wise H. The role of surface carbon in catalytic methanation. *Journal of Catalysis* **43**(1/3):363–366; 1976.

[32] Rabo JA, Risch AP, Poutsma ML. Reactions of carbon monoxide and hydrogen on Co, Ni, Ru, and Pd metals. *Journal of Catalysis* **53**(3):295–311; 1978.

[33] Goodman DW, Kelley RD, Madey TE, Yates JJT. Kinetics of the hydrogenation of CO over a single crystal nickel catalyst. *Journal of Catalysis* **63**(1):226–234; 1980.

[34] Alstrup I. On the kinetics of co methanation on nickel surfaces. *Journal of Catalysis* **151**(1):216–225; 1995.

[35] Inoue H, Funakoshi M. Carbon monoxide methanation in a tubewall reactor. *Int Chem Eng.* **21**(2):276–283; 1981.

[36] Underwood RP, Bennett CO. The CO/H₂ reaction over nickel-alumina studied by the transient method. *Journal of Catalysis* **86**(2):245–253; 1984.

[37] Kopyscinski J, Schildhauer TJ, Vogel F, Biollaz SMA, Wokaun A. Applying spatially resolved concentration and temperature measurements in a catalytic plate reactor for the kinetic study of CO methanation. *Journal of Catalysis* **271**(2):262–279; 2010.

[38] Andersson MP, Abild Pedersen F, Remediakis IN, Bligaard T, Jones G, Engbaek J, et al. Structure sensitivity of the methanation reaction: H₂-induced CO dissociation on nickel surfaces. *Journal of Catalysis* **255**(1):6–19; 2008.

[39] Coenen JWE, van Nisselrooy PFMT, de Croon MHJM, van Dooren PFHA, van Meerten RZC. The dynamics of methanation of carbon monoxide on nickel catalysts. *Applied Catalysis* **25**:1–8; 1986.

[40] Vannice MA. The catalytic synthesis of hydrocarbons from H₂/CO mixtures over the group VIII metals: II. The kinetics of the methanation reaction over supported metals. *Journal of Catalysis* **37**(3):462–473; 1975.

[41] Vlasenko VM, Yuzefovich GE. Mechanism of the Catalytic Hydrogenation of Oxides of Carbon to Methane. *Russ Chem Rev* **38**:728–739; 1969.

[42] Huang CP, Richardson JT. Alkali promotion of nickel catalysts for carbon monoxide methanation. *Journal of Catalysis* **51**(1):1–8; 1978.

[43] Golodets GI. Mechanism and kinetics of CO hydrogenation on metals. *Theor Exp Chem* **21**(5):525–529; 1985.

[44] Sanchez Escribano V, Larrubia Vargas MA, Finocchio E, Busca G. On the mechanisms and the selectivity determining steps in syngas conversion over supported metal catalysts: An IR study. *Applied Catalysis A* **316**(1):68–74; 2007.

[45] Yang CH, Soong Y, Biloen P. A comparison of nickel- and platinum-catalyzed methanation, utilizing transient-kinetic methods. *Journal of Catalysis* **94**(1):306–309; 1985.

[46] Zarfl J, Ferri D, Schildhauer TJ, Wambach J, Wokaun A. DRIFTS study of a commercial Ni/Al_2O_3 CO methanation catalyst. *Applied Catalysis A* **324**:8–14; 2015.

[47] Czekaj I, Loviat F, Raimondi F, Wambach J, Biollaz S, Wokaun A. Characterization of surface processes at the Ni-based catalyst during the methanation of biomass-derived synthesis gas: X-ray photoelectron spectroscopy (XPS). *Appl Catal A* **329**:68–78; 2007.

[48] Kopyscinski J, Schildhauer TJ, Biollaz SMA. Production of synthetic natural gas (SNG) from coal and dry biomass – A technology review from 1950 to 2009. *Fuel* **89**(8):1763–1783; 2010.

[49] Rhyner U. *Reactive Hot Gas Filter for Biomass Gasification*. Dissertation, ETH Zürich, Nr. 21102, 2013.

[50] Kaufman Rechulski MD. *Catalysts for High Temperature Gas Cleaning in the Production of Synthetic Natural Gas from Biomass*. Dissertation, EPF Lausanne, Nr. 5484, 2012

[51] Seemann MC, Schildhauer TJ, Biollaz SMA. Fluidized bed methanation of wood-derived producer gas for the production of synthetic natural gas. *Ind Eng Chem Res* **49**(15):7034–7038; 2010.

[52] Kopyscinski J, Seemann MC, Moergeli R, Biollaz SMA, Schildhauer TJ. Synthetic natural gas from wood: Reactions of ethylene in fluidised bed methanation. *Applied Catalysis A: General* **462/463**:150–156; 2013.

[53] Sinfelt JH. Catalytic hydrogenolysis on metals. *Catalysis Letters* **9**(3/4):159–171; 1991.

[54] Ken-Ichi Tanaka KA, Takahiro M. Intermediates and carbonaceous deposits in the hydrogenolysis of ethane on a Ni-Al_2O_3 catalyst. *Journal of Catalysis* **81**:328–334; 1983.

[55] Goodman DW. Structure/reactivity relationships for alkane dissociation and Hydrogenolysis using single crystal kinetics. *Catalysis Today* **12**:189–199; 1992.

[56] Goodman DW. Ethane hydrogenolysis over single crystals of nickel: Direct detection of structure sensitivity. *Surface Science Letters* **123**(1):L679–L685; 1982.

[57] Leach HF, Mirodatos C, Whan DA. The exchange of methane, ethane, and propane with deuterium on silica-supported nickel catalyst. *Journal of Catalysis* **63**(1):138–151; 1980.

[58] Zarfl J, Schildhauer TJ, Wambach J, Wokaun A. Conversion of ethane/ethylene/acetylene under methanation conditions. Manuscript in preparation, 2016.

[59] Zarfl J. *Methanation of Biomass-Derived-Synthesis Gas - In Situ DRIFTS Studies over an Alumina Supported Nickel Catalyst*. Dissertation, ETH Zürich Nr. 22183, 2015.

[60] Chuang SC, Pien SI. Infrared studies of reaction of ethylene with syngas on Ni/SiO_2. *Catalysis Letters* **3**(4):323–329; 1989.

[61] Zaera F, Hall RB. Low temperature decomposition of ethylene over Ni(100): Evidence for vinyl formation. *Surface Science* **180**(1):1–18; 1987.

[62] Zhu XY, White JM. Interaction of ethylene and acetylene with Ni(111): A SSIMS study. *Surface Science* **214**:240–256; 1989.

[63] Zuhr RA, Hudson JB. The adsorption and decomposition of ethylene on Ni(110). *Surface Science* **66**(2):405–422; 1977.

[64] Zhu XY, Castro ME, Akhter S, White JM, Houston JE. C-H bond cleavage for ethylene and acetylene on Ni(100). *Surface Science* **207**(1):1–16; 1988.

[65] Zaera F, Hall RB. High-resolution electron energy loss spectroscopy and thermal programmed desorption studies of the chemisorption and thermal decomposition of ethylene and acetylene on nickel(100) single-crystal surfaces. *Journal of Physical Chemistry* **91**:4318–4323; 1987.

[66] Mo YH, Kibria AKMF, Nahm KS. The growth mechanism of carbon nanotubes from thermal cracking of acetylene over nickel catalyst supported on alumina. *Synthetic Metals* **122**(2):443–447; 2001.

[67] Sheppard N, Ward JW. Infrared spectra of hydrocarbons chemisorbed on silica-supported metals: I. Experimental and interpretational methods; acetylene on nickel and platinum. *Journal of Catalysis* **15**(1):50–61; 1969.

[68] Bridier B, Lopez N, Perez-Ramirez J. Partial hydrogenation of propyne over copper-based catalysts and comparison with nickel-based analogues. *Journal of Catalysis* **269**(1):80–92; 2010.

[69] Horiuti I, Polanyi M. Exchange reactions of hydrogen on metallic catalysts. *Transactions of the Faraday Society* **30**:1164–1172; 1934.

[70] Wasylenko W, Frei H. Direct Observation of Surface Ethyl to Ethane Interconversion upon C_2H_4 Hydrogenation over Pt/Al_2O_3 Catalyst by Time-Resolved FT-IR Spectroscopy. *Journal of Physical Chemistry B* **109**(35):16873–16878; 2005.

[71] Cremer PS, Su X, Shen YR, Somorjai GA. Ethylene Hydrogenation on Pt(111) Monitored in Situ at High Pressures Using Sum Frequency Generation. *Journal of the American Chemical Society* **118**(12):2942–2949; 1996.

[72] Trimm DL, Liu IOY, Cant NW. The oligomerization of acetylene in hydrogen over Ni/SiO_2 catalysts: Product distribution and pathways. *Journal of Molecular Catalysis A: Chemical* **288**:63–74; 2008.

[73] Spanjers CS, Held JT, Jones MJ, Stanley DD, Sim RS, Janik MJ, et al. Zinc inclusion to heterogeneous nickel catalysts reduces oligomerization during the semi-hydrogenation of acetylene. *Journal of Catalysis* **316**:164–173; 2014.

[74] Bartholomew CH. Mechanisms of catalyst deactivation. *Applied Catalysis A* **212**(1/2): 17–60; 2001.

[75] Rechulski MDK, Schildhauer TJ, Biollaz SMA, Ludwig C. Sulfur containing organic compounds in the raw producer gas of wood and grass gasification. *Fuel* **128**:330–339; 2014.

[76] Rhyner U, Edinger P, Schildhauer TJ, Biollaz SMA. Applied kinetics for modeling of reactive hot gas filters. *Applied Energy* **113**:766–780; 2014.

[77] Struis RPWJ, Schildhauer TJ, Czekaj I, Janousch M, Ludwig C, Biollaz SMA. Sulphur poisoning of Ni catalysts in the SNG production from biomass: A TPO/XPS/XAS study. *Applied Catalysis A* **362**(1/2):121–128; 2009.

[78] Twigg MV (ed.) *Catalyst Handbook*. Wolfe, London, 1989.

[79] Bartholomew CH. Carbon deposition in steam reforming and methanation. *Catalysis Review – Science and Engineering* **24**(1):67–117; 1982.

[80] Rostrup Nielsen JR. Industrial relevance of coking. *Catalysis Today* **37**(3):225–232; 1997.

[81] Trimm DL. Catalyst design for reduced coking (review). *Applied Catalysis* **5**(3): 263–290; 1983.

[82] McCarty JG, Wise H. Hydrogenation of surface carbon on alumina-supported nickel. *Journal of Catalysis* **57**(3):406–416; 1979.

[83] Figueiredo JL, Bernardo CA. Filamentous Carbon Formation on Metals and Alloys. *Carbon Fibers Filaments and Composites, NATO ASI Series* **177**:441–457; 1990.

[84] McCarty JG, Sheridan DM, Wise H, Wood BJ. *Hydrocarbon Reforming for Hydrogen Fuel Cells: A Study of Carbon Formation on Autothermal Reforming Catalysts.* Final report prepared for U.S. Department of Energy, 1981.

[85] Wiltner A. *Untersuchungen zur Diffusion und Reaktion von Kohlenstoff auf Nickel- und Eisenoberflächen sowie von Beryllium auf Wolfram.* Dissertation, Universität Bayreuth, 2004.

[86] Helveg S, Sehested J, Rostrup-Nielsen JR. Whisker carbon in perspective. *Catalysis Today* **178**(1):42–46; 2011.

[87] Otsuka K, Kobayashi S, Takenaka S. Catalytic decomposition of light alkanes, alkenes and acetylene over Ni/SiO_2. *Applied Catalysis A: General* **210**:371–379; 2001.

[88] Baumhakl C. *Direct Conversion of Higher Hydrocarbons During Methanation and Impact of Impurities.* Presentation at the Second Nuremberg Workshop Methanation and Second Generation Fuels, Nuremberg, 2014.

[89] Gardner DC, Bartholomew CH. Kinetics of Carbon Deposition during Methanation of CO. *Ind Eng Chem Prod Res Dev* **20**(1):80–87; 1981.

[90] Schildhauer TJ, Struis RPWJ, Bachelin D, Seemann MC, Damsohn M, Ludwig C, Biollaz SMA, Abolhassani-Dadras S. Determination of carbon deposition on nickel catalysts by temperature programmed reaction with steam. Manuscript in preparation, 2016.

[91] Schildhauer TJ, Settino J, Teske SL. Modelling study of fixed and fluidized bed reactors for CO_2 methanation in Power-to-Gas applications. Manuscript in preparation, 2016.

[92] Schildhauer TJ, Newson E, Wokaun A. Closed cross flow structures – Improving the heat transfer in fixed bed reactors by enforcing radial convection. *Chemical Engineering and Processing: Process Intensification* **48**(1):321–328; 2009.

[93] Schildhauer TJ, Pangarkar K, van Ommen JR, Nijenhuis J, Moulijn JA, Kapteijn F. Heat transport in structured packings with two-phase co-current downflow. *Chemical Engineering Journal* **185/186**:250–266; 2012.

[94] Harms H, Höhlein B, Skov A. Methanisierung kohlenmonoxidreicher Gase beim Energie-Transport. *Chemie Ingenieur Technik* **52**(6):504–515; 1980.

[95] Rönsch S, Matthischke S, Müller M, Eichler P. Dynamische Simulation von Reaktoren zur Festbettmethanisierung. *Chemie Ingenieur Technik* **86**(8):1198–1204; 2014.

[96] Moeller FW, Ros H, Britz B. Methanation of coal gas for SNG. *Hydrocarbon Processing* 1974:69–74; **1974**.

[97] Eckle S. *SNG Technologies at Clariant.* Presentation at the 2nd Nuremberg Workshop Methanation and Second Generation Fuels,, Nuremberg, 2014.

[98] Eisenlohr KH, Moeller FW, Dry M. Influence of certain reaction parameters on methanation of coal gas to SNG. *ACS Fuels Division Preprints* **19**(3):1–9; 1974.

[99] GPGP. *Practical Experience Gained During the First Twenty Years of Operation of the Great Plains Gasification Plant and Implications for Future Projects.* US Department of Energy, Office of Fossil Energy, Washington, DC, 2006.

[100] Miller WR, Honea FI. *Great Plains Coal Gasification Plant Start-Up and Modification Report*. Fluor Technology Inc., Great Plains, 1986.

[101] Perry M, Eliason D. CO_2 *Recovery and Sequestration at Dakota Gasification Company*. Gasification Technology Conference, p. 35; 2004.

[102] Krier C, Hackel M, Hägele C, Urtel H, Querner C, Haas A. Improving the Methanation Process. *Chemie Ingenieur Technik* **85**(4):523–528; 2013.

[103] Pedersen K, Skov A, Rostrupnielsen J. Catalytic Aspects Of High-Temperature Methanation. *Abstracts Of Papers Of The American Chemical Society* **179**:60; 1980.

[104] Hoehlein B, Menzer R, Range J. High temperature methanation in the long-distance nuclear energy transport system. *Applied Catalysis* **1**:125–139; 1981.

[105] Rostrup-Nielsen JR, Skov A, Christiansen LJ. Deactivation in pseudo-adiabatic reactors. *Applied Catalysis* **22**(1):71–83; 1986.

[106] Rostrup Nielsen JR, Pedersen K, Sehested J. High temperature methanation: Sintering and structure sensitivity. *Applied Catalysis, A* **330**:134–138; 2007.

[107] Nguyen TTM. *Topsoe's Synthesis Technology for SNG with Focus on Methanation in General and Bio-SNG in Particular*. Presentation at the First International Conference on Renewable Energy Gas Technology (REGATEC). Malmö, 2014.

[108] Nguyen TTM, Wissing L, Skjoth-Rasmussen MS. High temperature methanation: Catalyst considerations. *Catalysis Today* **215**:233–238; 2013.

[109] Li C. *Current Development Situation of Coal to SNG in China*. Presentation at IEA-MOST Workshop: Advances in deployment of fossil fuel technologies. Beijing, 2014.

[110] Göteborg Energi. Press release, 18.12.2014. http://gobigas.goteborgenergi.se/En/News (accessed 15 December 2015).

[111] Lohmüller R. Methansynthese mit kombinierten isothermen und adiabaten Reaktoren. *Linde Berichte aus Technik und Wissenschaft* **41**:3–11; 1977.

[112] etogas, Home page. www.etogas.de, 2012 (accessed 15 December 2015).

[113] Dixon AG. An improved equation for the overall heat transfer coefficient in packed beds. *Chemical Engineering and Processing: Process Intensification* **35**(5):323–331; 1996.

[114] VDI. *VDI-Wärmeatlas. 6th edn*, VDI-Verlag, Dusseldorf; 2006.

[115] Pangarkar K, Schildhauer TJ, van Ommen JR, Nijenhuis J, Moulijn JA, Kapteijn F. Heat transport in structured packings with co-current downflow of gas and liquid. *Chemical Engineering Science* **65**(1):420–426; 2010.

[116] Cybulski A, Eigenberger G, Stankiewicz A. Operational and Structural Nonidealities in Modeling and Design of Multitubular Catalytic Reactors. *Industrial and Engineering Chemistry Research* **36**(8):3140–3148; 1997.

[117] Bey O, Eigenberger G. Gas flow and heat transfer through catalyst filled tubes. *International Journal of Thermal Sciences* **40**(2):152–164; 2001.

[118] Tsotsas E. Transportvorgänge in Festbetten Geschichte, Stand und Perspektiven der Forschung. *Chemie Ingenieur Technik* **64**(4):313–322; 1992.

[119] Tronconi E, Groppi G. A study on the thermal behavior of structured plate-type catalysts with metallic supports for gas/solid exothermic reactions. *Chemical Engineering Science* **55**(24):6021–6036; 2000.

[120] Gunn DJ, Khalid M. Thermal dispersion and wall heat transfer in packed beds. *Chemical Engineering Science* **30**(2):261–267; 1975.

[121] Eigenberger G, Kottke V, Daszkowski T, Gaiser G, Kern HJ. Regelmässige Katalysatorformkörper für technische Synthesen, Fortschritt-Berichte VDI, Reihe 15. *Umwelttechnik* **112**; 1993.

[122] Schlereth D, Hinrichsen O. A fixed-bed reactor modeling study on the methanation of CO_2. *Chemical Engineering Research and Design* **92**(4):702–712; 2014.

[123] Schaaf T, Grünig J, Schuster M, Orth A. Speicherung von elektrischer Energie im Erdgasnetz – Methanisierung von CO_2-haltigen Gasen. *Chemie Ingenieur Technik* **86**(4):476–485; 2014.

[124] Pangarkar K, Schildhauer TJ, van Ommen JR, Nijenhuis J, Moulijn JA, Kapteijn F. Experimental and numerical comparison of structured packings with a randomly packed bed reactor for Fischer-Tropsch synthesis. *Catalysis Today* **147**(Supplement):S2–S9; 2009.

[125] Groppi G, Tronconi E. Design of novel monolith catalyst supports for gas/solid reactions with heat exchange. *Chemical Engineering Science* **55**(12):2161–2171; 2000.

[126] Montebelli A, Visconti CG, Groppi G, Tronconi E, Kohler S. Optimization of compact multitubular fixed-bed reactors for the methanol synthesis loaded with highly conductive structured catalysts. *Chemical Engineering Journal* **255**:257–265; 2014.

[127] Tronconi E, Groppi G, Boger T, Heibel A. Monolithic catalysts with 'high conductivity' honeycomb supports for gas/solid exothermic reactions: characterization of the heat-transfer properties. *Chemical Engineering Science* **59**:4941–4949; 2004.

[128] Sudiro M, Bertucco A, Groppi G, Tronconi E. *Simulation of a Structured Catalytic Reactor for Exothermic Methanation Reactions Producing Synthetic Natural Gas*. In: Pierucci S, Ferraris GB (eds) Computer Aided Chemical Engineering. pp. 691–696, Elsevier, Amsterdam, 2010.

[129] Bajohr S, Schollenberger D, Götz M. *Methanation with Honeycomb Catalysts*. Presentation at the 2nd Nuremberg Workshop Methanation and Second Generation Fuels. Nuremberg, 2014.

[130] Schildhauer TJ, Geissler K. Reactor concept for improved heat integration in autothermal methanol reforming. *International Journal of Hydrogen Energy* **32**(12):1806–1810; 2007.

[131] Velocys, Home page. www.velocys.com; 2012 (accessed 15 December 2015).

[132] Liu Z, Chu B, Zhai X, Jin Y, Cheng Y. Total methanation of syngas to synthetic natural gas over Ni catalyst in a micro-channel reactor. *Fuel* **95**:599–605; 2012

[133] Greyson M, Demeter JJ, Schlesinger MD, Johnson GE, Jonakin J, Myers JW. *Synthesis of Methane*. Report of Investigation 5137, Department of the Interior, Bureau of Mines; 1955.

[134] Schlesinger MD, Demeter JJ, Greyson M. Catalyst for Producing Methane from Hydrogen and Carbon Monoxide. *Industrial Engineering Chemistry* **48**(1):68–70; 1956.

[135] Streeter RC, Anderson DA, Cobb JJT. Status of the Bi-Gas program – part II: evaluation of fluidized bed methanation catalysts. Proceedings of the Eigth Synthetic Pipeline Gas Symposium, p. 57; 1976.

[136] Graboski MS, Diehl EK. Design and operation of the BCR fluidized bed methanation PEDU. *Proceedings of the Fifth Synthetic Pipeline Gas Symposium*, p. 89; 1973.

[137] Streeter RC. Recent developments in fluidized-bed methanation research. *Proceedings of Ninth Synthetic Pipeline Gas Symposium. Proceedings of Ninth Synthetic Pipeline Gas Symposium*, pp. 153–165; 1977.

[138] Cobb JT Jr, Streeter RC. Evaluation of fluidized-bed methanation catalysts and reactor modeling. *Ind Eng Chem Process Des Dev* **18**(4):672–679; 1979.

[139] Alcorn WR, Cullo LA. *Nickel-Copper-Molybdenum Methanation Catalyst.* United States Patent 3962140; 1976.

[140] Graboski MS, Donath EE. *Combined Shift and Methanation Reaction Process for the Gasification of Carbonaceous Materials.* United States Patent 3904386; 1975.

[141] Friedrichs G, Proplesch P, Wismann G, Lommerzheim W. *Methanisierung von Kohlenvergasungsgasen im Wirbelbett Pilot Entwicklungsstufe, Technologische Forschung und Entwicklung – Nichtnukleare Energietechnik.* Prepared for Bundesministerium fuer Forschung und Technologie, Thyssengas GmbH, 1985.

[142] Hedden K, Anderlohr A, Becker J, Zeeb HP, Cheng YH. *Gleichzeitige Konvertierung und Methanisierung von CO-reichen Gasen.* Prepared for Bundesministerium für Forschung und Technologie, Forschungsbericht T 86-044, DVGW-Forschungsstelle Engler-Bunte-Institut, Universität Karlsruhe, 1986.

[143] Lommerzheim W, Flockenhaus C. One stage combined shift-conversion and partial methanation process for upgrading synthesis gas to pipeline quality. *Proceedings of the Tenth Synthetic Pipeline Gas Symposium*, pp. 439–451, 1978.

[144] Zeeb HB. *Desaktivierung von Nickelkatalysatoren bei der Methanisierung von Wasserstoff/Kohlenmonoxid-Gemischen unter Druck.* Dissertation, Universität Karlsruhe, 1979.

[145] Becker J. *Untersuchungen zur Desaktivierung von Nickelkatalysatoren und Kinetik der gleichzeitigen Methanisierung und Konvertierung CO-reicher Gase unter Druck.* Dissertation, Universität Karlsruhe, 1982.

[146] Anderlohr A. *Untersuchungen zur gleichzeitigen Methanisierung und Konvertierung CO-Reicher Gase in einer katalytischen Wirbelschicht.* Dissertation, Universität Karlsruhe, 1979.

[147] Cheng YH. *Untersuchungen zur gleichzeitigen Methanisierung und Konvertierung CO-reicher Synthesegase in Gegenwart von Schwefelwasserstoff.* Dissertation, Universität Karlsruhe, 1983.

[148] Seemann MC, Schildhauer TJ, Biollaz SMA, Stucki S, Wokaun A. The regenerative effect of catalyst fluidization under methanation conditions. *Applied Catalysis A: General* **313**(1):14–21; 2006.

[149] Kopyscinski J, Schildhauer TJ, Biollaz SMA. Methanation in a fluidized bed reactor with high initial CO partial pressure: Part I Experimental investigation of hydrodynamics, mass transfer effects, and carbon deposition. *Chemical Engineering Science* **66**(5):924–934; 2011.

[150] Kopyscinski J, Schildhauer TJ, Biollaz SMA. Methanation in a fluidized bed reactor with high initial CO partial pressure: Part 2 Modeling and sensitivity study. *Chemical Engineering Science* **66**(8):1612–1621; 2011.

[151] Biollaz SMA, Schildhauer TJ, Ulrich D, Tremmel H, Rauch R, Koch M. Status report of the demonstration of BioSNG production on a 1 MW SNG scale in Güssing. *Proceedings of the 17th European Biomass Conference and Exhibition*, p. 125, 2009.

[152] Galnares A. *The Gaya Project and GDF SUEZ's activities in the field of BioSNG.* Presentation at the Second Nuremberg Workshop Methanation and Second Generation Fuels. Nuremberg, 2014.

[153] Guerrini O, Perrin P, Marchand B, Prieur-Vernat A. Second Generation Gazeous Biofuels: from Biomass to Gas Grid. *Oil and Gas Science and Technology – Rev. IFP Energies Nouvelles* **68**(5): 925–934; 2013.

[154] Liu J, Shen W, Cui D, Yu J, Su F, Xu G. Syngas methanation for substitute natural gas over Ni-Mg/Al$_2$O$_3$ catalyst in fixed and fluidized bed reactors. *Catalysis Communications* **38**:35–39; 2013.

[155] Li J, Zhou L, Li P, Zhu Q, Gao J, Gu F, Su F. Enhanced fluidized bed methanation over a Ni/Al$_2$O$_3$ catalyst for production of synthetic natural gas. *Chemical Engineering Journal* **219**:183–189; 2013.

[156] Götz M, Bajohr S, Graf F, Reimert R, Kolb T. Einsatz eines Blasensäulenreaktors zur Methansynthese. *Chemie Ingenieur Technik* **85**(7):1146–1151; 2013.

[157] Götz M. *Recent Developments in Three Phase Methanation*. Presentation at the Second Nuremberg Workshop Methanation and Second Generation Fuels, Nuremberg, 2014.

[158] Frank ME, Sherwin MB, Blum DB, Mednick RL. Liquid phase methanation – shift PDU results and pilot plant status. *Proceedings of the Eighth Synthetic Pipeline Gas Symposium*. pp. 159–179, American Gas Association, Chicago, 1976.

[159] Frank ME, Mednick RL. Liquid phase methanation pilot plant results. *Proceedings of the Ninth Synthetic Pipeline Gas Symposium*. pp. 185–191, American Gas Association, Chigago, 1977.

[160] ChemSystem. *Liquid Phase Methanation/Shift*. Prepared for U.S. Energy Research and Development Administration, NO. E-(49-18)-1505, Chem System Inc., 1976.

[161] ChemSystem. *Liquid Phase Methanation/Shift – Pilot plant operation and laboratory support work*. Prepared for DoE (No. E4-75-C-01-2036), Chem System Inc., 1979.

[162] Teske SL, Couckuyt I, Schildhauer TJ, Biollaz S, Maréchal F. Integrating rate based models into multi-objective optimisation of process designs using surrogate models. *Proceedings of the 26th International Conference on Efficiency, Cost, Optimization, Simulation and Environmental Impact of Energy System*, p.143, 2013.

[163] Gassner M. *Process Design Methodology for Thermochemical Production of Fuels from Biomass – Application to the Production of Synthetic Natural Gas from Lignocellulosic Resources*. Dissertation, EPF Lausanne, 2010.

[164] Heyne S. *Bio-SNG from Thermal Gasification - Process Synthesis, Integration and Performance*. Dissertation, Chalmers University of Technology, Gothenburg, 2013.

[165] Teske SL. *Integrating Rate Based Models into a Multi-Objective Process Design and Optimisation Framework using Surrogate Models*. Dissertation, EPF Lausanne, 2014.

[166] Zhang J, Fatah N, Capela S, Kara Y, Guerrini O, Khodakov AY. Kinetic investigation of carbon monoxide hydrogenation under realistic conditions of methanation of biomass derived syngas. *Fuel* **111**:845–854; 2013.

[167] Hannoun H, Regalbuto JR. Mixing characteristics of a micro-Berty catalytic reactor. *Industrial and Engineering Chemistry Research* **31**(5):1288–1292; 1992.

[168] Horn R, Williams KA, Degenstein NJ, Schmidt LD. Syngas by catalytic partial oxidation of methane on rhodium: Mechanistic conclusions from spatially resolved measurements and numerical simulations. *Journal of Catalysis* **242**(1):92–102; 2006.

[169] Horn R, Degenstein NJ, Williams KA, Schmidt LD. Spatial and temporal profiles in millisecond partial oxidation processes. *Catalysis Letters* **110**(3/4):169–178; 2006.

[170] Horn R, Williams KA, Degenstein NJ, Schmidt LD. Mechanism of H_2 and CO formation in the catalytic partial oxidation of CH_4 on Rh probed by steady-state spatial profiles and spatially resolved transients. *Chemical Engineering Science* **62**(5):1298–1307; 2007.

[171] Dalle Nogare D, Degenstein NJ, Horn R, Canu P, Schmidt LD. Modeling spatially resolved profiles of methane partial oxidation on a Rh foam catalyst with detailed chemistry. *Journal of Catalysis* **258**(1):131–142; 2008.

[172] Michael BC, Donazzi A, Schmidt LD. Effects of H_2O and CO_2 addition in catalytic partial oxidation of methane on Rh. *Journal of Catalysis* **265**(1):117–129; 2009.

[173] Schildhauer T, Newson E, Müller S. The equilibrium constant for the methylcyclohexane–toluene system. *Journal of Catalysis* **198**(2):355–358; 2001.

[174] Bosco M, Vogel F. Optically accessible channel reactor for the kinetic investigation of hydrocarbon reforming reactions. *Catalysis Today* **116**(3):348–353; 2006.

[175] Bosco M. *Kinetic Studies of the Autothermal Gasoline Reforming for Hydrogen Production for Fuel Cell Applications.* Dissertation, ETH Zürich, 2006.

[176] Farrell RJ, Ziegler EN. Kinetics and mass transfer in a fluidized packed-bed: Catalytic hydrogenation of ethylene. *AIChE Journal* **25**(3):447–455, 1979.

[177] Tschedanoff V, Maurer S, Schildhauer TJ, Biollaz SMA. *Advanced two-phase model supporting the scale up of a fluidized bed methanation reactor.* Presentation at the Ninth International Symposium on Catalysis in Multiphase Reactors, December 2014, Valpré, Lyon, 2014.

[178] Levenspiel O. *Chemical Reaction Engineering*, 3rd edn. John Wiley & Sons, Inc. New York, 1999.

[179] Froment GF, Bischoff KB. *Chemical Reactor Analysis and Design.* John Wiley & Sons, Inc., New York, 1990.

[180] Baerns M, Hofmann H, Renken A. *Chemische Reaktionstechnik – Lehrbuch der Technischen Chemie.* Georg Thieme Verlag, Stuttgart, 1987.

[181] Parlikkad NR, Chambrey S, Fongarland P, Fatah N, Khodakov A, Capela S, Guerrini O. Modeling of fixed bed methanation reactor for syngas production: Operating window and performance characteristics. *Fuel* **107**:254–260; 2013.

[182] Xu J, Froment GF. Methane steam reforming, methanation and water-gas shift: I. Intrinsic kinetics. *AIChE Journal* **35**(1):88–96; 1989.

[183] Schildhauer TJ. *Untersuchungen zur Verbesserung des Wärmeübergangs in katalytischen Festbettreaktoren für Energiespeicheranwendungen.* Dissertation, ETH Zürich Nr. 14301, 2001.

[184] May WG. Fluidized-bed reactor studies. *Chemical Engineering Progress* **55**(12): 49–56; 1959.

[185] van Deemter JJ. Mixing and contacting in gas-solid fluidized beds. *Chemical Engineering Science* **13**(3):143–154; 1961.

[186] Bellagi A. *Zur Reaktionstechnik der Methanisierung von Kohlenmonoxid in der Wirbelschicht.* Dissertation, RWTH Aachen, 1979.

[187] Kai T, Furusaki S, Yamamoto K. Methanation of Carbon Monoxide by a fluidized catalyst bed. *Journal of Chemical Enginering Japan* **17**(3):280–285; 1984.

[188] Liu Y, Hinrichsen O. CFD simulation of hydrodynamics and methanation reactions in a fluidized-bed reactor for the production of synthetic natural gas. *Industrial and Engineering Chemistry Research* **53**(22):9348–9356; 2014.

[189] Rüdisüli M, Schildhauer TJ, Biollaz SMA, van Ommen JR. Bubble characterization in a fluidized bed with vertical tubes. *Industrial and Engineering Chemistry Research* **51**:4748–4758; 2012.

[190] Maurer S, Wagner STJ E C, Biollaz SMA, van Ommen JR. Bubble size in fluidized beds with and without vertical internals. *Proceedings of the 11th International Conference on Fluidized Bed Technology*. 2014:577, 2014.

[191] Glicksman LR. Scaling relationships for fluidized beds. *Chemical Engineering Sciences* **39**(9):1373–1379; 1984.

[192] Rüdisüli M, Schildhauer TJ, Biollaz SMA, van Ommen JR. Evaluation of a sectoral scaling approach for bubbling fluidized beds with vertical internals. *Chemical Engineering Journal* **197**:435–439; 2012.

[193] Maurer S, Schildhauer TJ, van Ommen JR, Biollaz SMA, Wokaun A. Scale-up of fluidized beds with vertical internals: Studying the sectoral approach by means of optical probes. *Chemical Engineering Journal* **252**:131–140; 2014.

[194] Kemmer H. Die Kohlenoxydreinigung (Entgiftung) des Stadtgases. *Angewandte Chemie* **49**(7):133–137; 1936.

5

SNG UPGRADING

Renato Baciocchi, Giulia Costa, and Lidia Lombardi

5.1 INTRODUCTION

The gas stream leaving a catalytic methanation reactor does not fulfil the quality specifications of the natural gas grid. Before injection into the natural gas grid, an up-grading step is necessary. In many cases where a nearly stoichiometric gas mixture is fed to the methanation reactor, this up-grading can be a simple condensation and drying step, because water is the only other reaction product besides methane. As discussed in chapter 4 of this book, stoichiometric mixtures of hydrogen and carbon oxides (H_2/CO slighty above three, H_2/CO_2 slighty above four, respectively) are typical for the coal to SNG processes by Lurgi, Davy Process Technology and Haldor Topsoe (TREMP® process), but also for some of the power to gas configurations.

In other processes, the water gas shift reaction to adapt the H_2/CO ratio is not conducted upstream, but together with methanation in one step, for example, in the fluidized bed methanation process developed by PSI. This results in a substantial content of carbon dioxide in the raw SNG leaving the methanation unit. Further processes use the CO_2 stemming from the gasification and/or the upstream water gas shift section as ballast gas in the methanation unit to dampen the temperature increase. This is the case for the VESTAS process by Foster–Wheeler/Clariant or the pre-methanator in the GoBiGas 20 MW_{SNG} demonstration plant (refer to chapters 4 and 6 of this book). In all these cases, the raw SNG will contain methane, carbon dioxide, water vapor, and hydrogen. Therefore, the up-grading can be discussed for one example, but the results will hold for all these processes.

Synthetic Natural Gas from Coal, Dry Biomass, and Power-to-Gas Applications, First Edition.
Edited by Tilman J. Schildhauer and Serge M.A. Biollaz.

This chapter aims at investigating and comparing the separation processes needed for upgrading synthetic natural gas (SNG) produced in second generation biomass conversion processes, which are run under operating conditions properly chosen in order to maximize the methane yield. Two process routes were considered as reference for this chapter: an autothermal dual fluidized bed gasification process followed by pressurized fluidized bed methanation, which is implemented in the Güssing pilot plant (confer also to chapter 8), and a hydrothermal gasification/methanation process [1], which is further discussed in chapter 10 within this book.

The typical expected compositions of the raw SNG obtained by each process route are reported in Table 5.1. SNG from biomass can be fed to the natural gas grid, provided that it complies with the pipeline feed-in specifications, also shown in Table 5.1. Methane and ethane are obviously the most valuable components in SNG and their recovery is also important in order to avoid uncontrolled emissions to the atmosphere, which could be harmful considering for instance the high greenhouse potential of methane (21 times that of carbon dioxide). In the framework of the actions aimed at climate change mitigation, rather pure carbon dioxide (95% minimum) may also represent a valuable product. In the near future, carbon dioxide storage sites are expected to be developed, and carbon dioxide could be transported to the storage sites through dedicated pipelines [2].

As shown in Table 5.1, besides the bulk separation of CO_2 from CH_4, the achievement of the natural gas pipeline specifications also requires the removal of minor constituents of the SNG. Among these, the most critical species are water, hydrogen, and nitrogen. In fact, provided that a separation process for removing CO_2 from the SNG is used, both the hydrogen and nitrogen concentrations of the SNG are expected to rise to around 4%, and the CH_4 purity will be at best 92%, that is well below the minimum specified purity of 96%. Additionally, for the hydrothermal route, after CO_2/CH_4 separation, the H_2 concentration is expected to increase to 8%, that is well above the maximum limit specified for pipeline natural gas. In addition, the SNG generated from both routes is basically saturated with water vapor at the relevant

TABLE 5.1   Specifications of: (i) SNG Produced via Pressurized Methanation (Güssing), (ii) SNG Produced via Hydrothermal Gasification, (iii) Pipeline Feed-In Natural Gas, (iv) Carbon Dioxide for Storage (Data taken from [1]).

Component	Specifications (% vol)			
	Güssing	Hydrothermal	Pipeline	Storage
CO_2	36	28–60	<6	>95
CH_4	57	40–70	>96	—
H_2	2.5	1–10	<5	—
CO	0.10	0.1–1.0	<1	—
Hydrocarbons	0.05	1–3	Not specified	—
NH_3	ppm	Traces	Not specified	—
N_2	2.50	0	—	—
H_2O	1.23	1–10	<60% R.H.	—

operating conditions (temperature and pressure); hence it needs to be dehydrated to comply with the water dew point requirements for pipeline transport (see Table 5.1). In fact, over long distance transport, if water were not removed, condensation could occur causing a reduction in the volumetric capacity of the system, as well as an increase of the operating pressure. Furthermore, particularly when CO_2 and H_2S are present in the gas, dehydration is necessary to prevent corrosion or erosion problems in pipelines and equipment [3].

5.2 SEPARATION PROCESSES FOR SNG UPGRADING

As discussed above, in order to meet pipeline specifications, SNG needs to be upgraded by removal of carbon dioxide, water, hydrogen, and eventually also nitrogen. This means that at least three binary separations need to be dealt with: CH_4/CO_2, CH_4/H_2, CH_4/H_2O. The separation processes available for these purposes and their technical and economic feasibility are briefly discussed in the following paragraphs. It should be noted that since currently there are only few industrial scale applications of SNG production where CO_2 removal downstream of methanation is needed, the description and evaluation of the most suitable upgrading techniques for SNG treatment are based on data available for other upgrading applications performed on gas streams presenting analogies to SNG such as biogas upgrading.

5.2.1 Bulk CO_2/CH_4 Separation

The bulk separation of CH_4 and CO_2 may be achieved by applying one of the following process options: physical or chemical absorption, membrane separation, adsorptive separation, or low temperature separation. Although the choice of the most effective separation option is the result of a detailed techno-economical optimization that considers the integration of the treatment in the whole SNG generation process (see for instance [4]), some general considerations are provided in this section that may help in performing at least a preliminary selection of potentially suitable separation processes to be further investigated.

5.2.1.1 Separation by Absorption The separation of carbon dioxide from process gas streams with solvents is already a commercial option for different industrial applications, especially in the petrochemical sector [2]. Absorption is typically performed using solvents characterized by high solubility of carbon dioxide (physical absorption) or alkaline solvents that can react with carbon dioxide (chemical absorption).

Chemical absorption using aqueous solution of alkanolamines is currently considered the benchmark separation process for natural gas sweetening: carbon dioxide is separated from the gas stream by contacting it with a solution of either monoethanolamine (MEA), diethanolamine (DEA), or methyl diethanolamine (MDEA) at relatively low temperature (40–60 °C) and at atmospheric pressure in a packed absorption column, where the amines react with CO_2 forming the corresponding carbamates.

The spent amine solution is then heated with steam up to typically 100–140 °C, allowing on the one hand to regenerate the amines and on the other hand to strip CO_2, that is commonly vented to the atmosphere but can possibly be used or stored, as it is the case of the CCS plants of Sleipner, Snohvit, and In-Salah [5]. Due to their high CO_2 absorption capacity also at low partial pressure, separation with amines is currently being proposed as the benchmark process also for carbon dioxide capture from flue gas, as a possible Greenhouse gases mitigation option [2]. Other feasible chemical solvents include hot potassium carbonate and ammonia. Potassium carbonate solutions absorb carbon dioxide at high pressure (typically 1 MPa) and relatively high temperature (>100 °C), leading to the formation of a solution rich in bicarbonate ions. The reaction is reversed by simply depressurizing the liquid solution, thus allowing to release pure CO_2, while regenerating the carbonates in the solution that are recycled to the CO_2 absorption step [6]. The main patented processes based on hot potassium carbonate are the UOP Benfield process and the Catacarb process [7]. These processes are commercially offered for applications at a minimum CO_2 partial pressure of 210–345 kPa [7]. UOP recommends the application of the Benfield process for feed conditions ranging between 10 and 120 bar total pressure, with acid gas concentration between 5 and 35% [8]. The comparative analysis of the CO_2 vapor–liquid equilibria with potassium carbonate resulting for different activators suggests to apply this process for CO_2 partial pressures above 700 kPa. Most of the applications of hot potassium carbonate are for acid gas separation in the ammonia, hydrogen and ethylene oxide plants [7]. Ammonia has been originally proposed for processes operating at ambient temperature, that is, 20–25 °C; more recently the so-called chilled ammonia process, operating between 0 and 10 °C, has been developed by Alstom [9]. The low temperature of the chilled ammonia process allows minimizing the loss of ammonia during the CO_2 absorption step, which is the main drawback of the ambient temperature process. Regeneration takes place at high temperature (50–200 °C) and pressure (2–136 bar), the latter required in order to avoid ammonia losses into the vapor phase [10]. If applied for post-combustion capture of flue gases, the chilled ammonia process allows for a reduction of the energy consumption in the desorber compared to the energy consumption of the process using aqueous alkanolamines, and it benefits from the use of potentially less harmful solvents [10].

Physical absorption is also used for CO_2 separation in many industrial applications and also in commercial coal to SNG plants to remove CO_2 together with impurities upstream of the methanation unit (see also previous chapters). As the CO_2 loading capacity of the solvent depends in this case on the CO_2 partial pressure in the gas stream to be treated, physical absorption is ideally suitable for feeds characterized by high total pressure and CO_2 concentration. There are a few patented processes based on physical absorption that differ mainly in the solvent employed. These are the Selexol process (Dow Chemical), based on dimethylethers of polyethylene glycol (DMPEG), the Purisol process (Lurgi), based on n-methyl-2-pirolydone (DMP), the Rectisol process (Lurgi and Linde), based on chilled methanol, and the Fluor Solvent one (Fluor), based on propylene carbonate [11]. The advantage of physical solvents with respect to chemical ones is that regeneration can be obtained with lower energy requirements, as there is no chemical bond to break. On the other hand, physical solvents typically require a higher pressure to achieve CO_2 loading rates as high as those of chemical solvents. For instance the typical operating pressure of the Selexol process is 3 MPa at approximately 313 K, whereas those of the Rectisol process are 8 MPa at 213–263 K [12]. Making reference to Selexol, this solvent has a very high CO_2/H_2 selectivity (around 45) and is suggested for acid gas partial pressures above 300–350 KPa, making it a very suitable candidate for CO_2 capture from Syngas obtained from gasification plants. Selexol also guarantees a fair CO_2/CH_4 selectivity (around 9), making it in principle also suitable for SNG upgrading [13].

5.2.1.2 Separation with Membranes
Membrane separation is based on the selective permeability property of membranes [14], which allows enrichment of the permeate leaving the membrane module in the more permeable component, whereas the retentate is enriched in the less permeable one.

Membrane materials can be classified according to two main criteria, these being the transport mechanism and the nature of the constituent material. Considering the first criteria, there are mainly three types of transport mechanisms: diffusion, solution–diffusion, and facilitated transport. In the last group, the so-called carriers (which are responsible for the facilitated transport of the target molecules) can be either fixed or mobile. According to the second criteria (i.e., materials), membranes are divided in three groups: organic polymers, inorganic, and mixed matrix membranes (MMMs). More details on the materials properties and selection criteria can be found in [15].

Good membrane materials must be characterized by both high selectivity and high permeability. High selectivity is needed to achieve a high purity product, while high permeability is desired to minimize membrane area and thus the capital cost. Unfortunately, as shown in Figure 5.1, polymeric materials are characterized by a rather general trade-off between permeability and selectivity, which is well described by the Robeson's upper limit [16].

Among the glassy polymeric membranes, commercial cellulose acetate (CA) is a type of CO_2 selective membrane widely applied on a large scale for the separation of CO_2 from crude natural gas as well as from mixtures with hydrocarbons in enhanced oil recovery operations; however, it should be noted that CA is characterized by both selectivity ($\alpha = 21$) and permeability values quite far from the Robeson upper limit [17]. A commercial polyimide membrane (Matrimid) offers slightly improved performance, with CO_2/CH_4 selectivity and CO_2 permeability up to 30 and 10 barrers (1 barrer $= 10^{-10}$ cm^3 (STP)·cm·cm^{-2}·s^{-1}·cm-Hg^{-1}), respectively (see again Figure 5.1). Nevertheless, both these materials lose their performance during operation. For instance, the actual selectivity of CA membranes usually drops to 12–15 as a result of swelling-induced plasticization of the material in contact with CO_2 and a similar loss of performance is reported for commercial polyimide membranes [16]. Cross-linking or the addition of functional groups has been found to be effective in improving the trade-off between selectivity and permeability [16], while increasing the stability of the membrane also in aggressive conditions [18].

A completely different alternative is provided by inorganic materials: zeolites, as SAPO-34 [19], carbon molecular sieves [20, 21], mixed matrix membranes [22], and

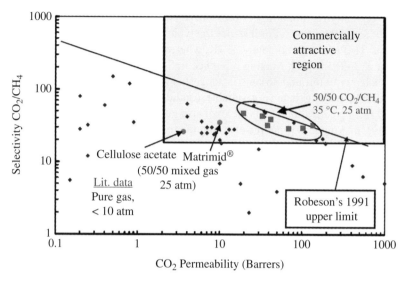

FIGURE 5.1 Trade-off between CO_2/CH_4 selectivity and CO_2 permeability. Reproduced from [16]. *Source*: Wind 2004 [16]. Reproduced with permission of Elsevier.

facilitated transport membranes [23, 24]. These may provide a way of overcoming Robeson's upper limit, although most of them are not yet ready for commercial application, due to low stability under real operating conditions or simply too high costs [25].

As far as the process configuration is concerned, a single membrane stage cannot typically achieve the required purity and recovery of methane in the retentate. For this reason, it is common practice to use a cascade of membrane stages that can be combined using different configurations [26]. Recently, three configurations were identified as the most promising ones for CO_2/CH_4 separation from SNG: a two-stage countercurrent recycling, a three-stage countercurrent recycling, and a three-stage with one common recycle loop (see Figure 5.2 for the summary of all investigated configurations) [4].

5.2.1.3 Bulk CO_2/CH_4 Separation by Adsorption

Adsorptive separation may also represent an effective tool for addressing the separation of a CH_4/CO_2 stream to obtain at the same time a pure CH_4 stream and a high purity CO_2 stream. The separation of CH_4/CO_2 mixtures was investigated using either equilibrium adsorbents, such as activated carbon, zeolite 13X, zeolite 5A, silica–gel, and metal–organic frameworks, or adsorbents that operate under the kinetic control regime, such as carbon molecular sieves (CMS), clinoptilolites, titanosilicates, DDR zeolites, and SAPO-34 [27]. In both cases, the most retained compound (heavy product, HP) is CO_2, whereas the less retained one (light product, LP) is CH_4.

Adsorptive separation is typically applied using the pressure swing adsorption (PSA) process, which is a cyclic process where adsorption is performed at higher pressure and desorption at lower pressure, possibly operating the latter step under

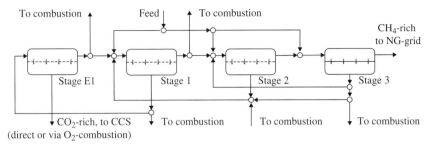

FIGURE 5.2   Cascade of membrane separation stages for CH_4/CO_2 separation from SNG. Reproduced from [4]. *Source*: Baciocchi 2009 [4]. Reproduced with permission of Elsevier.

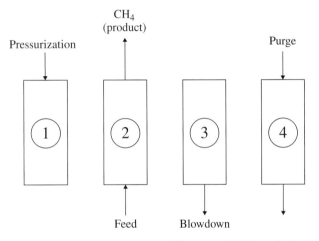

FIGURE 5.3   Schematic diagram of a standard Skarstrom's PSA cycle. Reproduced from [28]. *Source*: Cavenati 2006 [28]. Reproduced with permission of American Chemical Society.

vacuum conditions (vacuum swing adsorption, VSA) [14]. The PSA process in the standard stripping configuration (Skarstrom cycle), whose schematic diagram is reported in Figure 5.3, is a widely applied technology which typically allows to obtain a pure gas stream rich in the less retained compound (LP), whereas the other stream (HP) is usually characterized by low purity. This cycle comprises countercurrent pressurization with methane, feed (where purified methane is obtained as product at high pressure), countercurrent blowdown at low pressure to partially regenerate the adsorbent, and countercurrent purge at low pressure with methane to displace carbon dioxide from the product end [28].

A PSA Skarstrom cycle was applied for CH_4/CO_2 bulk separation using CMS at operating pressures between 0.37 and 0.04 MPa for the separation of a 50/50 feed mixture and achieved a purity above 90% in both product streams [29]. The comparison between CMS and zeolite-based PSA processes shows that a higher purity of the light product (CH_4) with higher recoveries can be obtained using CMS, with also a slightly higher productivity (feed h^{-1} kg^{-1} desorbent), despite this comparison being made considering operation of the CMS process at a slightly lower pressure (vacuum) than that of the zeolite process (0.03 MPa compared to 0.034 MPa) [29].

The Skarstrom cycle was also applied to the ternary separation of $CH_4/CO_2/N_2$ (60/20/20) in a bed packed with zeolite 13X [28]; in this case, the high pressure step was performed at 0.01–0.250 MPa, and the lower pressure step at 0.01 MPa, while two temperatures were studied, namely 300 and 323 K. The results showed rather low CH_4 purity and recovery values, between 73–85% and 88–27%, respectively [28].

In order to overcome the intrinsic limitation of the Skarstrom cycle in achieving high purity and recovery of methane, two co-current pressure equalization steps were added for the separation of a 50/50 CH_4/CO_2 gas mixture, thus leading to a six-step PSA cycle [30]; in this way, a 95.8% CH_4 purity with a 71.2% recovery and a productivity of $0.14 m^3$ $CH_4/(h kg$ adsorbent) were achieved, using CMS and operating between 0.4 MPa and atmospheric pressure. The Skarstrom cycle and its modifications may allow the achievement of a high purity of the light product (CH_4), but generally fail in achieving in the meantime also a high purity of the heavy component (CO_2).

The so-called Sircar's cycle (from the name of the inventor), that consists in a five-step PSA cycle (see Figure 5.4), overcomes this limitation [31]. This cycle is a derivation of the standard Skarstrom stripping cycle. The modification introduced, that is, the heavy product recycle, is basically a combination of the stripping PSA with a completely different PSA configuration, the enriching PSA. In this way, the light component is rinsed and also the pore volume of the adsorbent bed is enriched in the heavy component (CO_2). The configuration shown in Figure 5.4 using zeolite 13X achieves more than 99% purity of both CH_4 and CO_2, starting from a roughly equimolar CH_4/CO_2 mixture [31].

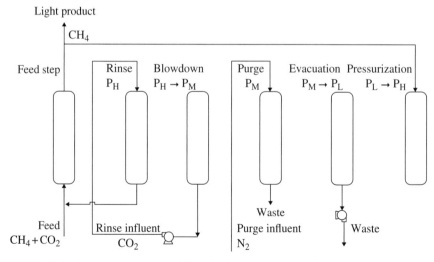

FIGURE 5.4 Schematic diagram of Sircar's five-step cycle for CH_4 and CO_2 production. Reproduced from [31]. *Source*: Knaebel 2003 [31]. Reproduced with permission of Springer.

5.2.1.4 *Low Temperature Separation*

This separation option, which relies on the different boiling points of CO_2 and CH_4, is commonly referred to as cryogenic separation. In this paragraph, we will rather refer to low temperature separation, considering that the term cryogenic strictly applies to processes operating below 120 K, whereas CO_2/CH_4 processes are typically performed above this limit [32].

There are no examples of low temperature separation for SNG upgrading. Nevertheless, this separation option has been applied for bio-methane production from biogas and for CO_2 capture from syngas in pre-combustion processes. In the former case, the biogas is compressed up to 8 MPa and cooled down to –45 °C, allowing the CO_2 to condense, conveying it to a further separation step for CH_4 recovery; the biogas, already enriched in methane, is further cooled to –55 °C and then expanded, allowing the separation of the carbon dioxide in solid form and achieving a very high CH_4 concentration (97%). A pilot scale application of this process is reported in the Netherlands [14].

Low temperature H_2/CO_2 separation for hydrogen production from syngas, generated from either coal gasification or natural gas reforming, also requires fairly high operating pressure. It can be shown that high CO_2 capture rates (let us say above 80%) can be achieved only for gas mixtures with a CO_2 concentration above at least 40% [32]. It is worth pointing out that these results are obtained for an operating temperature slightly above the CO_2 solidification temperature, which is close to the triple point temperature, that is 216.6 K or –56.6 °C [32]. Low temperature separation can be performed using either rotational and flash separation or distillation. In the former case, the gas is first cooled at –50 to –80 °C and/or expanded (2–3 MPa), thus forming a mixture which consists predominantly of gaseous methane containing a mist of small micron-sized droplets of liquid CO_2, that are separated by applying the patented apparatus of the rotational particle separator [33]. As an alternative, syngas can be separated after compression to around 5 MPa in a distillation column operating between 231 and 261 K [34].

5.2.2 Removal of other Compounds and Impurities

As reported previously, besides carbon dioxide and methane concentrations, critical parameters for synthetic natural gas in terms of pipeline injection specifications are its water and hydrogen content (particularly in the case of hydrothermal gasification). In addition, in order to increase the CH_4 content of the upgraded gas, also nitrogen removal may be required. The types of techniques that may be potentially applicable to dehydrate SNG correspond basically to those currently applied for water separation from natural gas, that is absorption by liquid desiccants, adsorption by solid desiccants, and membrane separation [4]. It should be noted that the type of process selected to perform the CO_2/CH_4 separation step may influence the selection of the position of the dehydration process. If adsorption were adopted for methane separation, water should be removed upstream this treatment, since it is strongly retained by most adsorbents and thus considerably reduces their capacity [35]. A similar limitation applies to low temperature CO_2/CH_4 separation.

Regarding the first type of process, typically employed absorbents include calcium chloride, lithium chloride, and glycol solutions. Glycol in particular has proved to be the most effective liquid desiccant since it exhibits high hygroscopicity, low vapor pressure, high boiling point, and low solubility in and of natural gas [36]. The four types of glycol that have been successfully used to dehydrate natural gas are ethylene glycol (EG), diethylene glycol (DEG), triethylene glycol (TEG), and tetraethylene glycol (T4EG); TEG is the preferred choice since it is considered to be the most cost-effective solvent [36]. As shown in Figure 5.5, the wet high pressure gas stream is contacted counter-currently with the dry and relatively cool TEG solution (termed strong) in a tray tower or packed column [3]. One of the main parameters that must be defined to design a TEG dehydration unit is the minimum concentration of TEG that is required for meeting outlet gas specifications in the strong solution entering the top of the dehydrator. This value is influenced by the operating pressure and temperature; specifically the higher the pressure, the lower the concentration of TEG required; whereas the higher the temperature, the higher the concentration of TEG needed to achieve specifications [3]. Working at higher pressure has also been shown to allow reducing the required inner diameter of the dehydrator, as well as the minimum TEG circulation rate, that then allows an adequate distribution of the glycol solution [3]. Operational disadvantages of dehydration processes based on absorption with glycol solutions include foaming and the formation of decomposition products and of high viscosity solutions at high glycol concentrations and low temperature which may hence prove difficult to pump. Furthermore, this process may cause detrimental environmental impacts in relation to fugitive emissions, soil contamination, and fluid disposal problems [36].

Solid desiccants reduce the moisture content of gases by chemical reaction, formation of hydrated compounds, or by adsorption. The most commonly employed types of adsorbents include alumina based adsorbents, activated carbon, molecular

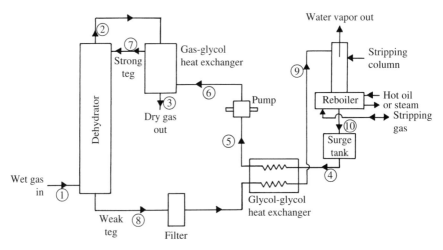

FIGURE 5.5 Scheme of the TEG dehydration process. Reproduced from [3]. *Source*: Gandhidasan 2003 [3]. Reproduced with permission of Taylor and Francis.

sieves, and silica based adsorbents [36]. Gas drying by using a PSA or a thermal swing adsorption (TSA) process is a common practice. Typically NaX zeolite (TSA, PSA) or activated alumina (PSA) are used as adsorbents [35]. Figure 5.6 reports water adsorption isotherms (amount adsorbed vs relative humidity) at 297 K for NaX zeolite (curve a) and alumina (curve b). The zeolite adsorbs water very strongly (isotherm type I), whereas alumina adsorbs water more moderately (type IV isotherm). Consequently, water is difficult to desorb from zeolite and the process is very energy intensive. Alumina, on the other hand, presents low water adsorption capacity at low partial pressures. The optimum water adsorbent for a PSA drier should present a type I isotherm with an adsorption affinity in between NaX and alumina (such as curve f in Figure 5.6). Modified activated carbon can fulfil that role. In fact, the hydrophobic nature of common activated carbon (CeCa) exhibited by curve c can be changed by introducing polar oxygen groups on the surface. Curve f shows a Type I water adsorption isotherm produced by oxidation of CeCa carbon by heating the carbon in a 45% HNO_3 solution at 353 K in the presence of a copper acetate catalyst [35]. Compared to the first type of analyzed dehydration process, adsorption with solid dessicants may allow to achieve lower dew points over a wide range of operating conditions and may prove more cost-effective to treat limited gas flow rates. In addition these systems are relatively free from corrosion and foaming problems [36]. The gas is flowing also in this case from the top to the bottom of a packed bed unit, which after a certain time of operation must be regenerated. Generally a portion of the entering wet gas (10%) is used for the regeneration step. This gas is sent through a heat exchanger and heated to 200–325 °C and piped to the unit requiring regeneration [37]. Concerning the effect of the operating conditions on design or processing parameters,

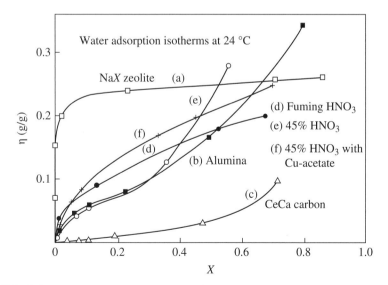

FIGURE 5.6 Pure water vapor adsorption isotherms for various types of adsorbents. Reproduced from [35]. *Source*: Sircar 1996 [35]. Reproduced with permission of Elsevier.

the operating temperature of the gas seems to have negligible effect on the dryer shell thickness but considerable impact on the desiccant mass required for dehydration [36]. As for the operating pressure of the gas, its increase has shown to decrease the mass of desiccant required for dehydration up to a pressure of 6 MPa.

Membrane technology is another attractive alternative process for natural gas dehydration. Membrane systems are often compact with a small footprint, passive, without moving parts, and do not exert the detrimental environmental impacts exhibited by glycol absorption systems [38]. Water is more condensable than methane (indicated by a much higher critical temperature or Lennard–Jones temperature), and thus, it presents higher solubility than methane in polymers. In addition, water molecules are smaller than methane ones, and thus, water also presents a much higher diffusivity. Hence, both solubility selectivity and diffusivity selectivity strongly favor the permeation of water over methane (as well as over nitrogen and carbon dioxide) in all polymers. However, the types of polymers that appear to present the highest water permeability as well as H_2O/CH_4 selectivity values are hydrophilic rubbery polymers [38]. In particular, block copolymers containing poly(ethylene oxide) (PEO; such as Pebax® and PEO-PBT) have attracted significant interest for dehydration of permanent gases such as nitrogen (e.g., [39]). These PEO-containing copolymers are characterized by a very high water vapor permeability (more than 50 000 Barrers) and H_2O/CH_4 selectivity (more than 5000). Furthermore, Pebax® copolymers are commercially available, which makes them good candidates to be fabricated into industrial membranes [38]. However, at present membrane systems are only used for niche dehydration applications [40]. The separation performance of membrane systems has shown to be restricted by the moderate feed to permeate pressure ratio encountered in this application, resulting in high membrane area requirements and high methane losses [41]. In order to overcome this issue and develop a competitive dehydration process based on membranes, Lin et al. [41] analyzed different process design configurations and found that a countercurrent design using a dry gas stream as sweep could prove to be the most promising process for natural gas dehydration. Specifically, the authors reported that, based on modeling results, a natural gas stream of $1.5\,m^3$ (STP) s^{-1} could be dehydrated from 1000 to 100 ppm with a membrane area as small as $120\,m^2$ and a potential methane loss of as little as 0.78%, which is even lower than what is currently achieved by glycol dehydrators (around 1% methane losses) [41].

Regarding the removal of other impurities such as hydrogen and nitrogen, the papers published so far on PSA application to natural gas separation report experimental results performed on model systems made up by two main components, that is, CH_4 and CO_2. One of the few exceptions discussing the issue of impurities reports the utilization of landfill gas for producing pure CH_4 and CO_2 [31]. The process selected for the bulk separation of CO_2 from CH_4 is Sircar's PSA cycle [42] (shown in Figure 5.4) and discussed in Subsection 5.2.1.3. The process was modified in order to deal with the removal of the main impurities in landfill gas, that is, water, nitrogen, and chlorinated hydrocarbons. In this study water and hydrocarbons are removed first in a TSA unit. Downstream from the TSA, a PSA unit is used for bulk CH_4/CO_2

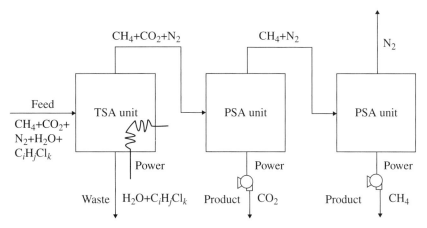

FIGURE 5.7 Schematic diagram of multistage adsorption process for splitting landfill gas. Reproduced from [31]. *Source*: Knaebel 2003 [31]. Reproduced with permission of Springer.

separation, and finally, a second PSA unit is introduced for separating nitrogen from methane. The scheme of the process is shown in Figure 5.7.

The purification of H_2 from mixtures is often addressed in the literature. Generally, H_2 is the main compound of the mixture. For instance a PSA process for a 80/20 H_2/CH_4 mixture in which activated carbon is used for selective adsorption of CH_4 over H_2 has been reported [43]. The separation of a 50/50 H_2/CH_4 gas mixture in activated carbon has been reported once in the literature [44]. In this work, several PSA configurations were modelled and their efficiency compared. The basic process considered is made up of the following cyclic steps:

1. Pressurization of the bed with the feed or with part of the H_2 generated from the next stage.
2. High pressure adsorption in which H_2 is produced.
3. High pressure CH_4 purge or co-current depressurization.
4. Counter-current blow down.
5. Low pressure H_2 purge.

In the two latter stages methane is produced and the bed is regenerated for the next cycle [44]. The results obtained for the two different configurations considered in stage 1 showed that H_2 purity decreases monotonically when H_2 recovery increases, whereas CH_4 purity increases (and its recovery decreases) in both cases. However, for the same H_2 recovery rate, the H_2 purity obtained from the PSA process using H_2 pressurization is always higher than that using feed pressurization. Finally, CH_4 purity showed not to be significantly affected by the gas used for the pressurization step. Regarding the two options tested for stage 3, the CH_4 purge showed to lead to higher purities for both H_2 and CH_4 products, with a CH_4 recovery in the range of approximately 35–95% [44].

A few studies have addressed so far the H_2/CH_4 and N_2/CH_4 separations by membrane technology. Hradil et al. [45] tested for H_2/CH_4 separation the application of heterogeneous membranes containing polymer adsorbent and polymeric binders specifically prepared in the laboratory. The H_2 permeability of these membranes was in the range 10^{-13} to 10^{-12} mol Pa^{-1} s^{-1} m^{-1}, whereas CH_4 permeability was about two orders lower. The evaluated selectivity of the heterogeneous membranes ranged from 3 to 120. The permeability to and selectivity for H_2 increased with increasing amounts of H_2 in the gas mixture and decreased with temperature. In conclusion, an increased diffusion flux and permeability were observed in heterogeneous membranes containing a polymer adsorbent and a higher selectivity was found compared to zeolite-filled membranes [45]. As for N_2/CH_4 separation, Lokhandwala et al [46] developed and tested the performance of membranes that selectively permeate methane and retain nitrogen (composite rubbery polymers) or selectively permeate nitrogen and retain the hydrocarbon (composite glassy polymers). Methane-selective membranes showed to be preferable for most applications. However, selectivities proved to be only moderate for either type of membrane, so multi-step or multi-stage membrane systems are needed to produce a low nitrogen product gas. However, for a feed gas containing 4–8 mol% nitrogen (such as SNG), a simple two-step bank of modules could be used to produce the separation yields required. In this system the pressurized feed gas passes across the surface of the membrane and the permeate, depleted in nitrogen, is re-pressurized. Over 85% methane recovery was shown to be achievable by this process [46].

5.3 TECHNO-ECONOMICAL COMPARISON OF SELECTED SEPARATION OPTIONS

As previously discussed, SNG upgrading requires different separation steps, involving bulk CO_2/CH_4 separation and several purification steps aimed at removing other components. The overview made in this chapter suggests that different processes are in principle applicable to the different separation steps required for SNG upgrading. To our knowledge, a comparative study of the different upgrading options is currently missing. There are only a few papers on the energy requirements related to the different available SNG upgrading technologies, although they are limited to bulk CO_2/CH_4 separation. Namely, Guo et al. [47] estimate an energy requirement of $566 \, MJ \, t^{-1}$ CO_2 for CO_2 removal from crude synthetic natural gas using physical absorption, whereas Gassner et al. [4] estimate $620 \, MJ \, t^{-1}$ CO_2 for membrane separation. To the best of our knowledge, there is no such figure for adsorptive separation. Nevertheless, it is worth mentioning that for CO_2 removal from biogas the energy requirement was $459 \, MJ \, t^{-1}$ CO_2 for PSA against $275 \, MJ \, t^{-1}$ CO_2 for chemical absorption with amines [47].

More data on the different separation options and the associated costs are instead available for the upgrading of biogas or synthetic gas obtained from thermal gasification processes. As to the former, a review of different candidate technologies for the upgrading of biogas from fermentation to natural gas grid quality is given in [14] while the economics are also discussed elsewhere [48, 49], but the conclusions cannot be transferred directly to SNG production because of large differences in plant scale and thermal integration opportunities.

As to thermal gasification, an application that in our view is closer to the second generation SNG discussed in this chapter, Chandel and Williams [50] report that, for a coal to SNG plant, the syngas cleanup system accounts for 15.8% of the total equipment capital cost for the components. They also calculate the SNG specific cost in the case of venting to the atmosphere the CO_2 separated by means of physical absorption (in this case physical absorption seems quite adequate, due to the high pressure of coal gasification), which results in the range of US\$ 8.42–9.53MBtu^{-1} (i.e., 21–24 Euro MWh^{-1}) depending on the type of coal. Even if these costs seem to be quite optimistic with respect to other literature sources, it is interesting how the same authors show the dependency of the SNG cost on the coal price: an increase in the coal price by 100% increases the cost of SNG by 12.5–18.8%. They also report that if the separated CO_2 is associated to CCS, the SNG production price becomes 23–26 Euro MWh^{-1}.

Calculation of SNG production cost from various coal ranks, with reference to a input to the gasifier of 500 MWth, using physical absorption for CO_2 separation, is also reported by NETL [51] and is in the range 48–53 Euro MWh^{-1}, in the case without CCS, and in the range 53–58 Euro MWh^{-1}, with CCS.

More details can be found in Alamia [52], who performs an interesting comparison of three different possibilities for the upgrading section, based on the use of membranes, monoethanolamine (MEA) absorption, both coupled with CCS, and PSA, without CCS, starting from the same entering syngas composition and considering a 100 MW thermal input gasifier. In this case, the PSA process has the lowest capital investment, with an annual capital investment of 5 MEuro a^{-1}, mainly due to the absence of the CCS, while the other two solutions are slightly more expensive, around 5.8 MEuro a^{-1}. Alamia (2010) [52] also calculates the total annual costs for the three upgrading sections, showing that, depending on the assumptions made on the electricity cost, the annual cost for membranes is 10.7–12.2 MEuro a^{-1}; the annual cost for PSA is 8.1–8.9 MEuro a^{-1}; the annual cost for MEA absorption is 9.1–9.8 MEuro a^{-1}.

Similarly, Heyne and Harvey [53] estimate the total investment and operation costs for SNG production from forest residues in a 100 MW thermal input atmospheric gasifier, considering for the upgrading section membranes, PSA and MEA. The cost contribution of the gas upgrading section varies from 13 to 22% of the total fixed capital investment and is the lowest for the membrane case. The SNG specific production cost varies from 104.4–105.5 Euro MWh^{-1} for MEA, respectively with and without CCS, to 108.1–110.5 Euro MWh^{-1} for membranes, respectively with and without CCS, to 112.9 for PSA, without CCS.

Also Gassner and Marechal [54] estimate the SNG production cost from lignocellulosic biomass, using different technologies and different ways of integrating into the plant some selected CO_2 removal processes as physical absorption, PSA, and membranes. They calculate a SNG specific production cost of about 76–107 Euro

MWh^{-1} for an input thermal capacity of 20 MW$_{th}$ whereas 59–97 Euro MWh^{-1} at scales of 150 MW$_{th}$ thermal input and above.

Gassner et al. [4] calculate the SNG production cost in case of membranes, which results in approximately 103 Euro MWh^{-1} without CCS, and 107 Euro MWh^{-1}, with CCS. In this case the authors also show that the additional cost of captured and avoided CO$_2$ is strongly dependent on the cost of electricity and lies in the range of 15–40 Euro t^{-1}.

However, when selecting one of the CO$_2$ removal techniques, one should consider that membranes and PSA operating costs are strongly dependent on the electricity market scenario, while in the case of absorption they are essentially independent of electricity prices.

In summary, it appears that producing SNG from coal is less expensive than producing it from biomass, and that the choice of the CO$_2$ removal technique influences the final cost of production of only a small percentage (4–8%). Also the decision of capturing the removed CO$_2$ has a limited effect on the final SNG production cost, with an increase of a few percentage points (1–4%) in case of biomass and slightly more (9–10%) in the case of coal. External conditions as the electricity price, CO$_2$ emission trading scheme and allowance price, feedstock price, or plant size, seem to be of larger influence on the final cost, rather than the selected CO$_2$ removal technique.

REFERENCES

[1] CCEM. *Second Generation biogas. New Pathways to efficient use of Biomass. Final Report*. Swiss Competence Center for Energy and Mobility, Zurich; 2012.

[2] IPCC. *IPCC Special Report on Carbon Dioxide Capture and Storage*. Intergovernmental Panel on Climate Change/Cambridge University Press, Cambridge, UK; 2005.

[3] Gandhidasan P. Parametric analysis of natural gas dehydration by a triethylene glycol solution. *Energy Sources* **25**(3): 189–201; 2003.

[4] Gassner M, Baciocchi R, Maréchal F, Mazzotti M. Integrated design of a gas separation system for the upgrade of crude SNG with membranes. *Chemical Engineering and Processing: Process Intensification* **48**(9): 1391–1404; 2009.

[5] Eiken O, Ringrose P, Hermanrud C. Lesson learned from 14 years of CCS operations: Sleipner, In-Salah and Snohvit. *Energy Procedia* **4**: 5541–5548; 2011.

[6] Kothandaraman A. *Carbon Dioxide Capture by Chemical Absorption: A Solvent Comparison Study*. Dissertation, Massachusetts Institute of Technology, Boston, USA; 2010.

[7] Chapel DG, Mariz CL, Ernest J. *Recovery of CO$_2$ from Flue Gases: Commercial Trends*. Paper 340 at the Annual Meeting of the Canadian Society of Chemical Engineering, Saskatoon, Canada; 1999.

[8] UOP. *UOP Benfield Process*. Benfield Process; 2008. www.virtu-media.com/what_we_do/multimedia/gasprocessing/Flash/pdfs/Techsheets/Benfield_Process.pdf (accessed 15 December 2015).

[9] Kozak F, Petig A, Morris E, Rhudy R, Thimsen D. Chilled ammonia process for CO$_2$ capture. *Energy Procedia* **1**: 1419–1426; 2009.

[10] Darde V, Thomsen K, van Well WJM, Stenby EH. Chilled ammonia process for CO_2 capture. *Energy Procedia* **1**: 1035–1042; 2009.

[11] Gupta M, Coyle I, Thambimuthu K. *CO_2 Capture Technologies and Opportunities in Canada. Strawman Document for CO_2 Capture and Storage (CC + S) Technology Roadmap.* First Canadian CC + S Technology Roadmap Workshop, 18–19 September 2003. CANMET Energy Technology Centre Natural Resources, Calgary, Alberta, Canada; 2003.

[12] Chen WH, Chen SM, Hung CI. Carbon dioxide capture by single droplet using Selexol, Rectisol and water as absorbents: A theoretical approach. *Applied Energy* **111**: 731–741; 2013.

[13] UOP. *UOP Selexol Technology for Acid Gas Removal. UOP*, New York; 2009. http://www.uop.com/?document=uop-selexol-technology-for-acid-gas-removal&download=1 (accessed 15 December 2015).

[14] Ryckebosch E, Drouillon M, Vervaeren H. Techniques for transformation of biogas to biomethane, *Biomass and Bioenergy* **35**: 1633–1645; 2011.

[15] Sridhar S, Smitha B, Aminabhavi TM. Separation of carbon dioxide from natural gas mixtures through polymeric membranes – A review. *Separation and Purification Reviews* **36**: 113–174; 2006.

[16] Wind JD, Paul DR, Koros WJ. Natural gas permeation in polyimide membranes. *Journal of Membrane Science* **228**: 227–236; 2004.

[17] Bhide B, Voskericyan A, Stern S. Hybrid processes for the removal of acid gases from natural gas. *Journal of Membrane Science* **140**(1): 27–49; 1998.

[18] Koros W. Mahajan R. Pushing the limits on possibilities for large scale gas separation: which strategies? *Journal of Membrane Science* **175**(2): 181–196; 2000.

[19] Li S, Martinek JG, Falconer JL, Noble RD. High-pressure CO_2/CH_4 separation using SAPO-34 membranes. *Industrial and Engineering Chemistry Research* **44**(9): 3220–3228; 2005.

[20] Ismail A, David L. A review on the latest development of carbon membranes for gas separation. *Journal of Membrane Science* **193**(1): 1–18; 2001.

[21] Hagg M, Lie J, Lindbrathen A. Carbon molecular sieve membranes – A promising alternative for selected industrial applications. *Advanced Membrane Technology* **984**: 329–345; 2003.

[22] Chung TS, Jiang, LY, Li Y, Kulprathipanja S. Mixed matrix membranes (MMMs) comprising organic polymers with dispersed inorganic fillers for gas separation. *Progress in Polymer Science* **32**: 483–507; 2007.

[23] Zou J, Ho WSW. CO_2-selective polymeric membranes containing amines in crosslinked poly(vinyl alcohol). *Journal of Membrane Science* **286**: 310–321; 2006.

[24] Li Y, Chung T, Kulprathipanja S. Novel Ag + -zeolite/polymer mixed matrix membranes with a high CO_2/CH_4 selectivity. *AICHE Journal* **53**(3): 610–616; 2007.

[25] Baker R. Future directions of membrane gas separation technology. *Industrial and Engineering Chemistry Research* **41**(6): 1393–1411; 2002.

[26] Qi R, Henson M. Optimization-based design of spiral-wound membrane systems for CO_2/CH_4 separations. *Separation And Purification Technology* **13**(3): 209–225; 1998.

[27] Santos MPS, Grande CA, Rodrigues AE. Pressure swing adsorption for biogas upgrading. Effect of recycling streams in pressure swing adsorption design. *Industrial Engineering and Chemistry Research* **50**: 974–985; 2011.

[28] Cavenati S, Grande C, Rodrigues A. Removal of carbon dioxide from natural gas by vacuum pressure swing adsorption. *Energy and Fuels* **20**(6): 2648–2659; 2006.

[29] Kapoor A, Yang R. Kinetic separation of methane carbon dioxide mixture by adsorption on molecular-sieve carbon. *Chemical Engineering Science* **44**(8): 1723–1733; 1989.

[30] Kim MB, Bae YS, Choi DK, Lee CH. Kinetic separation of landfill gas by a two-bed pressure swing adsorption process packed with carbon molecular sieve: Nonisothermal operation. *Industrial and Engineering Chemistry Research* **45**(14): 5050–5058; 2006.

[31] Knaebel K, Reinhold H. Landfill gas: From rubbish to resource. *Adsorption – Journal of the International Adsorption Society* **9**(1): 87–94; 2003.

[32] Berstad D, Anantharaman R, Neksa P. Low temperature CO_2 capture technologies – Applications and potential. *International Journal of Refrigeration* **26**: 1403–1416; 2013.

[33] Brouwers JJH, Kemenade JJP. *Condensed Rotational Separation to Upgrade Sour Gas.* Proceedings of the The Sixth Sour Oil and Gas Advanced Technology (SOGAT) Conference, Abu Dhabi, April 28–May 1. pp. 35–39; 2010.

[34] Berstad D, Neksa P, Giovag GA. Low-temperature syngas separation and CO_2 capture for enhanced efficiency of IGCC power plants. *Energy Procedia* **4**: 1260–1267; 2011.

[35] Sircar S, Golden T, Rao M. Activated carbon for gas separation and storage. *Carbon* **34**(1):1–12; 1996.

[36] Gandhidasan P, Al-Farayedhi A, Al-Mubarak A. Dehydration of natural gas using solid desiccants. *Energy* **26**(9): 855–868; 2001.

[37] Wunder JWJ. How to design a natural-gas drier. *Oil and Gas Journal* **60**(32): 137–148; 1962.

[38] Lin H, Thompson SM, Serbanescu-Martin A, Wijmans JG, Amo KD, Lokhandwala KA, Merkel TC. Dehydration of natural gas using membranes. Part I: Composite membranes. *Journal of Membrane Science* **413/414**: 70–81; 2012.

[39] Potreck J, Nijmeijer K, Kosinski T, Wessling M. Mixed water vapor/gas transport through the rubbery polymer PEBAX® 1074. *Journal of Membrane Science* **338**: 11–16; 2009.

[40] Baker RW, Lokhandwala K. Natural gas processing with membranes: an overview. *Industrial and Engineering Chemistry Research* **47**: 2109–2121; 2008.

[41] Lin H, Thompson SM, Serbanescu-Martin A, Wijmans JG, Amo KD, Lokhandwala KA, Low BT, Merkel TC. Dehydration of natural gas using membranes. *PartII: Sweep/countercurrent design and field test. Journal of Membrane Science* **432**: 106–114; 2013.

[42] Sircar S, Koch WR, Van Sloun J. *Recovery of Methane from Landfill Gas.* US Patent 4 770 676; 1988.

[43] Waldron W, Sircar S. Parametric study of a pressure swing adsorption process. *Adsorption – Journal of the International Adsorption Society* **6**(2): 179–188; 2000.

[44] Doong S, Yang R. A Comparison of gas separation performance by different pressure swing adsorption cycles. *Chemical Engineering Communications* **54**(1/6): 61–71; 1987.

[45] Hradil J, Krystl V, Hrabánek P, Bernauer B, Kočiřík M. Heterogeneous membranes based on polymeric adsorbents for separation of small molecules. *Reactive and Functional Polymers* **61**(3): 303–313; 2004.

[46] Lokhandwala KA, Pinnau I, He Z, Amo KD, DaCosta AR, Wijmans JG, Baker RW. Membrane separation of nitrogen from natural gas: A case study from membrane synthesis to commercial deployment. *Journal of Membrane Science* **346**: 270–279; 2010.

[47] Guo W, Feng F, Song G, Xiao J, Shen L. Simulation and energy performance assessment of CO2 removal from crude synthetic natural gas via physical absorption process. *Journal of Natural Gas Chemistry* **21**: 633–638; 2012.

[48] Patterson T, Esteves S, Dinsdale R, Guwy A. An evaluation of the policy and techno-economic factors affecting the potential for biogas upgrading for transport fuel use in the UK. *Energy Policy* **39**: 1806–1816; 2011.

[49] Browne J, Nizami AS. Thamsiriroj T, Murphy JD. Assessing the cost of biofuel production with increasing penetration of the transport fuel market: a case study of gaseous biomethane in Ireland. *Renewable and Sustainable Energy Reviews* **15**: 4537–4547; 2011.

[50] Chandel M, Williams E. *Synthetic Natural Gas (SNG): Technology, Environmental Implications, and Economics*. Climate Change Policy Partnership, Duke University; 2009.

[51] National Energy Technology Laboratory. *Cost and Performance Baseline for Fossil Energy Plants. Volume 2: Coal to Synthetic Natural Gas and Ammonia*. DOE/NETL, Washington, D.C.; 2011.

[52] Alamia A. Thermo-Economic Assessment of CO2 Separation Technologies in the Framework of Synthetic Natural Gas (SNG) Production. Dissertation, Chalmers University of Technology, Göteborg, Sweden; 2010.

[53] Heyne S, Harvey S. Impact of choice of CO_2 separation technology on thermo-economic performance of Bio-SNG production processes. *International Journal of Energy Research* **26**: 45–47; 2013. DOI: 10.1002/er.3038.

[54] Gassner M, Maréchal F, 2009. Thermo-economic process model for thermochemical production of synthetic natural gas (SNG) from lignocellulosic biomass. *Biomass and Bioenergy* **33**: 1587–1604; 2009.

6

SNG FROM WOOD – THE GOBIGAS PROJECT

JÖRGEN HELD

6.1 BIOMETHANE IN SWEDEN

The interest for biomethane as a transportation fuel in Sweden is strong. This is a result of the decarbonized power generation sector and a strong political will to replace the fossil fuels in the transportation sector. Electricity production [1] in Sweden 2012, dominated by hydro and nuclear power, is shown in Figure 6.1.

   With the natural gas grid limited to the southwest of Sweden, two parallel biomethane developments have taken place. In areas with access to the natural gas grid, biomethane has taken advantage of the natural gas infrastructure in terms of pipeline distribution and CNG filling stations. During 2012, biogas was injected into the natural gas grid at eight different sites in the southwest of Sweden [2]. In areas without access to the gas grid it has been logical to upgrade locally produced biogas and provide nearby CBG filling stations with biomethane. In total there are 55 biogas upgrading plants in Sweden [3]. The latter development has faced challenges in terms of balancing production and demand in an emerging and growing market. Most of the nongrid-connected filling stations have to rely on LNG as back-up, and in future on LBG. Contrary to many other countries the natural gas distributors in Sweden have a positive view on renewable methane, have actively supported the market build-up, and have taken responsibility for its development. However, it may be noted that more than 70% of the biogas in Sweden is produced at sewage water treatment plants and co-digestion plants.

Synthetic Natural Gas from Coal, Dry Biomass, and Power-to-Gas Applications, First Edition.
Edited by Tilman J. Schildhauer and Serge M.A. Biollaz.
© 2016 John Wiley & Sons, Inc. Published 2016 by John Wiley & Sons, Inc.

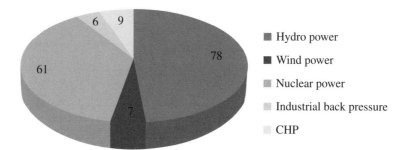

FIGURE 6.1   The electricity production in Sweden amounted to 163 TWh in 2012. Export minus import of electricity amounted to 20 TWh. The power generation is more or less decarbonized since the industrial back pressure and CHP are mainly biomass based.

All upgraded biogas in Sweden is utilized as vehicle fuel, directly or virtually through the green gas principle. The green gas principle allows a CNG/CBG vehicle to be refueled with natural gas as long as the same amount of biogas is injected into the natural gas grid. This is an important step for the development of the biomethane market, since the owner of a CNG/CBG vehicle can benefit from biogas tax exemption (no energy and no carbon dioxide tax) independent of whether the actual filling station is connected to a biogas plant or not.

Other market drivers are the 40% reduction of the company car benefit taxation for NGVs, support for the demonstration of new technology within the renewable energy gas area, and a new support scheme for manure based biogas production. Previously there was an investment support for actions reducing climate gas emission which was both favorable for biogas projects and an investment support for building refuelling stations for CNG/CBG vehicles. Several municipalities offered free parking for biogas powered vehicles and the government supported private individuals with 10 000 SEK (approx. 1100 euros) if they bought a car that could use biogas. During several decades the government has supported extensive biogas research and development through different national initiatives and programs.

All these actions have resulted in a growing biomethane market within the transportation sectors as can be seen in Figures 6.2 and 6.3.

While the amount of upgraded biogas used as vehicle fuel has increased steadily, biogas production has been fairly stable at 1.4–1.6 TWh/year for the last few years. This results in a larger share of the produced biogas being upgraded and used as vehicle fuel. Note that landfill gas is not cleaned, upgraded, and used as vehicle fuel in Sweden even if it is technically feasible. The main reason is that the deposit of organic material has been banned since 1 January 2005 and production is declining, as seen in Table 6.1. In addition, landfill gas may contain elevated and varying levels of different impurities normally not present in traditional biogas.

In fact it seems like the share of biogas being upgraded has hit a ceiling and there is an urgent need for increased biomethane production capacity to feed the growing market. New substrates and new production technologies, such as gasification and methanation of lignocellulose feedstock, are of major interest [8], especially for a

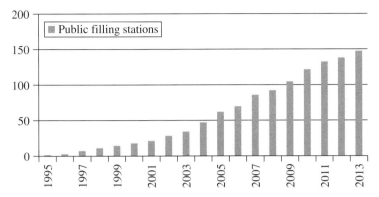

FIGURE 6.2 Number of public filling stations for natural gas and biogas in Sweden. In addition there were 58 nonpublic filling stations for buses and heavy duty trucks, such as garbage truck, at the end of 2013.

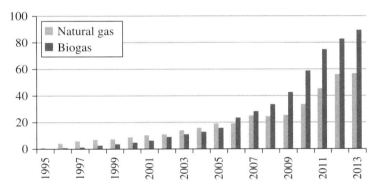

FIGURE 6.3 Amount of sold biogas and natural gas as vehicle fuel in million Nm^3. It is notable that biogas surpassed natural gas as vehicle fuel in 2006 and has continued to stay ahead since then.

TABLE 6.1 The Share of Biogas used as Vehicle Fuel [2, 4–7].

	2008	2009	2010	2011	2012
Total biogas production [GWh]	1359	1363	1387	1473	1589
Landfill gas [GWh]	369	335	296	270	254
Flaring [GWh]	195	135	112	115	165
Total biogas production excluding landfill gas and flaring [GWh]	795	893	979	1088	1170
Biogas sold as vehicle fuel [GWh][a]	327	410	574	729[b]	808[b]
Share of biogas (excluding landfill gas and flaring) used as vehicle fuel [%]	41	46	59	67	69

[a] Amount of sold biogas is taken from Statistics Sweden, www.scb.se.
[b] In these numbers small amounts of imported biogas used as vehicle fuel are included.

country like Sweden, with almost 70% of its area covered with forest. Due to the strong pulp and paper industry the logistics for transporting large quantities of biomass is well developed in Sweden and biomass fueled CHP plants of more than 100 MW$_{th}$ are not uncommon.

6.2 CONDITIONS AND BACKGROUND FOR THE GOBIGAS PROJECT IN GOTHENBURG

Gothenburg Energy, fully owned by the Gothenburg municipality, envisions bio-methane as one of the most important renewable fuels of the future and points out that it is a great advantage that distribution can take place through the existing (natural) gas grid.

The aim of the GoBiGas project is twofold:

1. To demonstrate the possibilities to produce bio-SNG from forest residues through gasification and methanation.
2. To build a plant which can provide for the growing need of renewable and CO_2-neutral biomethane.

The local conditions for the chosen site are advantageous in many aspects.

- There are good logistics for feedstock through the nearby harbor and railroad. The harbor can receive feedstock transported over the sea as well as on Göta älv, a river that connects to Vänern, the largest lake in Sweden. Vänern is situated in a forestry landscape and several pulp and paper industries and saw mills are situated on the shores of the lake.
- Large amounts of biofuels are already handled at the site through the existing hot water central.
- The site has access to the natural gas grid (4 and 35 bar) as well as the district heating system.
- The produced biomethane is targeted for the transportation sector and the Gothenburg area is well developed in terms of filling stations; and both the regional and the national CNG/CBG market is rapidly growing.

The history of the GoBiGas project started with prestudies of gasification technology for bio-SNG production in 2005. The ambition was to build a large plant with a capacity of 100 MW that could provide 100 000 vehicles with fuel. A first pilot study to gain an understanding of the technology followed by an evaluation of possible solutions were undertaken before the specific solution was selected. In 2007 Gothenburg Energy financed the rebuilding of the Chalmers CFB boiler which was supplemented with a 2–4 MW indirect gasifier at a cost of 13 million SEK (approx. 1.4 million euros). The gasifier was delivered by Metso Power and the erection was finalized in record time. The first measuring campaign was conducted only six

FIGURE 6.4 Aerial view of the recently erected GoBiGas plant (green building) adjacent to the existing hot water central. Picture taken 4 June 2014 by webcam and downloaded from the GoBiGas website, www.gobigas.se. With permission from Gothenburg Energy.

months after the start of construction. To minimize the risk it was decided to split the GoBiGas project into two parts, first a smaller demonstration plant (20 $MW_{bio\text{-}SNG}$), to be followed by a second plant of commercial size. The first part of the project was awarded funding from the Swedish Energy Agency on 25 September 2009 and the state aid approval from the European Commission came on 14 December 2010. The Gothenburg Energy Board took the decision to implement the project two days later on 16 December 2010. The plant, shown in Figure 6.4, was inaugurated by the Swedish energy minister, Anna-Karin Hatt on 12 March 2014.

6.3 TECHNICAL DESCRIPTION

The GoBiGas gasification section is supplied by Metso Power and licensed by Repotec. Haldor Topsoe is the technology provider of the gas cleaning and methanation. The building and installation work was managed by Jacobs as the EPCM (Engineering, Procurement, Construction Management). The gasification technology is based on a concept developed by Prof. Hermann Hofbauer, Technical University of Vienna, and commercialized through Repotec. The gasification technology was proven in an 8 MW_{th} plant in Güssing, Austria. The Güssing plant has been in

**TABLE 6.2   Technical Data [9]. Data from Held J. (editor)
Conference Proceedings 1st International Conference
on Renewable Energy Gas Technology. ISBN 978-91-981149-0-4.
Renewable Energy Technology International AB, 2014.**

GoBiGas – the first part	
Fuel capacity [MW_{th}]	32
Biomethane capacity [MW_{bioSNG}]	>20
District heating [GWh/year]	50
Electricity consumption [MW]	3
RME consumption [MW]	0.5
Operating hours [h/year]	8000

operation since 2002 and has accumulated more than 90 000 h of operation. In 2009 bio-SNG production was demonstrated in the Güssing plant through the European Bio-SNG project. A slipstream was diverted to a 1 MW fluidized bed methanation reactor developed by PSI and commercialized through CTU. The produced bio-SNG was successfully used as vehicle fuel in passenger cars. A crew from Gothenburg Energy participated in the operation of the fluidized bed methanation unit and this newly demonstrated technology was initially the favourite to be selected for the GoBiGas plant. However, during the evaluation of different technologies, Gothenburg Energy decided to go for technologies that were well proven and had strong suppliers. Different options were evaluated and the TREMP methanation technology by Haldor Topsoe was finally chosen. The technical data of the first part of the GoBiGas project is shown in Table 6.2.

Initially wood pellets will be used as feedstock but the ambition is to switch to forest residues. During the start-up nitrogen is used as the inert gas during the fuel feeding but during normal operation the separated carbon dioxide will be used. A simplified principal layout of the GoBiGas plant is shown in Figure 6.5. The gasifier (1) is operated at 850 °C and the combustor (2) at 930 °C. Hot bed material is transferred from the combustion reactor to the gasifier where the feedstock is converted to gaseous products and char. The char, together with bed material, is transferred to the combustion reactor where the char is combusted. The raw synthesis gas is cooled down in the syngas cooler (3) to a temperature of approximately 180 °C before entering the baghouse filter (4). Tars are separated in the RME scrubber (5) and transferred, together with spent RME, to the combustion chamber. In this way the energy in the tars and the spent RME is recovered as heat. Remaining tars are captured by the activated carbon bed (6). The four beds are operated alternately and when saturated they are regenerated by steam. The raw synthesis gas is pressurized to 16 bar in a six-stage intercooled compressor (7). Unsaturated hydrocarbons, such as ethylene may cause carbon deposition and deactivation of the downstream fixed bed catalysts. As much as 2.5 vol% of ethylene has been observed in the synthesis gas and this has to be addressed in order to ensure a good operation. In the GoBiGas plant the unsaturated hydrocarbons, primarily ethylene, and organic sulfur-containing compunds such as mercaptanes and

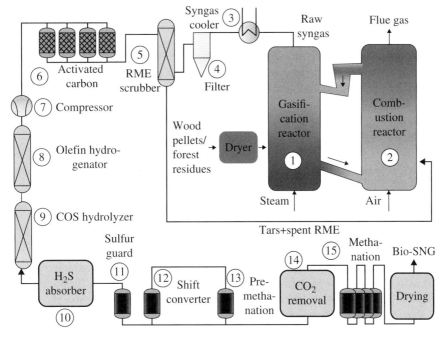

FIGURE 6.5 Simplified principal layout of the GoBiGas plant.

thiophenes are hydrogenated in the olefin hydrogenator (8). Shown below are the hydrogenation reactions for ethylene, methyl mercaptan, and ethyl mercaptan, respectively:

$$C_2H_4 + H_2 \rightarrow C_2H_6$$

$$CH_3SH + H_2 \rightarrow CH_4 + H_2S$$

$$C_2H_5SH + H_2 \rightarrow C_2H_6 + H_2S$$

The carbonyl sulfide, COS, is hydrolyzed in the COS hydrolyzer (9) according to

$$COS + H_2O \rightarrow H_2S + CO_2$$

The COS hydrolyzer is supplemented with a chlorine absorption catalyst based on potassium carbonate on an aluminium oxide support.

The H_2S is removed in an amine scrubber (10) together with approximately 50% of the carbon dioxide. The amine scrubber is from BASF. The remaining sulfur is removed down to 0.1 ppm through a ZnO bed (11). Part of the synthesis gas is shifted (12) to adjust the H_2/CO ratio. The synthesis gas then undergoes a pre-methanation step (13) before the remaining carbon dioxide is removed down to 0.1% in a two-stage amine scrubber (14). The final methanation takes place in four consecutive fixed bed methanation reactors (15). The gas is then dried through zeolites and delivered at 5 bar to Swedegas' compressor station where the bio SNG is pressurized to 35 bar before injection into the transmission grid.

6.4   TECHNICAL ISSUES AND LESSONS LEARNED

A large-scale demonstration of new technology is by default associated with technical uncertainties and GoBiGas is no exception. There have been initial problems with high levels of tar clogging the syngas cooler, which in turn has resulted in repeatedly shut downs. At first it was suspected that the bed material, olivine, needed additional time to become more catalytically active but the problem remained. The problem has now been solved by adding additives to the bed material.

6.5   STATUS

During spring 2014 the problems with the high tar levels was solved and the gasifier has successfully been producing synthesis gas of good quality (see Table 6.3). Till August 2014 the total time of operation of the gasifier is approximately 1000 h. Methanation is expected to be up and running during autumn 2014, including the injection of the produced bio-SNG into the transmission grid.

6.6   EFFICIENCY

In the first part of the GoBiGas project the aim is to prove the technology in the 20 MW_{bioSNG} scale and verify the performance goals. These goals include a biomass to bio-SNG efficiency of ~ 65%, an overall efficiency of ~ 90%, and the ability to operate the plant 8000 h/year.

6.7   ECONOMICS

The cost for the GoBiGas plant, including all the expenses related to the build-up of the project organization since 2005, pre-studies, evaluations, and so on, amounts to approximately 165 million euro. The GoBiGas plant is not expected to be commercial at this scale and extra costs have been spent on redundant or over-sized systems in order to gain experience and have a larger operating margin. Since the plant is

TABLE 6.3   **Gas Composition on a Dry Basis After the Gasifier.**

Component	Vol%
CH_4	8–9
H_2	38–40
CO	22–23
CO_2	23–25

considered as a demonstration platform it is heavily equipped with measuring devices at a numerous of measuring points. Extra safety precautions have been taken. During the building phase more than 400 people were simultaneously working at the site and no major accident was reported.

6.8 OUTLOOK

The investment in an industrial scale facility such as the first stage of the GoBiGas project is an important and necessary step to build confidence, get experience, and provide a platform for further development and optimization in order to bridge the gap from pilot and demonstration scale to commercial facilities. A potential investor basically needs to know if the technology works, if it is reliable, if the expected efficiency is obtained, and if there is a need for further improvements before realizing plants on a commercial scale. Taking into account that such investments are in excess of 100 million euro, it is obvious that the whole bio-SNG community benefits from the Gothenburg Energy effort.

In this first step of the GoBiGas project the aim is to prove and verify the technology. The second step of the project involving a commercial plant with a capacity of 80–100 $MW_{bio-SNG}$ has already been granted funding through the European NER 300 program.

By proving the technology at the 20 $MW_{bio-SNG}$ scale the foundation is laid to reach a fully commercial plant. The chosen technology will benefit from the scale-up effect and hence the specific investment and operating costs for a second 80–100 $MW_{bio-SNG}$ plant are expected to be significantly reduced.

In parallel to the GoBiGas project several development projects are being conducted in collaboration with Chalmers University of Technology. Tests are done in laboratory scale facilities as well as in the Chalmers 2–4 MW indirect gasifier. A lot of effort is spent on improved gas cleaning and one of the main R&D paths is related to chemical looping reforming. Gothenburg Energy has in collaboration with Chalmers University of Technology, the Technical University of Berlin, and Renewable Energy Technology International AB been granted funding through the ERANET BESTF (Bio Energy Sustaining the Future) program to develop such a gas cleaning concept.

ACKNOWLEDGEMENTS

The author wants to thank Ingemar Gunnarsson and Lars Andersson, Gothenburg Energy, for the informative study tour at the GoBiGas plant and for valuable input, feedback, and comments.

REFERENCES

[1] Swedish Energy Agency. *Energy in Sweden 2013* (in Swedish). ET 2013:22, Swedish Energy Agency, Sweden; 2013.

[2] Swedish Energy Agency. *Production and Utilization of Biogas, Year 2012* (in Swedish). ES 2013:07, Swedish Energy Agency, Sweden; 2013.

[3] IEA Bioenergy. *Foz do Iguacu*. Country Reports, IEA Bioenergy Task 37 – Energy from Biogas. IEA Bioenergy, Brazil; 2014.

[4] Swedish Energy Agency. *Production and Utilization of Biogas, Year 2008* (in Swedish). ES 2010:01, Swedish Energy Agency, Sweden; 2010.

[5] Swedish Energy Agency. *Production and Utilization of Biogas, Year 2009* (in Swedish). ES 2010:05, Swedish Energy Agency, Sweden; 2010.

[6] Swedish Energy Agency. *Production and Utilization of Biogas, Year 2010* (in Swedish). ES 2011:07, Swedish Energy Agency, Sweden; 2011.

[7] Swedish Energy Agency. Production and Utilization of Biogas, Year 2011 (in Swedish). ES 2012:08, Swedish Energy Agency, Sweden; 2012.

[8] Held J. *Small and Medium Scale bioSNG Production Technology*. Renewtec Report 001:2013, ISSN 2001-6255, Renewable Energy Technology International AB, Lund, Sweden; 2013.

[9] Held J. (ed.) *Conference Proceedings of the First International Conference on Renewable Energy Gas Technology*. ISBN 978-91-981149-0-4. Renewable Energy Technology International AB, Lund, Sweden; 2014.

7

THE POWER TO GAS PROCESS: STORAGE OF RENEWABLE ENERGY IN THE NATURAL GAS GRID VIA FIXED BED METHANATION OF CO_2/H_2

MICHAEL SPECHT, JOCHEN BRELLOCHS, VOLKMAR FRICK, BERND STÜRMER, AND ULRICH ZUBERBÜHLER

7.1 MOTIVATION

7.1.1 History "Renewable Fuel Paths at ZSW"

The Centre for Solar Energy and Hydrogen Research (ZSW) has been working in the area of renewable fuel generation – especially methanol, methane, and hydrogen – since the late 1980s. Pilot plants were built for these fuel generation pathways:

1. Power to methanol: CH_3OH from electrolytic H_2 and CO_2;
2. Biogas to methanol: CH_3OH from CO based syngas via bio-methane steam reforming;
3. Biogas to gas: CH_4 from biogas via membrane separation;
4. Biomass to gas: CH_4 from CO based syngas via biomass carbonate looping gasification;
5. Power to gas (P2G®): CH_4 from electrolytic H_2 and CO_2.

Synthetic Natural Gas from Coal, Dry Biomass, and Power-to-Gas Applications, First Edition.
Edited by Tilman J. Schildhauer and Serge M.A. Biollaz.
© 2016 John Wiley & Sons, Inc. Published 2016 by John Wiley & Sons, Inc.

The common goal of the different routes is to produce and utilize renewable carbon based fuels for energy storage, energy transmission, power supply and transport. The focus of the present article is on the P2G® process.

7.1.2 Goal "Energiewende"

The German term "Energiewende" refers not only to the generation of renewable electricity but also to the sustained provision of energy for the heat and mobility consumer sectors. The goal of the transformation of our energy system is a sustainable and secure full supply with renewable energies (RE). Although wind and solar energy occur fluctuatingly, the supply of final energy – electricity, heat, and fuels – must take place at all times without restriction of use and without the waste of surplus RE. This means that, on the one hand, the demand during times of a limited supply of RE must be covered and, on the other hand, a surplus of RE during times of lower demand must not go unused. A secure and sustainable full supply of RE will therefore not be successful without adequate storage of energy and without the provision of regenerative fuels (hydrogen and carbon based fuels).

7.1.3 Goal "Power Based, Carbon Based Fuels"

The sustainable generation of electricity based and carbon based fuels represents a major challenge. The energy system of the future will also require carbon based fuels (C fuels) for certain areas of mobility (e.g., long-distance truck haulage, air traffic, ship traffic), for the chemical industry (e.g., plastics), and for nonsubstitutable applications in the area of electricity generation (e.g., natural gas substitute for seasonal energy storage). Therefore, the goal is a sustainable, CO_2 neutral energy system and not a decarbonization of the energy system. A promising opportunity is the utilization of electrolytic hydrogen to maximize the yield of carbon based fuels from biomass as a carbon resource.

7.1.4 Goal "P2G®"

The original definition of power to gas was oriented towards for the conversion of (renewable) electricity to a (renewable) gas, such as methane and/or hydrogen, in order to store this in the existing natural gas infrastructure and make it available to different use sectors at different times [1, 7, 17]. The most important applications of P2G® in a RE based energy system are:

1. Long-term storage for RE – e.g., from fluctuatingly acquired photovoltaic/wind power energy surpluses;
2. Stabilization of the power grid by grid-driven electricity consumption and grid-driven power feed-in via reconversion of the energy carrier produced;
3. Partial transfer of the energy transport from the power grid to the gas grid;

4. Production of sustainable fuels for mobility which cannot be substituted elsewhere (with the short-term realisable entry market "CH_4 mobility", in the mid term with "H_2 mobility" and in the long term with the provision of liquid hydrocarbons, e.g., for air traffic).

The opportunity for the bidirectional coupling of the power grid and the gas grid lies in converging the systems into a sustained energy supply with electricity, heat, and fuels using the existing grids for energy distribution and storage.

7.1.5 Goal "Methanation"

The most important goals with the design of the reactor concept for methane synthesis are temperature control in the reactor and maximization of the conversion rate of the educts CO_2 and electrolytic H_2 for the production of substitute natural gas. Normally, CO based syngas is used for methanation and not a CO-free educt gas. Here, the focus was on the rapid industrial implementation in a 6000 kW_{el} plant by the year 2013. This involves the construction of the so-called "e-gas plant" for the car manufacturer Audi AG by the plant construction firm ETOGAS GmbH. The goal is sustainable mobility using methane-powered vehicles.

7.2 THE POWER TO FUEL CONCEPT: CO-UTILIZATION OF (BIOGENIC) CARBON AND HYDROGEN

Regenerative sources of carbon available over the long term for the production of renewable, C based fuels are biomass and CO_2 from the air and bodies of water. Assuming a density of 500 kg/m^3 and a carbon content of 50% by weight, 1 m^3 of dry biomass yields 250 kg of concentrated carbon. By comparison, airborne carbon is highly diluted; assuming a carbon content of 400 ppm$_v$ one standard cubic meter of air contains only around 0.0002 kg of carbon as diluted, gaseous CO_2. Due to the dilution, the extraction of concentrated CO_2 from the air and water bodies requires a high energy input.

Consequently, biomass is of vital importance for the transformation of our energy system. Biomass is the only renewable energy which contains carbon. Nevertheless, the sustained availability of bioenergy potentials is limited and, in the case of cultivated biomass, directly competes with the cultivation of foodstuffs. If a sustained and secure full supply with RE is to succeed, biomass must be regarded in future as a carbon resource and not as an energy source. Biomass is primarily required to cover future demands for carbon based fuels.

Because of its elemental composition, biomass is not suitable for converting 100% of the carbon to carbon based fuels using conventional conversion processes. A highly efficient C conversion is, however, essential in order to obtain the maximum yield of fuel. For the highly efficient production of carbon based fuels, not only new production processes are required, but also an "upgrade" of the limited biomass resources (extension of the energy coverage by raising the rate of C conversion).

A highly promising approach is the production of biogas and the thermochemical conversion of biomass (e.g., gasification) in combination with externally supplied hydrogen from electrolysis. This enables electrolytic hydrogen to be generated directly from water and renewable electricity, and therefore RE to be stored chemically. This innovative combination of processes can decisively enhance the limited bioenergy potential.

As the following reaction equation shows, for the complete and therefore efficient conversion of carbon in the biomass[1] to methane (CH_4) it is necessary to supply additional hydrogen:

$$CH_{1.431}O_{0.661} + 1.946\,H_2 \rightarrow CH_4 + 0.661\,H_2O \tag{7.1}$$

Here, at standard conditions the biomass has a lower heating value (LHV) of 439.67 kJ/mol [3]. Under standard conditions, the lower heating values of hydrogen and methane are 241.82 kJ/mol$_{H2}$ and 802.27 kJ/mol$_{CH4}$, respectively. Comparing the products and educts, Equation (7.1) shows that 802.27 kJ methane can be produced from 439.67 kJ biomass and 470.58 kJ H$_2$ with 100% carbon conversion. This corresponds to an energy efficiency of 88% for the production of SNG (substitute natural gas). Assuming an efficiency of 70% for the electrolytic provision of hydrogen from regenerative surplus power, the overall energy efficiency is then 72%.

With the "upgrade" of biomass using electrolytic hydrogen, the yield of the carbon based energy carrier SNG can be maximized. It follows from Equation (7.1) that up to 802.27 kJ methane can be produced from 439.67 kJ biomass, corresponding to an energetic ratio of biomass input to SNG output of 182.47%.

Through the use of externally supplied hydrogen, a highly efficient utilization of the carbon stored in the biomass is achieved. In this way, the specific requirements for agricultural areas used for the provision of fuels can be curtailed and a sustainable energetic utilization of biomass is made possible. Figure 7.1 shows the future-oriented process for the conversion of biomass and renewable power to the required carbon based fuels [4].

The biomass is used as a carbon resource and can be converted in the anaerobic biogas process or in thermochemical biomass conversion by biomass to gas (BtG) – possibly in combination with gas to liquids (GtL) – to carbon based energy carriers, such as methane (CH_4), higher hydrocarbons (C_xH_y) and alcohols or ethers ($C_xH_yO_z$). Electrolytic H$_2$ can be utilized either for the conversion of carbon (CO_2) to methane or directly combined with the respective production process.

Figure 7.2 illustrates the degree to which the fuel yield from biomass can be increased by the proposed approach [4]. The comparison shows that first-generation biofuels (bioethanol and biodiesel) have relatively poor conversion efficiencies. For the production of SNG from biomass without regenerative surplus power, a more efficient conversion is possible compared with biodiesel and bioethanol. Depending

---

[1] Assuming that biomass contains 50% carbon, 6% hydrogen, and 44% oxygen by weight, a molar composition of 1.0 part carbon, 1.431 parts hydrogen, and 0.661 parts oxygen is characteristic for the biomass.

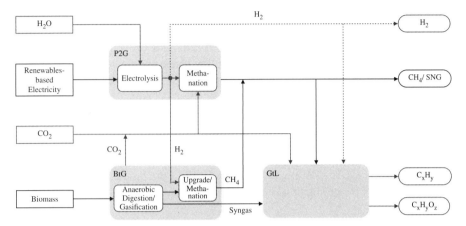

FIGURE 7.1 Approach for the highly efficient utilization of biogenic carbon in the production of regenerative, carbon based fuels. BtG: biomass to gas; GtL: gas to liquids.

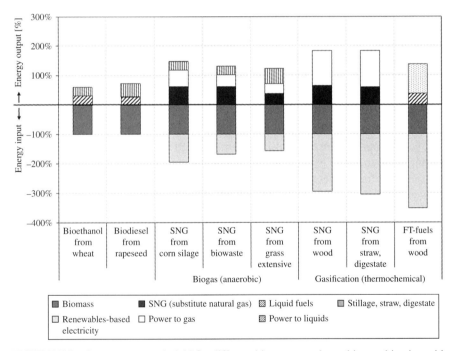

FIGURE 7.2 Input energy and yield for different bioenergy paths and in combination with renewable power for the electrolytic provision of hydrogen in the production of C fuels from biomass. With the use of biomass and electricity the "power to gas" and "power to liquids" stages add to "SNG" and to "Liquid fuels" in the case of FT fuels.

on the educt and the bioenergy path, energy conversion efficiencies of more than 60% can be achieved. Furthermore, in the case of anaerobic biogas production the yield of SNG can be nearly doubled when electrolytic hydrogen is fed in (i.e., the CO_2 fraction of the biogas is also converted to methane). In addition, digested residues remaining from the production of biogas can be recycled either as humus or nutrition suppliers or efficiently converted by thermochemical conversion. In the case of gasification, a maximum of three times the energy yield is possible with the production of SNG, as here no digested residues remain behind. An alternative to biodiesel and bioethanol is the production of liquid fuels by Fischer–Tropsch synthesis (FT fuels). From an energy standpoint, due to the lower hydrogen to carbon ratio of FT fuels ($-CH_2-$) compared with SNG (CH_4) the production of FT fuels is less efficient than the production of SNG by anaerobic digestion and thermochemical conversion.

In order to meet the future requirements for carbon based fuels in a sustainable energy system, it is advantageous to utilize the available regenerative potential of bioenergy preferentially for the production of carbon based fuels. For the generation of power and heat, renewable technologies not based on biogenic resources can be used. With the described approach, pressure on the limited biomass resources – in combination with electrolytic hydrogen produced by renewable power – can be sustainably relieved. Compared with biodiesel and bioethanol, the fuel yield per input energy unit of biomass and per utilized agricultural area is increased by up to a factor of six.

7.3 P2G® TECHNOLOGY

P2G® is an approach for the (seasonal) storage of RE. In the P2G® process, the fluctuatingly acquired electricity from RE, particularly from wind power and photovoltaics, is used for the electrolytic production of hydrogen which together with CO_2 is converted in a synthesis reactor to methane, the main constituent of natural gas, and this is fed as SNG into the natural gas grid. In the existing natural gas infrastructure, the chemical energy carrier methane produced from RE is efficiently stored, distributed, and made available for need based use. A particular advantage compared with other storage options is the use of the natural gas grid with its high storage and transport capacity. The P2G® process offers the option of merging the power grid and the gas grid into an integrated complete system for the on-demand use, distribution, and storage of energy.

For the long-term storage and the seasonal equalization of RE, only chemical secondary energy carriers such as hydrogen and carbon based fuels (e.g., natural gas substitute) which can be produced from different RE come into question today. These currently represent the only foreseeable options for the seasonal storage of RE with a capacity in the TWh range (using the example of Germany). SNG can be reconverted in modern gas and steam power stations and in decentralized CHP plants when electricity is in demand, find uses in industry or as a fuel – e.g., as "e-gas" in mobility.

The main process components of a P2G® plant are water electrolysis and methanation. Electrolysis is available on the market in the required MW_{el} power class as an alkaline system. However, this is to a limited extend the case for polymer electrolyte membrane (PEM) electrolysis. The solid oxide electrolyte (SOE) electrolysis is currently under development. At the present time, because of the still lacking market, the available electrolyzers are, however, not standard products and must be specially designed and manufactured for the MW_{el} power class, with the high specific costs which this entails. Beside the goal of improving efficiency, the focus of the electrolysis lies particularly in the development of new electrolysis block concepts and the reduction of costs by the modularization of hydrogen-producing systems.

7.3.1   Methanation Characteristics for CO_2 Based Syngas

Methanation in combination with the P2G® process is distinguished by the following features:

1. The methanation of CO based synthetic gases has been technically realized and is available on the market. However, this is not true for H_2/CO_2 based educt gases.
2. The CO_2 based synthetic make-up gas fed to the SNG process is stoichiometrically adjusted with the goal of nearly total C conversion.
3. The full conversion of the "inert" CO_2 poses far greater demands on the reactor system than the methanation of CO.

Technologically, methane synthesis from largely stoichiometrically adjusted carbon oxides and hydrogen mixtures represents a special challenge. On the one hand, the highest possible reaction yield must take place in order to fulfil the regionally specific feed requirements for synthetic natural gas. On the other hand, due to the exothermic heat effect of the reaction, this conversion level can result in a pronounced rise in temperature, with negative consequences for the catalyst.

Particularly the educt gases for methanation resulting from the absorption-supported gasification of biomass (absorption enhanced reforming (AER) process [5, 11, 12]) and for the P2G® process are characterized by nearly stoichiometric carbon oxides – hydrogen concentrations (see, e.g., [11] and [13]). While this enables the direct production of SNG without downstream gas conditioning (CO conversion, CO_2 separation), at the same time it requires a considerable reduction of the CO_2 and H_2 concentrations in the product gas to comply with the specifications for the natural gas substitute.

A major challenge is the increased heat development due to the high reaction yield compared with non-stoichiometrically adjusted gases (e.g., dual fluidized bed (DFB) gasification [6, 14]). Table 7.1 gives the results of the reactor heat calculation, assuming maximum conversion of hydrogen ($X_{H2} = 100\%$) relative to the educt gas stream [10].

TABLE 7.1   Specific Heat of Reaction of Methanation with Full H_2 Conversion from Different Educt Gases (With/Without Steam).

		CO_2/H_2	DFB	Biogas/H^2	AER
H_2	vol%$_{db}$	80	39	64.3	67.5
CO_2	vol%$_{db}$	20	22.5	16.1	12
CO	vol%$_{db}$	0	23	0	8.5
CH_4	vol%$_{db}$	0	15.5	19.6	12
H_2O	vol%	0	50	20	20
Specific heat of reaction, wet/db	kWh/m$^3_{STP}$	0.41/0.41	0.22/0.33	0.27/0.33	0.36/0.43

Compared with a biogas/hydrogen mixture or a gasification gas obtained from DFB biomass gasification, the conversion of a CO_2/H_2 gas mixture and of an AER educt gas to methane releases a greater amount of heat. This can be explained on the basis of the high proportion of convertible educts in the gas stream fed in, in particular as methane no longer contributes to the heat of reaction in the course of the reaction.

For the choice of reaction conditions chemical equilibrium calculations can be used, as these allow a conclusion on the product composition for the maximum achievable rate of conversion relative to the operating parameters chosen. The goal is to control the process so that after condensing out the water, a product gas is obtained which can be fed into the grid. Figure 7.3 plots the required pressures for the synthesis of H gas (95 vol%$_{db}$ CH_4) and L gas (90 vol%$_{db}$ CH_4) according to the currently applicable grid input regulations in Germany [8, 9]. By way of example, equilibrium calculations were performed with different stoichiometric numbers SN (see Equation (7.2)) for an H_2/CO_2 mixture at temperatures of 200–300 °C, assuming a specified methane content in the product [10]. y_i is the volume fraction of the respective component i (with $i = H_2$, CO, CO_2).

$$SN = \frac{y_{H_2}}{3\,y_{CO} + 4\,y_{CO_2}}\;[-] \tag{7.2}$$

where γ_i is the volume fraction of the respective component i (with $i = H_2$, CO, CO_2).

It becomes clear that the minimum required pressure for synthesis increases disproportionately with temperature. Accordingly, the goal of the process design is the lowest possible reactor temperature in the regions with high reaction conversions (in the direction of the reactor outlet). This temperature is limited by the start-up temperature of the catalyst used and its conversion characteristics. The choice of catalyst thus clearly influences the process pressure and consequently the required energy input for the compression of the educt gas mixture. Beside the temperature, the educt stoichiometry also decisively influences the pressure. In the region under consideration, stoichiometric numbers SN approximating 1.00 require low process pressures. If the resulting product gas is to be fed into an L gas grid, significantly

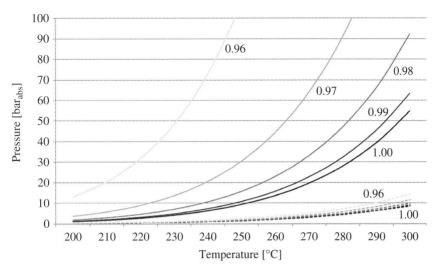

FIGURE 7.3 Required operating pressure for equilibrium conversion as a function of reactor temperature with different stoichiometric numbers (SN) for a CO_2/H_2 mixture. H gas = solid lines, L gas = dashed lines.

lower pressures are required for synthesis than for the production of an H gas. Thus, for example, if a catalyst has a minimum operating temperature of 240 °C, a pressure of at least 6.3 bar$_{abs}$ is required in order to comply with the limit value of CH_4 for H gas. For L gas, the respective value can already be complied with at a pressure upwards of 1.0 bar$_{abs}$. It should be noted that these considerations are fully valid only under chemical equilibrium. These figures represent the minimum requirement for the operating pressure. In practice, this value will be higher than the values calculated here [10].

Various reactor systems can be used for the methanation of carbon oxides. These must satisfy the requirements of sufficient temperature control in the main reaction zone and good reaction conversion. Within the scope of the P2G® process, the ZSW is investigating wall-cooled, catalyst-filled fixed bed reactors in tubular reactor and plate reactor designs (see below). These systems are designed so that sufficient heat dissipation is possible through the reactor walls. An important parameter here is the ratio of the cooling surface to the volume of the catalyst, as this directly influences the maximum temperature in the catalyst bed. Accordingly, in the case of a tubular reactor, the reaction heat released determines the maximum tube diameter. Possible cooling media are pressurized water, heat transfer oil, and molten salt.

An important advantage of wall-cooled reactors compared with adiabatic or isothermal systems is the possibility to adjust the temperature profile of the catalyst bed as required, which in exothermal reactions is characterized by very high temperatures in the region of the gas inlet and low temperatures at the end of the catalyst bed (≈ cooling medium temperature). This imposes a temperature gradient on the reactor. Due to the high temperatures, initially very high conversion rates are obtained, while the low temperatures simultaneously allow a favourable adjustment of the equilibrium position. This constellation allows nearly total conversion in a single reactor stage [15].

The basic prerequisite for such operation is efficient control of the heat of reaction, as otherwise a reduced catalyst life must be expected. A suitable measure for influencing the temperature profile is a graded educt gas feed, allowing the distribution of the heat of reaction over several fill regions. Reference [19] investigated this reactor design using a diluted H_2/CO stream. The authors showed that, for a constant educt gas rate, an educt gas stream fed in downstream considerably reduced the upstream temperature maximum.

In industrial applications tubular reactors are usually designed as tube bundle reactors. With this design, the catalyst is located inside the tube and the cooling medium (e.g., molten salt) in the jacket. Upscaling is achieved by increasing the number of tubes. The results of experimental investigations with a single tube can then be transferred directly to the tube bundle, provided that there is a uniform gas distribution over the tubes in the stream and therefore a uniform pressure loss over the individual catalyst beds.

An alternative reactor design is the plate reactor. With this reactor type, pressurized-water cooled plates are installed at regular intervals throughout the catalyst bed. In the hot spot region of the reactor, partial evaporation of the pressurized water occurs in the cooling plates, enabling good heat transfer.

The essential features of the reactors used by the ZSW are briefly described. The tubular reactors used in the 250 kW_{el} and the 6000 kW_{el} P2G® plants (see Subsection 7.3.2) are characterized by a molten cooling medium (molten salt reactors). A pre-defined temperature profile can be imposed on the reactor by at least two separate cooling circuits. An additional feature is the graded educt feed by a displacement pipe at the reactor inlet, which serves to limit the hot spot temperature. The design is shown schematically in Figure 7.4a.

The plate reactor is designed according to a cost-effective principle of construction, in which the cushion-shaped thermal plates are manufactured by expanding spot-welded plates under high pressure. The thermal plates are joined to form a heat exchanger package containing the pressurized water/steam cooling medium. The catalyst fill is located between the thermal plates. The heat of reaction is conducted away by the partial evaporation of the pressurized water/steam cooling medium. The cooling circuit pressure is adjusted so that the required temperature-dependent educt gas conversion is achieved at the outlet of the reactor. The design is shown schematically in Figure 7.4b.

If the product gas composition at the outlet of the reactor does not have the required quality, it is possible to compensate for this with an additional separation stage. Membrane gas separation methods are suitable in this connection. If, for example, the operation of the methane synthesis is chosen with overstoichiometric H_2, the excess hydrogen can be efficiently separated from the methane and recycled to the educt mixture [20]. At the same time the quantity of methane produced is increased, as the higher content of hydrogen favours the conversion of CO_2.

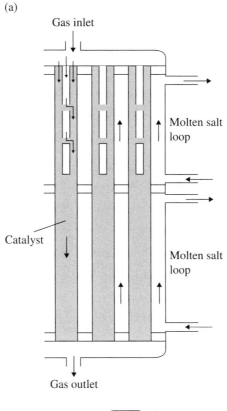

(a) Gas inlet

Molten salt loop

Catalyst

Molten salt loop

Gas outlet

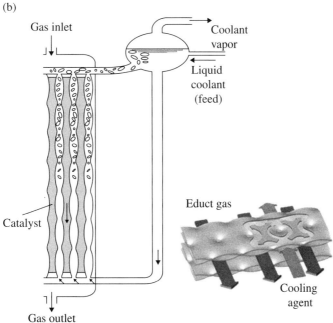

(b) Gas inlet

Coolant vapor

Liquid coolant (feed)

Educt gas

Catalyst

Cooling agent

Gas outlet

FIGURE 7.4 (a) Schematic representation of the molten salt-cooled tube bundle reactor design with two separate cooling circuits and graded educt gas feed by a displacement pipe. (b) Schematic representation of the water/steam-cooled plate reactor design, consisting of thermal plates joined to a heat exchanger package (grafic on right [16]).

7.3.2   P2G® Plant Layout of 25 kW$_{el}$, 250 kW$_{el}$, and 6000 kW$_{el}$ Plants

Two P2G® plants were constructed at the ZSW in the power classes 25 kW$_{el}$ and 250 kW$_{el}$. The ZSW is also participating in a 6000 kW$_{el}$ plant within the scope of basic engineering, commissioning and plant monitoring. These power figures refer to the DC electrical input power of the electrolyzers as a design basis.

The three plants are characterized by the following features: the 25 kW$_{el}$ plant is a test plant with two fixed bed methanation reactors connected in series with inter-stage condensation and a product gas recycle loop. The 250 kW$_{el}$ plant is a test plant of technically relevant size with variable configuration. Two different reactor systems (tube bundle and plate reactors) can be operated alone or in combination. A membrane gas processing stage enriches the methane in the product gas and allows the recycling of a hydrogen-rich gas (permeate) as an educt. The 6000 kW$_{el}$ plant is the world's first commercial power to gas plant which feeds methane into the natural gas grid. It was designed as a "once through" concept in the form of a tube bundle reactor with feed into the L gas grid. The designs of all three plants are shown schematically in Figures 7.5 to 7.7.

7.3.2.1   25 kW$_{el}$ P2G® Plant Layout
The ETOGAS GmbH, formerly SolarFuel GmbH, commissioned ZSW with the construction of the 25 kW$_{el}$ plant. The container-integrated design allows operation directly from a biogas plant using real gases as educt. The plant was designed for the conversion of CO_2 to methane from the "off gas" of biogas processing plants (for biomethane production, e.g., by pressure swing adsorption or amine scrubbing) or for the direct conversion of biogas with the

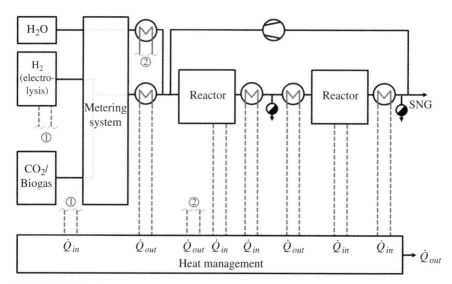

FIGURE 7.5   Schematic representation of the container-integrated 25 kW$_{el}$ P2G® test plant design.

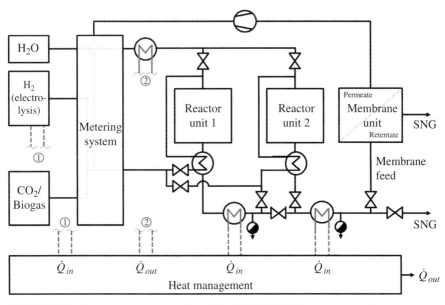

FIGURE 7.6 Schematic representation of the 250 kW$_{el}$ P2G® plant design; Reactor unit 1: plate reactor, Reactor unit 2: tube bundle reactor.

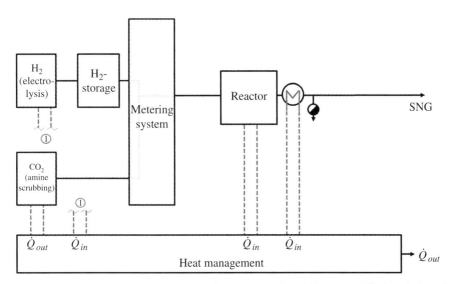

FIGURE 7.7 Schematic representation of the commercial 6000 kW$_{el}$ P2G® plant design of Audi AG in Werlte, Lower Saxony, Germany.

addition of hydrogen. It was completed in 2009. With this plant, the feasibility of the technology was first demonstrated using real gases from biogas plants.

The production of hydrogen takes place by alkaline high-pressure electrolysis with 25 kW_{el} input power. The educt gas streams are fed by means of metering devices to a two-stage reactor system with partial inter-stage condensation of water. The design of the synthesis unit with inter-stage condensation for the reduction of the moisture content increases the proportion of methane in the product gas. In order to limit the temperature in the hot spot zone of the first reactor, a part of the gas stream from the product gas is recycled via a "recycle loop". Deactivation of the catalyst due to the precipitation of carbon is prevented by feeding steam into the educt gas.

The reactors for the synthesis of methane are designed as tubular reactors with double jackets for tempering the reaction chamber. A heat transfer oil serves as the tempering medium. Beside pressure electrolysis and methanation, the container-integrated plant also houses the control electronics, including a filling module for natural gas vehicles consisting of gas drying, compressor, storage rack, and filling device.

7.3.2.2   250 kW_{el} P2G® Plant Layout   The 250 kW_{el} plant was constructed within the scope of a publicly funded project [2] and commissioned in 2012. The goal is to optimize the technology in intermittent and dynamic operation, as well as to identify cost-reduction potentials. The design is shown schematically in Figure 7.6.

The plant has a 250 kW_{el} alkaline high-pressure electrolyzer and features two different reactor systems which can be operated alone or in combination. Beside the tube bundle reactor, which has proved itself in the chemical industry, a pressurized water-cooled plate reactor is used. For tempering the hot spot region, different approaches have been followed. For the plate reactor, cooling of the main reaction zone takes place by the evaporation of water in the thermal plates, while for the multi-zone tube bundle reactor, the graded educt gas feed and molten salt cooling limit the temperature in the hot spot zone. Here, the multi-zone design of the molten salt cooling system allows the possibility to adjust the temperature profile of the methanation reactor as required.

For increasing the methane content it is possible to operate the plant as a two-stage synthesis process with partial inter-stage condensation, or as a single-stage methanation with downstream membrane gas upgrade. Processing of the methane-rich product gas to a substitute natural gas which can be fed into the grid takes place by means of membrane technology. Here, the separated hydrogen-rich gas (permeate) is recycled as educt gas. This interconnection allows overstoichiometric H_2 production in the methanation reactors with nearly total conversion of CO_2 [20].

7.3.2.3   6000 kW_{el} P2G® Plant Layout   The client for the 6000 kW_{el} plant in Werlte, Lower Saxony (Germany), is the car manufacturer Audi AG. The plant was designed and built by the plant construction firm ETOGAS GmbH. Commissioning took place in the fourth quarter of 2013. The goal of the plant is the production of an electricity based natural gas substitute to be fed into the natural gas grid and used as a sustainable fuel for mobility. The plant fed SNG with a methane content of >93 vol%$_{db}$ into the natural gas grid for the first time at the end of 2013 (Figure 7.8).

FIGURE 7.8   The 6000 kW$_{el}$ P2G® methanation plant of Audi AG in Werlte, Lower Saxony, Germany (molten salt-cooled tube bundle reactor); photo: Audi AG [18]. See Figure 7.4a for a schematic representation of the design.

The world's first commercial power to gas plant feeding SNG into the natural gas grid is characterized by a simple basic process technology concept in which no steam is dosed into the educt gas and the dried product gas is fed as a replacement gas into the existing on-site L gas grid without further processing steps (except air dosing to the product gas). Methanation is carried out in a molten salt-cooled tube bundle reactor with imposed temperature profile and simple reactor throughput ("once through"). The temperature control in the hot spot zone takes place in the individual tubes through a graded educt gas feed as well as through the heat transfer medium.

The CO_2 required for the process is separated from the biogas of a digester plant via amine scrubbing. Under full load, the P2G® plant produces a replacement gas volume flow of around 325 m³$_{STP}$/h (without the biomethane fraction, which is also fed into the gas grid). The replacement gas is fed into the local gas distribution grid (1.8 bar$_{abs}$). When the capacity of the distribution grid is exhausted (e.g., in the summer, with low gas consumption) the gas is fed into the transport network (35–45 bar$_{abs}$).

A special feature is the extraction of heat from the methanation reactor to supply the amine scrubbing. Heat is made available to the CO_2 stripper during operation of the P2G® plant. When the P2G® plant is in standby mode, the heat required for the

CO_2 stripper is supplied from combustion of part of the biogas stream produced. This means a vastly higher biomethane yield during methanation operation, as there is no combustion of the biogas.

A further feature of the 6000 kW_{el} plant is the intermediate storage of hydrogen. The hydrogen produced in the alkaline electrolysis with three electrolytic stacks is temporarily stored in a hydrogen pressure tank at approximately 10 bar_{abs}. This allows a temporary decoupling of the intermittent operation of the electrolyzer from methanation, reducing the number of start-up and shut-off ramps compared with electrolysis operation.

7.4 EXPERIMENTAL RESULTS

In addition to the building and operation of the P2G® plants in the power classes 25 kW_{el} and 250 kW_{el}, experimental investigations are being carried out at the ZSW on the hydrogeneration of gases containing oxides of carbon. In particular, these include screening of the catalysts available on the market, investigation on their deactivation and the processing of the product gases from methanation to a replacement gas using downstream membrane gas upgrade technology. This chapter discusses the most important results of these experiments.

7.4.1 Methanation Catalysts: Screening, Cycle Resistance, Contamination by Sulfur Components

7.4.1.1 Catalyst: Screening   Before use in the reactors, the catalysts available on the market are investigated with regard to their suitability for intermittent operation in methanation. The investigations include the characterization of catalysts with regard to start-up temperature, conversion behavior and catalyst activation, for example, and the identification of deactivation mechanisms (thermal stability, poisoning, and deposition of carbon) on different reactors and with different analytical methods, such as thermogravimetric analysis (TGA).

The investigations on start-up temperature and conversion behavior are carried out in a wall-cooled, oil-tempered tubular reactor with approximately 80 ml catalyst volume. Figure 7.9 shows typical results for the commercially available Ni based catalysts regarding the start-up temperature ($T_{start\text{-}up}$) and the temperature measured in the hot spot of the reactor ($T_{hot\,spot}$). The start-up temperature is determined by increasing the oil temperature in 10 K steps. The temperature above which the catalyst temperature in the hot spot zone of the reactor is higher than the oil temperature (due to exothermic reaction enthalpy) is referred to as the start-up temperature.

On the basis of the measured data in Figure 7.9 ($T_{Start\text{-}up}$, $T_{hot\,spot}$, y_{CH_4}) it can be said that catalysts 1, 3, and 5 are very well suited for CO_2 methanation. Besides their low start-up temperatures, these catalysts have high temperature gradients in the hot spot zone. Due to the exothermic character of the methanation reaction, this rise in

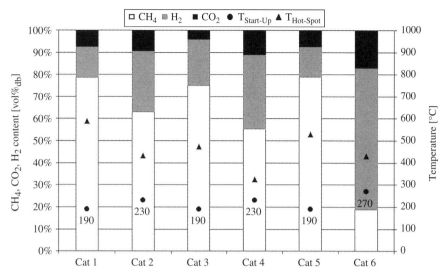

FIGURE 7.9   Start-up temperature ($T_{\text{start-up}}$), temperature in the hot spot zone ($T_{\text{hot spot}}$) and conversion characteristics (resulting gas composition) of Ni based catalysts on the market (Cat x) for CO_2 methanation. Educt gas: 80 vol%$_{\text{H2, db}}$/20 vol%$_{\text{CO2, db}}$. Database $T_{\text{start-up}}$ and $T_{\text{hot spot}}$: SV = 2000 l$_{\text{educt}}$/(l$_{\text{cat}}$·h) and p = 7 bar$_{\text{abs}}$. Database gas composition: SV = 3000 l$_{\text{educt}}$/ (l$_{\text{cat}}$·h); p = 1.5 bar$_{\text{abs}}$; $T_{\text{cooling medium}}$ = 300 °C.

temperature is an indicator for the high activity of the catalysts. These results were confirmed with regard to the methane concentrations obtained in the product gas. Here, the test parameters do not reflect the optimal operating points of the catalysts, however in view of the uniform operating conditions they allow comparison.

7.4.1.2   Catalyst: Cycle Resistance for Intermittent Operation   The catalysts for methanation in the power to gas plants are subject to intermittent and dynamic operation and therefore to frequent start-up and shut-down processes. The deactivation of the catalysts due to sinter effects, which reduce the catalytically active surface, are investigated in an oil-tempered reactor [V = 80 ml; space velocity, SV = 4000 l$_{\text{educt}}$/ (l$_{\text{cat}}$·h)]. The cycle resistance of the catalyst is determined in automatic, cyclical methanation operation with starting conditions equivalent to those of the standby mode (temperature and pressure maintained in an H$_2$ atmosphere) and the methanation, including the addition of H$_2$/CO$_2$. Figure 7.10 illustrates typical results for the course of the hot spot temperatures and the product gas content over a few cycles.

One cycle in Figure 7.10 comprises the start-up of methanation synthesis from the standby mode and the shut-down of synthesis to the standby mode, while the tempering medium maintains the temperature. Before beginning with the addition of CO$_2$, the temperature in the reaction chamber is identical to that of the tempering medium. The temperature of the tempering medium is adjusted to a level above the start-up temperature of the catalyst. In the standby mode, H$_2$ streams through the reactor. As soon as CO$_2$ enters the reactor, the exothermic methanation reaction begins. This results in a pronounced rise in temperature in the hot spot zone together

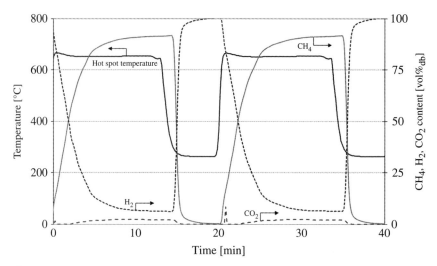

FIGURE 7.10   Typical test run in cyclic methanation operation for investigation of the cycle resistance of the catalysts. Educt gas composition: 80 vol%$_{H2,\ db}$/20 vol%$_{CO2,\ db}$, $T_{cooling\ medium} = 260\,°C$, SV $= 4000\ l_{educt}/(l_{cat}\cdot h)$, $p = 7\ bar_{abs}$.

with the conversion of the educt gases. After reaching stationary methanation operation, CO_2 dosing is discontinued and the reactor is flushed with hydrogen alone. As a result, the temperatures fall to the level of the tempering medium (around 260 °C). The process described above is automatically repeated following a brief holding time until reaching the pre-defined total number of cycles. Figure 7.11 shows the result of investigations over 900 cycles.

As Figure 7.11 shows, deactivation of the catalyst, caused by pronounced thermal loading, was not observed over 900 cycles. The temperatures in the reaction chamber and the quality of the product gas remain nearly constant. While thermal deactivation in the hot spot zone due to sinter effects of the catalyst cannot be entirely excluded, the results indicate that a long catalyst life can be expected. Furthermore, it can be said that, with intermediate H_2 storage, the number of cycles for methanation can be significantly reduced, extending the catalyst life still further (see Figure 7.7). Frequent start-up and shut-down processes, which are typical in P2G® plants because of the intermittent power draw, therefore do not significantly limit the catalyst life. Depending on the requirements of the power grid and the intermediate hydrogen storage capacity and the resulting number of start-up and shut-down processes during methanation, a technical catalyst life time of several years can be expected.

7.4.1.3   Catalyst: Contamination by Sulfur Components   Depending on the CO_2 source, minor components, particularly sulfur compounds, can contaminate the reaction chamber and cause deactivation of the catalyst. The effects of sulfur poisoning on a nickel based catalyst are measured by thermogravimetric analysis in a forced-flow tube reactor. The product gas composition and the catalyst mass are determined as a function of time for a high space velocity (SV) of 25 000 $l_{educt}/(l_{cat}\cdot h)$.

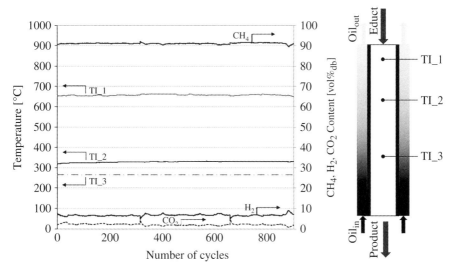

FIGURE 7.11 Temperature course and product gas quality as a function of the number of cycles (educt gas composition: 80 vol%$_{H2,\,db}$/20 vol%$_{CO2,\,db}$; $T_{cooling\,medium}$ (Oil)=260 °C, SV=4000 l$_{educt}$/(l$_{cat}$·h); p=7 bar$_{abs}$.

The influence on the course of concentration of the product gas during synthesis operation as a result of the declining reaction conversion and on the catalyst mass can be clearly seen in Figure 7.12. Small amounts of hydrogen sulfide lead to a deactivation of the catalyst. While the comparative measurement under the same reaction conditions without the addition of H_2S gives a methane content of 59 vol%$_{db}$ in the product gas, with the addition of 46 ppm$_{v,db}$ H_2S to the feed gas stream, a maximum of 53 vol%$_{db}$ methane is obtained. Thereafter the methane fraction falls off significantly. Simultaneously, the concentrations of hydrogen and carbon oxides increase significantly.

The mass of the catalyst over time is a further indicator for sulfur-related contamination. The initial mass decrease results from the reduction of nickel oxides by the hydrogen content of the educt gas (post hoc activation). The weight gain due to the formation of nickel sulfide is superimposed on this weight loss (catalyst deactivation). Comparative measurements prove that the lower the sulfur load, the longer the life of the catalyst. It follows from the results that, depending on the CO_2 source, high demands are placed on the gas production process for the removal of minor components in order to ensure a long life of the catalyst. The authors recommend << 0.1 ppm$_{v,db}$ H_2S sulfur load in the educt gas stream.

7.4.2 Results with the 25 kW$_{el}$ P2G® Plant

Experimental investigations were carried out for the conversion of gas mixtures containing H_2/CO_2 and $H_2/CO_2/CH_4$, respectively, with the 25 kW$_{el}$ P2G® plant. Both variants were investigated with "real gases" at different biogas plants. The educt

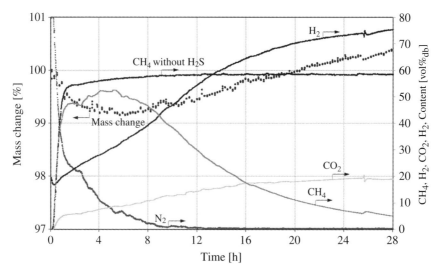

FIGURE 7.12 Degradation behavior of a Ni catalyst during the synthesis of methane in the presence of H_2S. At $t=0$ switch from 100 vol%$_{N2,db}$ to H_2/CO_2 educt gas (80 vol%$_{H2,db}$/20 vol%$_{CO2,db}$) with additional dosage of 46 ppm$_v$ H_2S in the educt gas stream. $T=260\,°C$, $p=7$ bar$_{abs}$, SV$=25\,000$ l$_{educt}$/(l$_{cat}$·h). The CH_4 concentration profile without H_2S in the educt is shown for comparison.

gases were biogas and "CO_2 off gas" taken from a processing plant for the production of biomethane from biogas. The educt gases were subjected to fine gas cleaning. The results of the SNG production from the educt gases "CO_2 off gas/H_2" and "biogas/H_2" are shown in Figure 7.13a, b. In both cases, the reactor was operated using the same operating parameters [T: 250–550 $°C$, $p=7$ bar$_{abs}$, Ni based catalyst, gas load SV: 2000–5000 l$_{educt}$/(l$_{cat}$·h)].

Figure 7.13a, b show that, with the use of both real gases – CO_2 from a pressure swing absorption plant and for purified biogas – a uniform gas quality can be achieved over a longer period. The synthesis operation remained stable during the entire test run. Because of the gas composition and the calorific properties, the gas mixture produced was a replacement gas (L gas) in accordance with the German DVGW G 260 and DVGW G 262 sets of regulations [8, 9]. After drying, further gas conditioning for feeding into the L gas grid is therefore not required.

7.4.3 Results with the 250 kW$_{el}$ P2G® Plant

The molten salt-cooled tube bundle reactors installed in the 250 kW$_{el}$ and the 6 MW$_{el}$ P2G® plants were designed based on experimental test series with a single tube. The results show that a scale-up from a single-tube reactor to a tube bundle reactor poses no problems. The modular design of this reactor type allows it to be adapted to the required gas rate by varying the number of reactor tubes.

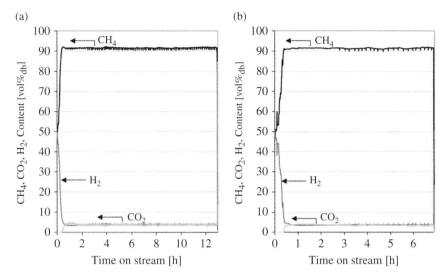

FIGURE 7.13 (a) Product gas composition for the conversion of CO_2/H_2 in the 25 kW$_{el}$ P2G® plant. As educt, the "off gas" from a pressure swing adsorption (PSA) was used, which processes biogas to biomethane in a biogas plant. (b) Product gas composition for the conversion of cleaned biogas with H_2 in the 25 kW$_{el}$ P2G® plant without previous separation of CO_2.

The pressurized water-cooled plate reactor was originally built for CO_2 methanation and operated under nearly realistic conditions. Due to the design of the plate reactor, the scaling of the size, spacing and number of the cooling plates is more difficult than with the tube bundle reactor. The goal with the plate reactor is to identify the operating characteristics for use in methanation, and in particular, to examine the potentials for cost reduction compared with other reactor concepts.

Figure 7.14a, b illustrate the operating characteristics of the two 250 kW$_{el}$ P2G® plant reactors with regard to repeated start-up and shut-down processes and load change. A stoichiometrically adjusted H_2/CO_2 mixture was added to each of the reactor systems, starting in the standby state under H_2 atmosphere. After reaching a stationary state with constant product gas quality, the investigation of the load change behaviour between 100 and 70% was carried out on the basis of stoichiometrically adjusted ramps for the educt gas streams. The system then reverted to the standby state under H_2 flushing and the entire sequence repeated.

Compared with shut-down in the standby mode, start-up from the standby mode requires a longer time due to the thermal inertia of the reactor systems and the increased flushing time with the mole number-reducing methanation. However, start-up and shut-down are in the order of only a few minutes. Following repeated start-up and shut-down processes, the original gas composition is obtained again. The results furthermore show that with the tube bundle reactor and the plate reactor,

(a)

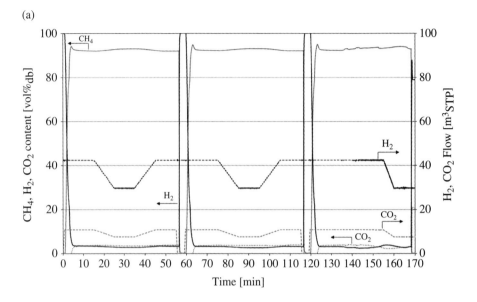

(b)

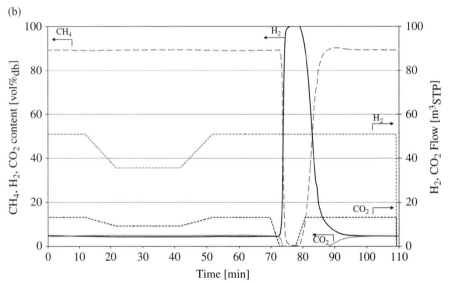

FIGURE 7.14 (a) Gas composition during start up/shut down and during load change with the tube bundle reactor of the 250 kW$_{el}$ P2G® plant as a function of the educt gas stream [$T = 200$–600 °C, $p = 7$ bar$_{abs}$, SV = 1365 l$_{educt}$/(l$_{cat}$ · h)]. (b) Gas composition during start-up/shutdown and during load change with the plate reactor of the 250 kW$_{el}$ P2G® plant as a function of the educt gas stream [$T = 200$–600 °C, $p = 7$ bar$_{abs}$, SV = 1365 l$_{educt}$/(l$_{cat}$ · h)].

load changes between 100 and 70% have no significant influence on the product gas quality.

The quality of the product gas – using catalysts of the latest generation and with optimal operation of the reactor with simple reactor throughput (once through) – is suitable for feeding a replacement gas into the German natural gas grid for the gas group L in accordance with the German guidelines [8, 9] without requiring a further downstream gas upgrade.

7.4.4 Results with the 250 kW$_{el}$ P2G® Plant in Combination with Membrane Gas Upgrade

In addition to the determination of operating concepts and options for cost reduction for P2G® plants, the production of a replacement gas for feeding into the H (*high* calorific gas) gas grid with CH$_4$ fractions typically > 95 vol%$_{db}$ is a major challenge. For the moderate operating pressures for methanation of < 10 bar$_{abs}$ as in Figure 7.3 with a "once through process", this is only possible to a limited extent. A highly promising option of downstream gas processing is the membrane gas separation process, in which the retentate is extracted as a replacement gas and the permeate is completely recycled to the educt gas stream for the methanation (recycle loop).

The results obtained from the process interconnection shown in Figure 7.6 with the two "once through" reactor concepts are displayed in Figure 7.15a (tube bundle reactor) and Figure 7.15b (plate reactor) as a function of the membrane area (unit area as *n* times the area of the smallest membrane module). The gas volume flow and the gas composition of the retentate are illustrated. In addition, the gas volume flows of the permeate stream and the CH$_4$ fraction of the product gas from methanation (= membrane feed) are plotted. The permeate stream increases with increasing membrane surface, while the retentate stream decreases. With the addition of a stoichiometrically adjusted H$_2$/CO$_2$ educt gas and the complete recycling of the permeate gas, the methane content in the retentate increases with increasing membrane area. The CH$_4$ fraction at the outlet of the methanation reactor (= membrane feed) remains nearly unchanged in spite of recycling.

For both reactor types, the gas qualities of the retentate gases with recycling of the permeate gases fulfil the requirements for feeding into the natural gas grid as a replacement gas, also for feeding into a group H gas grid.

Comparing the process chains with and without gas upgrade by membrane technology shows that, with a constant educt gas dosage with downstream membrane gas processing, not only is a high-calorific gas produced, but also that the operating pressure for the synthesis of methane is reduced (Figure 7.16). The interconnection of the methanation reactor and membrane realized at the ZSW enables simple regulation of the overall process pressure using the retentate-side pressure regulating valve of the membrane unit. The retentate gas produced requires only limited processing effort in order to fulfil the highest demands for feeding into the gas grid as a replacement gas.

Figure 7.17 shows the operating characteristics with regard to a load change in the methanation reactors with downstream membrane gas upgrade using the example of the tube bundle reactor. The product gas from methanation is fed to the membrane. With the tube bundle reactor, the retentate gas reaches a methane content of > 98 vol%$_{db}$ after 5 min. The fast start-up time is made possible by the time-delayed recycling of the permeate gas.

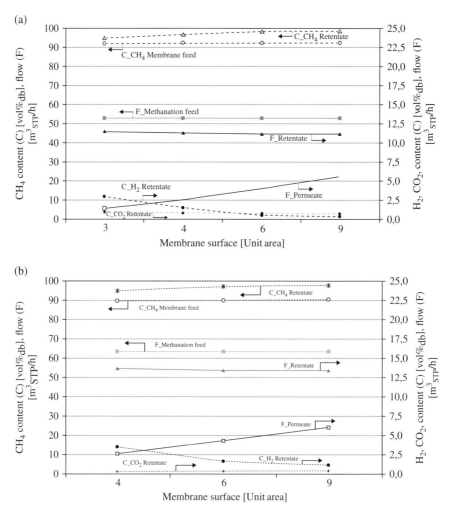

FIGURE 7.15 (a) Operating data for the tube bundle reactor ($p=5$ bar$_{abs}$) with downstream membrane gas upgrade ($p=5$ bar$_{abs}$) as a function of the membrane area: gas volume flow and gas composition of the retentate; gas volume flows of the permeate and methanation feed; CH$_4$ volume fraction of the methanation product gas (= membrane feed). (b) Operating data for the plate reactor ($p=5$ bar$_{abs}$) with downstream membrane gas upgrade ($p=5$ bar$_{abs}$) as a function of the membrane area: gas volume flow and gas composition of the retentate; gas volume flows of the permeate and methanation feed, CH$_4$ volume fraction of the methanation product gas (– membrane feed).

7.5 P2G® PROCESS EFFICIENCY

To determine the process efficiency, the P2G® process was modelled at the ZSW using the commercially available IPSEpro simulation software. The core of IPSEpro is an equation solver (Newton–Raphson algorithm) which is able to solve implicit

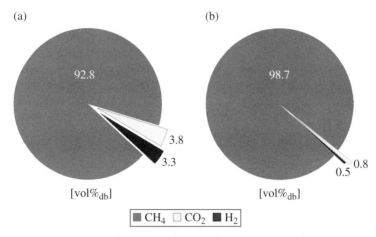

FIGURE 7.16 Gas composition with the tube bundle reactor: (a) without downstream membrane gas upgrade (SV = 1365 $l_{educt}/(l_{cat} \cdot h)$, $p_{methanation}$ = 7 bar_{abs}); and (b) with downstream membrane gas upgrade (SV = 1365 $l_{make\,up\,gas}/(l_{cat} \cdot h)$, $p_{methanation}$ = 5 bar_{abs}, $p_{membrane}$ = 5 bar_{abs}).

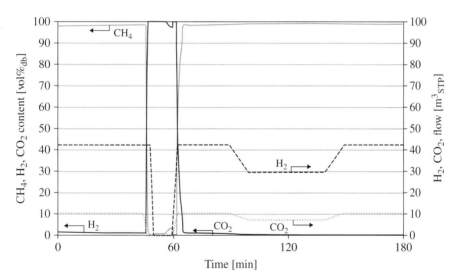

FIGURE 7.17 Gas composition during start-up/shut-down and during load change with the tube bundle reactor of the 250 kW_{el} P2G® plant as a function of the make-up gas volume flow with downstream membrane gas upgrade (T = 200–600 °C, p = 7 bar_{abs}, SV = 1365 $l_{make\text{-}up\,gas}/(l_{cat} \cdot h)$, membrane surface = 6 [unit area]).

equation systems. Taking the conservation of mass, chemical reactions and physical relationships into account, element and material balances can be simulated with individual models and complex process chains.

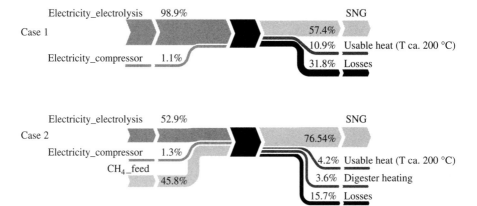

FIGURE 7.18 Energy balances for the P2G® process for a stoichiometric CO_2/H_2 gas stream (Case 1) and a stoichiometric biogas/H_2 stream (Case 2; biogas: 60 vol%$_{db}$ CH_4, 40 vol%$_{db}$ CO_2).

Figure 7.18 shows the energy balance for the P2G® process for two selected cases (once through process) without downstream membrane gas separation. The following cases were simulated:

- *Case 1*
  Methanation of a stoichiometric CO_2/H_2 gas stream.
- *Case 2*
  Methanation of a stoichiometric biogas/H_2 stream
  (biogas: 60 vol%$_{db}$ CH_4, 40 vol%$_{db}$ CO_2).

The results of the process simulation are based on the data shown in Table 7.2. For calculation of the conversion rate of the methanation reaction, the thermodynamic equilibrium was assumed at 8 bar$_{abs}$ and 260 °C.

As shown in Table 7.2, for Case 1 (stoichiometric CO_2/H_2 gas stream), the product gas from methanation fulfils the specifications for feeding into an L gas grid. For Case 2 the methane fraction of 95.8 vol%$_{db}$ is significantly higher, producing an H gas-equivalent replacement gas [8, 9].

For calculation of the chemical and thermal efficiency, in addition to the power of the applied electrolysis ($P_{el,electrolysis}$) and the chemical energy of the biogas ($P_{chem,educt}$), the most important electrical consumer loads, such as the CO_2/biogas compression and compression of the product gas to the feeding pressure following methanation ($P_{el,compression}$) were considered:

$$\eta_{chem} = \frac{P_{chem,product\ gas}}{P_{el,electrolysis} + P_{el,compression} + P_{chem,educt}} [-] \qquad (7.3)$$

$$\eta_{th} = \frac{\dot{Q}_{th,usable}}{P_{el,electrolysis} + P_{el,compression} + P_{chem,educt}} [-] \qquad (7.4)$$

TABLE 7.2   Basic Data for Determining the Energy Balances of the P2G® process.

		CH_4	H_2	CO_2	H_2O
		[vol%$_{db}$]	[vol%$_{db}$]	[vol%$_{db}$]	[vol%]
Case 1	Educt gas	0.0	80.0	20.0	0.0
	Product gas	93.9	4.9	1.2	65.2
Case 2	Educt gas	23.1	61.5	15.4	10.0
	Product gas	95.8	3.4	0.8	50.9
Electrolysis					
Energy demand of electrolysis			4.3	kWhel/m³H_2,STP	
Pressure			8.5	bar$_{abs}$	
Methanation					
Pressure educt gas			8.5	bar$_{abs}$	
Pressure product gas			8	bar$_{abs}$	
Temperature			260	°C	
Further energy and substance streams					
Pressure CO_2/biogas			1	bar$_{abs}$	
Temperature usable heat			ca. 200	°C	
Pressure gas grid feed-in			16	bar$_{abs}$	

In Case 1, the conversion of a stoichiometric CO_2/H_2 gas stream to an L gas suitable for feeding, a chemical efficiency of 57.4% is achieved (Figure 7.18). A quarter of the waste heat is produced at a temperature level of approximately 200 °C and can be used, for example, to provide heat to an amine scrubbing unit for CO_2 removal from biogas. The resulting thermal efficiency is 10.9%. In the case of direct methanation of biogas (without previous CO_2 separation in Case 2), a chemical efficiency of 76.5% results, with a thermal efficiency of 4.2% for the utilization of the waste heat at a temperature level of approximately 200 °C. One advantage of the combination P2G®-biogas plant is that part of the heat flow with $T < 200$ °C can be utilized for digester heating.

The losses resulting from the conversion of electrical power to chemically bound energy in the form of a natural gas substitute occur both in the electrolysis ($\eta = 70\%$, corresponding to 4.3 kWh$_{el}$/m³$_{H2,STP}$) and in the methanation unit ($\eta = 83.2\%$). As the efficiency of the methanation is determined by the heat of reaction of the CO_2 hydrogenation, the optimization potential with regard to the overall plant efficiency primarily concerns the electrolysis. An increase in the electrolytic efficiency from 70 to 80% (corresponding to 3.75 kWh$_{el}$/m³$_{H2,STP}$) results in an improvement of the overall plant efficiency to $> 65\%$ (Case 1).

7.6   CONCLUSION AND OUTLOOK

The P2G® process was first proposed in 2009 for the storage of renewable electricity in the form of a natural gas substitute for feeding into the natural gas grid. The resulting secondary energy carriers methane and/or hydrogen can be used on demand for reconversion or as a final energy carrier for the mobility and heating consumer sectors.

For a prototype proof of concept a 25 kW_{el} P2G® test facility was built at the ZSW in 2009. For process optimization a 250 kW_{el} plant was completed in 2012. A technical plant for the automotive sector with 6000 kW_{el} has been in operation since the end of 2013.

- The container-integrated 25 kW_{el} test facility comprises electrolysis, methanation, and control electronics, including a filling module for natural gas vehicles. The 25 kW_{el} P2G® system has been successfully operated at different anaerobic digester plants with CO_2 off gas (from a biomethane plant) and directly with biogas (without previous CO_2 separation). In both operation modes a replacement gas was produced with gas grid feed-in quality. The replacement gas thus produced – so-called e-gas (electricity based compressed SNG) – corresponds to the quality and calorific value of the natural gas present in the gas network.

- The main goals of the 250 kW_{el} P2G® project are the process upscaling to the energy relevant MW range, load dynamic operation concepts regarding the flexibility demands of the electricity market, and the identification of cost reduction potentials.

- In 2013, for the first time natural gas substitute was fed on an industrial scale from the 6 MW_{el} plant into the natural gas grid (e-gas plant of the car manufacturer Audi AG). This shows that – in spite of the still existing potential for optimization – the technology is already feasible in the MW power class today.

The following key arguments follow from the results of the experiments carried out with the P2G® plants in the power classes 25 kW_{el} and 250 kW_{el}:

- Methane production from CO based syngas is state of the art. This is not the case for H_2/CO_2 feed gases due to the poorly reactive CO_2 molecule. Nevertheless, synthesis can be realized with modified catalysts having high reaction selectivity and conversion rates.

- The renewable fuel methane can be produced decentrally in the MW power class in P2G® plants with significantly less processing effort than by the Fischer–Tropsch process or by methanol synthesis.

- The cycle resistance of the methanation catalyst was demonstrated for up to 900 cycles (standby/full load) without activity loss.

- A replacement gas can be produced by a "once through" process followed by drying of the gas without further processing (option for low-calorific gas networks).

- In combination with membrane gas upgrade, the methane content of the replacement gas can be increased up to 99 $vol\%_{db}$ (option for high-calorific gas networks).

- The start-up time of the methanation reactors from standby to full-load operation is in the order of only a few minutes. The same applies for the shut-down ramps.

- Due to the short start-up/shut-down ramps, the P2G® concept can provide positive and negative balancing energy to stabilize the electricity grid.

- For the conversion of a stoichiometrically adjusted CO_2/H_2 gas stream, the chemical efficiency of the P2G® process relative to the electrical energy input today is $< 60\%$. The optimization potential for the production of hydrogen by alkaline electrolysis and PEM electrolysis allows the conclusion that an increase of the overall plant efficiency to $> 65\%$ is possible (with the solid oxide electrolysis efficiencies $> 80\%$ appear realistic in the future).

For today's almost completely crude oil based fuel market, a diversification of the resource basis to include an increasing fraction of renewable energy is urgently necessary. By maximizing carbon utilization in the conversion of biogenic resources, the P2G® concept opens up a highly efficient path for the production of fuel.

In order to meet the future requirements for carbon based fuels, this provides the opportunity to utilize the sustained accessibility of bioenergy potential in combination with renewable electrical power for the production of carbon based fuels and not for the generation of electricity and heat, for which other renewable energies can be substituted. The fuel yield per utilized agricultural area is increased by up to a factor of six compared with first generation biofuels (biodiesel/bioethanol).

ACKNOWLEDGEMENTS

The authors express their gratitude for the financial support with the development of P2G® technology. The construction and operation of the 25 kW_{el} P2G® plant was financed by SolarFuel GmbH, which has traded under the name ETOGAS GmbH since 2013. The construction and operation of 250 kW_{el} P2G® plant was supported by the Federal Ministry for the Environment, Nature Conservation and Reactor Safety (BMU) on the basis of a resolution of the German Parliament (project number: 0325275). The construction of the 6000 kW_{el} P2G® plant was an investment of the car manufacturer Audi AG. The concept described in Chapter 7.2 for the use of biogenic resources with a high rate of carbon conversion was supported by the Baden-Württemberg Ministry for Rural Areas and Consumer Protection (MLR).

REFERENCES

[1] Specht M, Baumgart F, Feigl B, Frick V, Sturmer B, Zuberbuhler U, Sterner M, Waldstein G. Storing bioenergy and renewable electricity in the natural gas grid. *AEE Topics 2009. Forschungsverbund Erneuerbare Energien* **2010**:69; 2010.

[2] BMU. *Public Fund of the Federal Ministry for the Environment, Nature Conservation and Nuclear Safety in Germany*. BMU, Funding Code: 0325275 A-C, BMU, Bonn, Germany; 2011.

[3] Boie W. *Vom Brennstoff zum Rauchgas – Feuerungstechnisches Rechnen mit Brennstoffkenngrößen und seine Vereinfachung mit Mitteln der Statistik*. Teubner, Leipzig; 1957.

[4] Brellochs J, Specht M, Oechsner H, Schüle R, Eltrop L, Härdtlein M, Henßler M. *Konzeption für die: (Neu-)Ausrichtung der energetischen Verwertung von Biomasse und der Bioenergie-Forschung in Baden-Württemberg*. Bioenergieforschungsplattform

Baden-Württemberg; Studie im Auftrag des Ministeriums für Ländlichen Raum und Verbraucherschutz Baden-Württemberg; 2013. http://www.bioenergieforschungsplatt form-bw.de/pb/,Lde/1133469 (accessed 15 December 2015).

[5] Brellochs J, Marquard-Möllenstedt T, Zuberbühler U, Specht M, Koppatz S, Pfeifer C, Hofbauer H. *Stoichiometry Adjustment of Biomass Steam Gasification in DFB Process by In Situ CO_2 Absorption*. Proceedings of the International Conference on Poly-Generation Strategies, Vienna; 2009.

[6] Corella J, Toledo J, Molina G. A review on dual fluidized-bed biomass gasifiers. *Industrial and Engineering Chemistry Research* **46**:6831–6839; 2007.

[7] Specht M, Zuberbuhler U, Baumgart F, Feigl B, Frick V, Sturmer B, Sterner M, Waldstein G. *Storing Renewable Energy in the Natural Gas Grid - Methane via Power-to-Gas (P2G): A Renewable Fuel for Mobility*. In: Proceedings of the Sixth Conference "Gas Powered Vehicles - The Real and Economical CO2 Alternative", Stuttgart, Germany, p. 98; 2011.

[8] DVGW. *Arbeitsblatt DVGW G 260 – Gasbeschaffenheit*. Deutscher Verein des Gas- und Wasserfaches e.V. (DVGW). Wirtschafts- und Verlagsgesellschaft Gas und Wasser mbH, Bonn; 2012.

[9] DVGW. *Arbeitsblatt DVGW G 262 – Nutzung von Gasen aus regenerativen Quellen in der öffentlichen Gasversorgung*. Deutscher Verein des Gas- und Wasserfaches e.V. (DVGW). Wirtschafts- und Verlagsgesellschaft Gas und Wasser mbH, Bonn; 2011.

[10] Frick V. *Erzeugung von Erdgassubstitut unter Einsatz kohlenoxid-haltiger Eduktgase – Experimentelle Untersuchung und simulationsgestützte Einbindung in Gesamtprozessketten*. Dissertation, University of Stuttgart, Logos Verlag GmbH, Berlin; 2013.

[11] Marquard-Möllenstedt T, Stürmer B, Zuberbühler U, Specht M. *Fuels – Hydrogen Production – Absorption Enhanced Reforming*. In: Anon (ed.) Encyclopedia of Electrochemical Power Sources. Elsevier, Amsterdam, p. 249; 2009.

[12] Koppatz S, Pfeifer C, Rauch R, Hofbauer H, Marquard-Möllenstedt T, Specht M. H_2 rich gas by steam gasification of biomass with in situ CO_2 absorption in a dual fluidized bed system of 8 MW fuel input. *Fuel Processing Technology* **90**:914; 2009.

[13] Pearce B, Twigg M and Woodward C. *Methanation*. In: M. V. Twigg (ed.) Catalyst Handbook. Wolfe Publishing, Frome, England, p. 340; 1989.

[14] Pfeifer C, Koppatz S, Hofbauer H. Steam gasification of various feedstocks at a dual fluidised bed gasifier: Impacts of operation conditions and bed materials. *Biomass Conversion and Biorefinery* **1**:39; 2011.

[15] Seglin L, Geosits R, Franko B, Gruber G. *Survey of Methanation Chemistry and Processes. Methanation of Synthesis Gas*. In: L. Seglin (ed.) Advances in Chemistry, No. 146, p. 1. American Chemical Society, Washington, D.C.; 1974.

[16] DEG. *Home page*. DEG-Engineering GmbH, Gelsenkirchen, Germany; 2012. http://www.deg.de/(accessed 15 December 2015).

[17] Specht M, Brellochs J, Frick V, Stürmer B, Zuberbühler U, Sterner M, Waldstein G. Speicherung von Bioenergie und erneuerbarem Strom im Erdgasnetz. *Erdöl Erdgas Kohle* **126**(10):342; 2010.

[18] Audi. *Home page*. Audi AG, Ingolstadt, Germany; 2012. http://www.audi.de/(accessed 15 December 2015).

[19] Wollmann A, Benker B, Keich O, Bank R. Potenzialermittlung eines modifizierten Festbett-Rohrreaktors. *Chemie Ingenieur Technik* **81**(7):941.

[20] ZSW. European Patent Application PCT/EP2013/071095; 2013.

8

FLUIDIZED BED METHANATION FOR SNG PRODUCTION – PROCESS DEVELOPMENT AT THE PAUL-SCHERRER INSTITUT

TILMAN J. SCHILDHAUER AND SERGE M.A. BIOLLAZ

8.1 INTRODUCTION TO PROCESS DEVELOPMENT

The development of a process from idea to broad commercial application proceeds in several consecutive phases. When a technological development is concerned, these phases are often referred to as Technical Readiness Levels (TRL), a concept originally developed by NASA and now widely used, for example, by the European Commission [1]; see Table 8.1.

These TRLs can also be applied to the development of a chemical or energy conversion process such as the production of Synthetic Natural Gas (SNG). While the first three levels comprise literature research, thermodynamic analyses, and basic experiments, for example, to identify potential catalysts and promising operation conditions, TRL 4 asks for a clear concept of the complete process chain. At this level, the choice of gasifier, gas cleaning technologies, and methanation reactor type should be confined to a few promising combinations. The critical steps should be tested in bench-scale experiments, for example, simulating the gas compositions expected at the respective process steps with gas from bottles. Based on the results, which validate previous predictions, a first process chain model can be generated to compare the performance of the chosen process step combination to competing process variants or technologies.

Synthetic Natural Gas from Coal, Dry Biomass, and Power-to-Gas Applications, First Edition.
Edited by Tilman J. Schildhauer and Serge M.A. Biollaz.
© 2016 John Wiley & Sons, Inc. Published 2016 by John Wiley & Sons, Inc.

TABLE 8.1 Definition of Technical Readiness Levels (TRL) as used by the European Commission [1]. © European Union, 1995–2015.

Technology Readiness Level	Description
TRL 1	Basic principles observed
TRL 2	Technology concept formulated
TRL 3	Experimental proof of concept
TRL 4	Technology validated in the laboratory
TRL 5	Technology validated in relevant environment (industrially relevant environment in the case of key enabling technologies)
TRL 6	Technology demonstrated in relevant environment (industrially relevant environment in the case of key enabling technologies)
TRL 7	System prototype demonstration in operational environment
TRL 8	System complete and qualified
TRL 9	Actual system proven in operational environment (competitive manufacturing in the case of key enabling technologies; or in space)

TRL 5 signifies an important and also expensive step, costly in terms of time and funding to be invested. Here, the most important process steps should be integrated and tested in the relevant environment. In the case of SNG processes, this implies that gas cleaning, conditioning, and methanation should be tested with real gasification producer gas. If the gas up-grading comprises a critical or novel step, it should be tested here as well. The goal of the work at TRL 5 is to demonstrate in relatively small units the robustness of the process and the long-term stability of the materials used, such as catalysts. Typical test durations are several 100 to a few 1000 h, during which previously unknown issues easily will be detected. Successful tests at TRL 5 are the prerequisite for further up-scaling.

TRL 6 covers the up-scaling of the critical steps to pilot scale. While chemistry is more or less scale-independent, the hydrodynamic situation and therefore the residence time distribution, heat, and/or mass transfer in the up-scaled unit may change. In the worst case, the performance of the process step (e.g., selectivity or catalyst stability in the case of chemical reactors) may change as well. Therefore, up-scaling the process unit to pilot scale is favorably supported by modeling/simulation. To correctly predict which subprocess (chemical reaction kinetics, mass/heat transfer) will be limiting in the up-scaled unit, a so-called rate based model is necessary that includes an appropriate physico-chemical description of the relevant subprocesses in the unit. In order to validate the rate based models in pilot scale, it is advantageous to equip the pilot scale unit with diagnostic tools that generate data in the necessary quality. In case of, for example, methanation reactors, these diagnostic tools should, at a minimum, include measurement of inlet and outlet gas composition and temperature profile; further in some cases, also a concentration profile may be helpful (refer also to the respective sections in Chapter 4 in this book). A model that is successfully validated with pilot scale data allows for realistic optimization of the operating conditions and is a very helpful tool to de-risk up-scaling to commercial scale and to increase the trust of potential investors. Due to the complexity of TRL 6 units, the significant operation risk (given the amount of burnable gas in the pilot scale unit in the case of SNG processes), the funding situation, and to facilitate the transfer of knowhow, industry has to be involved in the development and construction of TRL 6 units.

The next step at TRL 7 is to integrate a pilot scale unit in the real environment to demonstrate the complete process chain. In the case of SNG processes, all process units from gasification to gas-upgrading should be realized at least in pilot scale and the applied equipment (reactors, pumps, etc.) should be further scalable to commercial scale. The aim at this stage is to demonstrate the technical feasibility of the complete process in pilot scale and to generate data that allow further improvement in process chain analysis, both with respect to efficiency and economics. Consequently, also TRL 7 units should comprise a number of relevant diagnostic tools to generate these data as far as necessary.

TRL 8 units then are demonstration plants in close to commercial scale that additionally also allow the verification of the economic predictions and cost calculations before the last step, the commercial applicability (TRL 9). The further market implementation can be described with the so-called Commercial Readiness Index (CRI), introduced by the Australian Reneable Energy Agency [2], whereby CRI 1 (Hypothetical commercial proposition) covers the technology development from TRL 2 to TRL 7, and CRI 2 (Commercial trial) corresponds to the end of the technology development, that is, TRL 8 and TRL 9. The next steps into the market are then Commercial scale-up (CRI 3), Multiple commercial application (CRI 4), Market competition driving widespread deployment (CRI 5) and lead finally to a "Bankable" grade asset class (CRI 6).

8.2   METHANE FROM WOOD – PROCESS DEVELOPMENT AT PSI

At the Paul-Scherrer Institut (PSI) in Switzerland, research on converting (dry) biomass to SNG started in the year 2000. The idea was promoted by the company Gazobois SA (Switzerland) since the early 1990s. In 1999, PSI joined with Gazobois SA and the Ecole Polytechnique Fédérale de Lausanne (EPFL) to conduct until 2002 preliminary studies that recommended specific gasification and methanation technologies [3]. Based on thermodynamic analyses, literature research, basic experiments, that is, work at TRL 1–3, the fast internally circulating fluidized bed (FICFB) gasification process developed by Prof. Hofbauer's group at TU Vienna (Austria) [4] and built by Repotec (Austria) was identified as most promising biomass gasification technology suitable for SNG production. The reasons were not only that the FICFB had already operated under commercial conditions in Güssing (Austria) since 2002 [5] and that a slip stream was available for long duration tests under industrially relevant conditions.

The main advantage of this allothermal gasification technology is the quality of the producer gas that is nearly nitrogen-free and contains a very high amount of hydrocarbons, especially methane. The reasons are the separation of gasification zone and combustion zone, on the one hand, and the relatively low gasification temperature of around 850 °C, on the other hand.

As shown in Figure 8.1, the wood chips are gasified in a bubbling fluidized bed with steam as gasification agent. This leads to a high hydrogen content (about 40% in the dry gas) and the absence of nitrogen. The bed material and unreacted char move through a siphon to the combustion zone which is realized as a riser. The bed material and the char are transported upwards by air whereby the char is burned and heats the bed material. The hot bed material is separated by a cyclone from the flue gas and drops back into the gasification zone where the sensible heat is used to run the endothermic gasification reactions. Meanwhile, the Güssing gasifier has more than 60 000 operating hours; the technology can be considered as very reliable (more than 7000 operation hours/year) and was chosen for several further projects [6], for example, within the GoBiGas project where it is up-scaled to 32 MW$_{th}$ (refer to Chapter 6 in this book).

The producer gas from the FICFB gasifier contains, besides 35–40% hydrogen, also CO and CO$_2$ (20–24% each) and methane (around 10%). This is an important advantage for the overall process chain efficiency, because methane produced in the gasification step was not further converted by endothermic steam reforming at 850 °C and does not need to be formed by exothermic methanation at 300–400 °C. In consequence, a higher fraction of the chemical energy contained in the wood is transferred to the SNG.

A second consequence of the relatively low gasification temperature is the significant amount of unsaturated hydrocarbons (up to 3 vol% of ethylene and acetylene), which can cause massive carbon formation in adiabatic fixed bed methanation reactors at the inherently high temperatures [7]. It was considered economically not viable to remove the olefins by physical scrubbing or to convert it chemically. Therefore, the Comflux fluidized bed methanation technology (see Chapter 4 in this

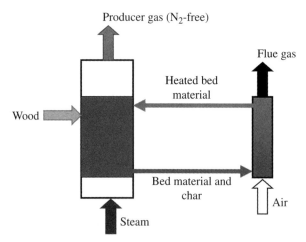

FIGURE 8.1 Allothermal dual fluidized bed gasification: the FICFB gasifier developed by TU Vienna and built by repotec GmbH.

book) was considered as potential technology due to its very good temperature control and its proven ability to combine methanation and water gas shift reaction (necessary to adjust the H_2/CO ratio).

Based on work from the 1970s reported in the literature and PSI's own bench scale experiments (TRL 4), suitable operation conditions for the fluidized bed methanation were chosen. Further, active carbon, an ammonia scrubber, and a zinc based sorbent bed were chosen as gas cleaning steps to remove impurities that could deactivate the catalyst. In 2003, a bench scale fluidized bed reactor including the gas cleaning section was connected to a slip stream of the FICFB gasifier during 120 h which can be considered as the first production of bio-SNG [8]. Based on this encouraging result, a fully automated 10 kW_{SNG} scale reactor system ("COSYMA") was designed and built at PSI and then transferred and connected to the gasifier in Güssing (Austria) in 2004 [9].

Several long duration tests at TRL 5 level were conducted which showed that the gas cleaning section was not sufficient to completely remove sulfur species causing catalyst deactivation after around 200 h. Thorough analysis of the producer gas elucidated the presence of several tens ppm of thiophene in the producer gas. Meanwhile, with improved gas sampling and analytics, a large number of organic sulfur species were identified (thiophene, benzo-thiophene, dibenzo-thiophene, and their derivatives) [10, 11]. In 2005, experiments with commercial molybdenum sulfide based hydrodesulfurization catalysts were conducted both in the laboratory and in the COSYMA set-up in Güssing, showing that conversion of thiophene is the key to higher catalyst lifetime. During these campaigns, similar results were obtained, as reported by the Energy Center of the Netherlands (ECN): relatively high temperatures and low space velocities in the HDS reactor are necessary to obtain sufficient thiophene conversion [12]. In consequence, a scrubber based thiophene removal step was integrated that enabled demonstrating catalyst stability for more than 1000 h in summer 2007; see Figure 8.2. The achieved high methane content of about 40% and the very low CO amount are close to the thermodynamic equilibrium [13].

These successful results obtained on the TRL 5 level showed that critical parts of the technology were stable and robust. Therefore the scale-up could be started. The broad experience gained with the reactor type during the Comflux project allowed the omission of work at TRL 6 and directly building a 1 MW_{SNG} process development unit (PDU; TRL 7). This was designed and erected by CTU AG (Winterthur, Switzerland) and Repotec Umwelttechnik GmbH (Vienna, Austria) within the European Union project BioSNG [14] which was substantially supported by funding from Swiss electricity producers and Austrian public funds. Mechanical completion was achieved in 2008. The plant was commissioned with the support of TU Vienna and PSI using producer gas from the FICFB gasifier (Biomassekraftwerk Güssing). The PDU includes the complete process chain from wood to SNG, including gasification, gas cleaning steps, conditioning, methanation, and gas purification in demo scale (TRL 7); see Figure 8.3 [15]. In December 2008, for the first time FICFB producer gas was converted to methane-rich gas in the PDU; the composition of the raw SNG was very similar to that obtained by COSYMA (see Figure 8.2) which proved the successful scale-up from 10 kW to 1 MW scale.

(a)

(b)

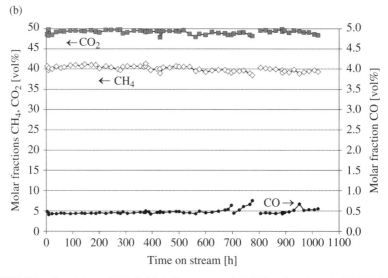

FIGURE 8.2   Activities at TRL 5: (a) fully automated 10 kW$_{SNG}$ bench scale COSYMA unit and (b) results of the long duration test in a slip stream from the FICFB gasifier in Güssing.

(a)

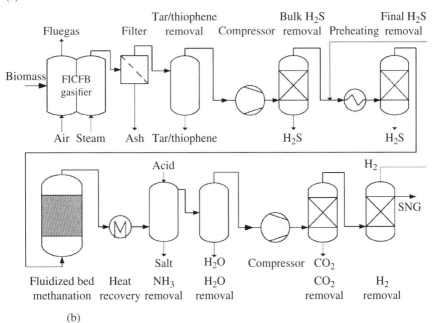

(b)

FIGURE 8.3 Block flow diagram (a) and photograph (b) of the 1 MW$_{SNG}$ process development unit (PDU, TRL 7) in Güssing, Austria, converting wood derived producer gas to SNG, from [13, 15]. Source: Kopyscinski 2010 [13]. Reproduced with permission from Elsevier.

In April 2009, the first operation of the full process chain was achieved, producing $100\,m^3/h$ of SNG in hydrogen gas quality (Wobbe index = 14.0, HHV = 10.67 kWh/m^3N) [16] that was fed to a CNG fueling station. A chemical or cold gas efficiency of around 61% was achieved, that is, the ratio of the lower heating value (LHV) of the SNG times the mass flow of SNG, divided by the LHV of the wood multiplied by the mass flow of the wood.

To support the commissioning and to generate data for process concept validation, continuous gas samples were taken by PSI at several sampling points between all major process units. For this purpose, several hundred meters of sampling lines were integrated to the PDU. Based on this gas sampling, valuable information on the performance of the gas cleaning and gas upgrading was obtained. Further, a gas sampling device with several sampling points and moveable thermocouple in a protective tube was introduced inside the main reactor, allowing measurement of the axial gas phase concentration profile and the temperature profile. Non-intrusive pressure sensors mounted to radial ports at several heights and circumference positions of the reactor were used to measure high frequent pressure fluctuations which are caused by rising gas bubbles and allow the derivation of conclusions on the hydrodynamic pattern inside the reactor. This information can be used to further develop a rate based model of the methanation reactor; the axial concentration profiles can be used for model validation. Like in the COSYMA, a special sampling device enabled catalyst samples to be taken during operation without them contacting the air. By appropriate characterization methods [10], it is possible to control the state of the catalyst inside the reactor with respect to carbon deposition and sulfur poisoning.

Ongoing research at PSI focuses on deepening the knowhow, on the one hand, and on broadening the applicability of fluidized bed methanation, on the other hand. The aim is to optimize the operating conditions for converting cleaned producer gas from different combinations of gasifiers and gas cleaning steps, to investigate the application of fluidized bed methanation for power to gas applications, and to support and de-risk scale-up. For this, a number of activities have been conducted in recent years which are partly described in more detail in Chapter 4 in this book. In situ IR spectroscopy experiments and systematic experiments in a micro-fluidized bed reactor helped to understand the reaction mechanisms of main and side reactions and of catalyst deactivation by carbon deposition. Suited set-ups with axial temperature and concentration measurements were developed to determine the kinetics of main and side reactions. A rate-based computer model has been developed that describes the interaction of kinetics, hydrodynamics, and mass transfer. The goal is to simulate fluidized bed methanation reactors and to predict reactor performance during scale-up and for varying input gases including CO_2- and H_2-rich gases in power to gas applications. A new pilot scale set-up ("GanyMeth", TRL 6) has been designed and constructed that allows hydrodynamic investigations and reactive tests for model validation at pressures up to 12 bar. For this reactor, a number of diagnostic tools and respective methods were developed to enable model validation with high quality data on temperature and concentration profiles.

Further research activities are investigating hot gas cleaning steps with the aim to further improve the efficiency of the process chain by avoiding any scrubbers and therewith the condensation and re-evaporation of water upstream of the methanation. These activities are partly described in Chapters 3 and 12 in this book. Finally, integrating rate based models of fluidized bed methanation and gas cleaning steps into a tool for thermo-economic analysis and multi-objective optimization shall allow the identification of pareto-optimal process configurations and operation conditions for each process unit. This way, further research shall be focused on the most promising options.

REFERENCES

[1] European Commission. *Technology Readiness Levels (TRL), Horizon 2020.* Work Programme 2014–2015 General Annexes, Extract from Part 19 – Commission Decision C(2014)4995; 2014.

[2] ARENA. *Home page.* 2008. Available at: http://arena.gov.au/resources/readiness-tools/(accessed on 12 December 2015).

[3] Biollaz SMA, Thees O. *Ecogas – Teilprojet: Methan aus Holz, Energieholzpotential Schweiz.* Technical Report prepared for Novatlantis ETH. Paul Scherrer Institut Villigen, Switzerland; 2003.

[4] Hofbauer H, Veronik G, Fleck T, Rauch R. *The FICFB Gasification Process.* In: Bridgwater AV, Boocock DGB (eds.) Developments in Thermochemical Biomass Conversion. Blackie Academic and Professional, London, pp. 1016–1025; 1997.

[5] Hofbauer H, Rauch R, Bosch K, Koch R, Aichernig C. *Biomass CHP Plant Güssing – A Success Story.* Expert Meeting on Pyrolysis and Gasification of Biomass and Waste. Strasbourg, France; 2002.

[6] FICFB. *Home page.* 2006. Available at: http://www.ficfb.at/(accessed 12 December 2015).

[7] Czekaj I, Loviat F, Raimondi F, Wambach J, Biollaz S, Wokaun A. Characterization of surface processes at the Ni-based catalyst during the methanation of biomass-derived synthesis gas: X-ray photoelectron spectroscopy (XPS). *Applied Catalysis A* **329**:68–78; 2007.

[8] Seemann MC, Biollaz SMA, Aichernig C, Rauch R, Hofbauer H, Koch R. *Methanation of Biosyngas in a Bench Scale Reactor Using a Slip Stream of the FICFB Gasifier in Güssing.* Proceedings of the Second World Conference and Technology Exhibition – Biomass for Energy, Industry and Climate Protection. Rome, Italy; 2004.

[9] Seemann MC, Schildhauer TJ, Biollaz SMA. Fluidized bed methanation of wood-derived producer gas for the production of synthetic natural gas. *Industrial Engineering and Chemical Research* **49**(15):7034–7038; 2010.

[10] Struis RPWJ, Schildhauer TJ, Czekaj I, Janousch M, Ludwig C, Biollaz SMA. Sulphur poisoning of Ni catalysts in the SNG production from biomass: A TPO/XPS/XAS study. *Applied Catalysis A* **362**(1/2):121–128; 2009.

[11] Rechulski MDK, Schildhauer TJ, Biollaz SMA, Ludwig C. Sulphur containing organic compounds in the raw producer gas of wood and grass gasification. *Fuel* **128**:330–339; 2014.

[12] Rabou LPLM, Bos L. High efficiency production of substitute natural gas from biomass. *Applied Catalysis B: Environmental* **111/112**:456–460; 2012.

[13] Kopyscinski J, Schildhauer TJ, Biollaz SMA. Production of synthetic natural gas (SNG) from coal and dry biomass - A technology review from 1950 to 2009. *Fuel* **89**(8):1763–1783; 2010.

[14] Bio-SNG. *European Union Project – Demonstration of the Production and Utilization of Synthetic Natural Gas (SNG) from Solid Biofuels (Bio-SNG)*, Project No. TREN/05/ FP6EN/S07.56632/019895; 2006. Available at: http://www.bio-sng.com (accessed 12 December 2015).

[15] Möller S. *The Güssing Methanation Plant*. Presentation at the BioSNG 09 – International Conference on Advanced Biomass-to-SNG technologies and Their Market Implementation. Zürich, Switzerland; 2009.

[16] Biollaz SMA, Schildhauer TJ, Ulrich D, Tremmel H, Rauch R, Koch M. *Status Report of the Demonstration of BioSNG Production on a 1 MW SNG Scale in Güssing*. Proceedings of the 17th European Biomass Conference and Exhibition. Hamburg, Germany; 2009.

9

MILENA INDIRECT GASIFICATION, OLGA TAR REMOVAL, AND ECN PROCESS FOR METHANATION

Luc P.L.M. Rabou, Bram Van der Drift, Eric H.A.J. Van Dijk, Christiaan M. Van der Meijden, and Berend J. Vreugdenhil

9.1 INTRODUCTION

Natural gas provides nearly 50% of the Dutch primary energy consumption. A substitute from renewable sources is needed to meet the goals for CO_2 reduction and the share of renewables [1]. The upgrading of biogas from digesters is well developed, but the potential is limited to a small percentage of the primary energy consumption. Substitute natural gas (bioSNG) by the gasification of solid biomass and the methanation of producer gas has large potential, but the technology is less mature. However, as biomass import will be needed, production plants at GW scale may be considered near a harbor.

Feedstock costs can account for up to 50% of bioSNG production costs if biomass is used from sustainably managed sources. That makes high efficiency in the conversion chain of paramount importance. In 2000, ECN evaluated a number of gasification technologies, selected indirect gasification as the best option, and started development of MILENA gasification technology. MILENA technology allows significantly higher efficiency in bioSNG production [2, 3]. The design allows operation

Synthetic Natural Gas from Coal, Dry Biomass, and Power-to-Gas Applications, First Edition.
Edited by Tilman J. Schildhauer and Serge M.A. Biollaz.
© 2016 John Wiley & Sons, Inc. Published 2016 by John Wiley & Sons, Inc.

at large scale and pressure. Present MILENA gasifiers operate at atmospheric pressure. According to our system analysis, that is the best choice for systems smaller than about 50 MW.

The ECN approach is to start with hydrocarbon-rich producer gas, obtained by indirect gasification using little steam and moderate temperatures (~800 °C). The producer gas is cleaned and converted into methane, using commercially available catalysts. The conversion involves a water gas shift to increase the amount of H_2, hydrogenation of higher and unsaturated hydrocarbons, reforming of aromatic hydrocarbons, and methane production from CO and H_2. Heat production per mole of methane is reduced when compared to production from CO and H_2 only, as the hydrogenation of hydrocarbons is less exothermic and reforming endothermic.

Moderate gasification temperatures bring the advantage that fuels can be considered which produce low-melting ash. These fuels can be comparatively cheap, as they are difficult to use in combustion or high-temperature (>900 °C) gasification processes. A disadvantage of the applied conditions is the larger production of tar (i.e., heavy aromatic compounds). For bioSNG production, deep tar removal is needed anyhow to prevent catalyst coking. The OLGA tar removal technology developed by ECN is able to remove completely all but the most volatile tar compounds. Tar separated from MILENA producer gas can be used to meet the heat demand of the gasification process, instead of additional fuel or recycled producer gas.

ECN bioSNG research focuses on these two technologies: MILENA indirect gasification and OLGA tar removal. Other important topics are catalytic conversion of organic sulfur compounds, especially thiophene (C_4H_4S), and of hydrocarbons such as benzene, toluene, and xylene (BTX) or ethylene (C_2H_4). BTX and ethylene may be separated from the producer gas instead of converted, if process and market conditions make separation attractive [4]. The aim is to obtain a biosyngas mixture which can be handled by any commercial fixed bed methanation technology. Final upgrading to gas grid quality is given limited attention. An exception is made for CO_2 removal, for which ECN has developed a technology using solid sorbents.

The SNG production process considered by ECN involves a number of steps at different pressure levels. Figure 9.1 shows the main process steps. Steam addition or gas recycling may be needed, but are not shown. Processes in the upper line will operate at ambient or mild pressure (<10 bar). Processes in the middle line will operate at mild pressure. The bottom line involves methanation at high pressure (>20 bar).

In small systems, the gasifier operates at atmospheric pressure. In future large systems, the gasifier will operate at mild pressure, obviating the need of the first compressor shown in Figure 9.1. The outlet pressure of the second compressor will depend on the SNG product specifications to be met. The compression duty is limited by upstream removal of CO_2 and water (not shown).

The main steps in the ECN bioSNG process are described in Section 9.2. The expected process efficiency and economy are treated in Section 9.3. Actual results, status, and plans are described in Section 9.4. The outlook for further research and future developments are discussed in Section 9.5.

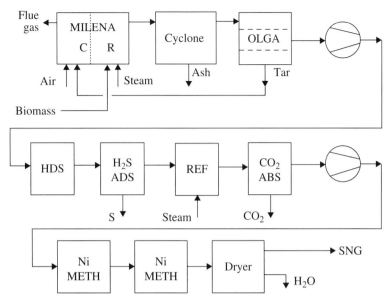

FIGURE 9.1 Main steps in the ECN bioSNG process.

9.2 MAIN PROCESS STEPS

9.2.1 MILENA Indirect Gasification

The MILENA indirect gasification process is designed to use little steam and to yield producer gas rich in hydrocarbons. Table 9.1 shows the producer gas composition on a dry basis for a large MILENA gasifier after tar removal. The H_2/CO ratio in the producer gas depends on process conditions, but is usually close to one. The water content is about 35%, depending on gasifier conditions and fuel moisture content. Table 9.1 also shows the contribution of each component to the producer gas heating value, the heating value loss on reaction with H_2 to methane, and the expected composition of the bioSNG product after methanation and upgrading. Not shown in Table 9.1 are contaminants, such as tar, NH_3, H_2S, HCl, and other compounds containing O, N, S, and Cl.

The MILENA gasifier (see Figure 9.2) consists of three zones: the gasifier section, the combustion section, and the settling chamber. It relies on the use of a bed material such as sand or olivine. The gasifier section operates in fast fluidization mode, with gas velocities around 6 m/s. The combustion section operates in bubbling bed mode, with gas velocities around 0.5–1.0 m/s. The settling chamber is part of the gasifier section where the gas velocity is reduced significantly. From here on, the gasifier and combustion sections will be referred to as riser and combustor.

Biomass is fed by a screw to the bottom part of the riser, where the bed material is fluidized by a small supply of steam or air. Biomass is heated quickly by contact with hot bed material. This produces pyrolysis gas and char and cools the bed material. The pyrolysis gas entrains bed material to the top of the riser, making room for hot bed material to flow from the combustor to the bottom of the riser.

TABLE 9.1   Composition of Tar-Free MILENA Producer Gas (Dry Basis), Contribution to Producer Gas Heating Value, Loss on Reaction with H_2 to CH_4, and Final BioSNG Composition.

	MILENA [vol% dry]	MILENA [% LHV]	Conversion loss [%]	bioSNG [vol% dry]
H_2	27	18		<0.5
CO	32	25	20	<0.1
CO_2	20			2
CH_4	14	30		95
C_2H_4	4	14	12	
C_6H_6	1	8	10	
$C_xH_y{}^a$	1	5	12	
N_2	1			3

^{a}Mainly C_2H_2, C_2H_6, C_7H_8, and small quantities of hydrocarbons with 3–5 C atoms.

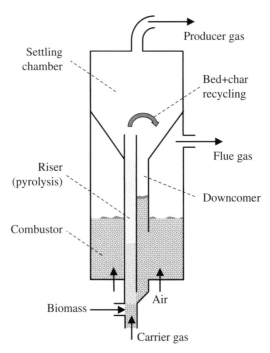

FIGURE 9.2   MILENA indirect gasifier with riser, combustor, and settling chamber in a single vessel.

At the top of the riser, gas and entrained solids enter the settling chamber. Here, the reduced gas velocity allows bed material and char to separate from the gas stream. Some dust will be entrained to the gas cooling and cleaning systems. Bed material and char are collected in a funnel and transported via a downcomer into the combustor. More than one downcomer can be applied if required.

In the combustor, the bed material is heated by combustion of char with air. Close to stoichiometric conditions are applied to obtain a high flue gas temperature for maximum heat transfer to the bed material. Secondary air can be injected above the bubbling bed to assure complete combustion and limit NO_x emissions. The low gas velocity leads to a long gas residence time. The combination of high temperature and long residence time makes it possible to use waste-derived fuels.

Transport of bed material and char between riser and combustor provides the main contributions to the heat balance. Further contributions may come from, for example, preheated combustor air, additional combustor fuel, recycling part of the producer gas or tar to the combustor, superheated steam or air/oxygen to the riser.

The gasifier operating pressure is limited by the balance between heat and mass transport. Higher pressure reduces the gas velocity and increases the gas density in the riser. The net result is a reduced drag force acting on the bed material, which leads to lower bed material transport to the combustion section. In order to maintain heat transport from the combustor to the riser with less bed material available to carry heat, the bed material has to be heated to a higher temperature in the combustor and cooled to a lower temperature in the riser. This will limit the operating pressure to about 7 bars.

The principles described above are also used in the Battelle [5] and FICFB [6] processes. The actual implementation of the principles in the MILENA process is quite different. Some of the distinguishing properties of the MILENA process are (see also Figure 9.2):

- The gasification section is a fast fluidized bed (riser), with fluidization mainly by producer gas. Only a small amount of steam, inert gas or air is needed for initial fluidization.
- The combustion section is a bubbling fluidized bed, with fluidization by combustion air.
- The gasification and combustion sections are placed within a single hull, which makes gasification easier at moderate pressure (up to about 7 bar).
- Char and bed material are separated from producer gas in a settling chamber and recycled to the combustion section via a downcomer (or several downcomers at large scale).
- Particles which are too small to be separated in the settling chamber are collected by a cyclone or dust filter and can be recycled to the combustion section.
- The process heat demand can usually be covered by combustion of the char and tar separated from the producer gas with preheated air. If more heat is required, some producer gas can be "leaked" to the combustion section by manipulating the pressure difference between the gasifier and combustion sections.
- Low steam use leads to producer gas with high (20–60 g/Nm^3) tar content, but that is not a problem for OLGA tar removal technology.

• Low steam use leads to producer gas with comparatively low (~35%) moisture content, reducing the cooling demand required to condense water before gas compression and reducing the volume of condensate produced.

9.2.2   OLGA Tar Removal

OLGA tar removal is a technology specially developed for the deep removal of tar. The goal is to be able to use the gas in downstream processes such as gas engines or catalytic reactors. Figure 9.3 shows a typical OLGA layout.

The technology is based on the idea that a liquid or solid mixture of tar, dust, and water should be avoided at all times. That can be realized if dust is removed at a temperature above the tar dew point and tar at a temperature above the water dew point. In the case of MILENA producer gas at atmospheric pressure, the tar dew point is around 450 °C and the water dew point around 75 °C. In the current system, a cyclone is used to remove most of the dust from the raw producer gas. Part of the chloride will form KCl, condense onto dust particles, and be removed by the cyclone.

OLGA technology comprises several process steps. In the first step, gas is cooled in a counter-current by a washing liquid to a temperature above the water dew point. Most of the heavy tar compounds condense and are collected in the liquid, together with most of the dust still present. To avoid plugging, a washing liquid is used which is able to dissolve the condensable tar compounds. Due to the rapid cooling of the gas, aerosols are formed from oversaturated tar vapour and fine dust particles. These aerosols are captured in a wet electrostatic precipitator (ESP).

In the second step, the remaining tar compounds are absorbed in a liquid with an affinity for light tar compounds. Even though the column operates above the water dew point, a tar dew point below 5 °C can be achieved. Volatile compounds such as benzene

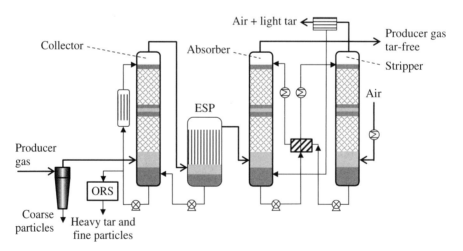

FIGURE 9.3 Typical OLGA lay-out with collector, absorber and stripper columns, electrostatic precipitator (ESP), and oil recovery system (ORS).

and toluene are hardly absorbed and remain in the gas. Water remains in the gas too. It can be removed downstream of OLGA by a water scrubber or water condenser.

In the third step, the tar-rich absorption liquid is regenerated. Tar compounds are desorbed by heating and stripping with air, nitrogen, or steam. Which gas is used depends on the location and the integration with the gasifier. The columns used in the three steps are referred to as collector, absorber, and stripper.

Key components in the absorption and desorption steps are phenol and naphthalene. The design of the absorber and stripper is such that 99.9% of both is removed and water can be condensed without phenol pollution. The stripper off-gas can be recycled to the MILENA combustor to provide additional heat input for the gasification.

For the first separation step, an oil recovery system (ORS) has been developed. The function of this device is to separate heavy tar and dust from the lighter tar fraction. This lighter tar fraction is used to maintain adequate viscosity in the collector. Heavy tar and dust are sent to the MILENA combustor to provide additional energy for gasification.

OLGA effectively removes all tar compounds heavier than toluene. It makes it possible to reuse the energy in the tar for the gasification process and it avoids a wastewater problem by separating tar above the water dew point.

9.2.3 HDS and Deep S Removal

The total sulfur content of producer gas from biomass gasification may vary from about 100 ppmv for clean fresh wood to more than 1000 ppmv for waste-derived fuels. The main sulfur compounds in producer gas are H_2S and COS. Other compounds found in significant concentrations (1–100 ppmv) are thiophene (C_4H_4S), thiols (e.g., CH_3SH, C_2H_5SH), CS_2, and thiophene derivatives which contain one or more (m)ethyl and/or benzyl groups, such as (di)methyl-thiophene, (di)benzo-thiophene, and so on.

Sulfur binds to nickel methanation catalysts and renders them inactive. The sulfur concentration must be reduced to 0.1 ppmv to obtain a catalyst life of several years. Deep sulfur removal by technology applied in chemical or coal to SNG plants is not economic for bioSNG plants smaller than 100 MW, unless they use biomass with high sulfur content. In the latter case, high sulfur removal costs should be compensated by low fuel costs.

Non-aromatic organic sulfur compounds are relatively easy to remove. They are readily hydrolyzed to H_2S and then captured by commercially available solid adsorbents or liquid absorbents. Iron oxide adsorbent or scrubbing with amine or a transition metal solution can be used to reach a ppmv sulfur concentration. Sub-ppmv levels can be reached with zinc or noble metal oxides. Heavy aromatic sulfur compounds can be removed by the OLGA tar removal system. The lighter aromatic sulfur compounds can be bound to metal impregnated active carbon, but the BTX bind to active carbon too. That leads to a considerable loss of potential methane output and quick saturation of the active carbon, as BTX are present in much larger concentrations than thiophene.

The ECN approach is to use a commercial hydrodesulfurization (HDS) catalyst for conversion of organic sulfur compounds into H_2S, in series with commercial technology for H_2S removal. ECN research focuses on the operating conditions and performance of the HDS catalyst. If sulfided CoMoO is used, a temperature above 350 °C is required to obtain sufficient catalytic activity. As the OLGA exit temperature is around 100 °C, heat must be supplied.

If gasification is performed at atmospheric pressure, compression of the gas may provide part of the required heat input. However, prior to compression the gas needs to be dried by cooling in order to prevent compressor damage by condensate formation and to reduce the compressor power demand. The relatively low moisture content of MILENA producer gas reduces the cooling duty involved. The cooling and drying step removes some of the producer gas contaminants, notably NH_3.

The HDS catalyst also functions as a hydrogenation catalyst for unsaturated aliphatic compounds, such as acetylene (C_2H_2), ethylene (C_2H_4), and propene (C_3H_6). Furthermore, the HDS catalyst promotes the water gas shift and methanation reactions, given the right conditions. All these reactions are exothermic. In case of MILENA producer gas, they raise the producer gas temperature by about 200 °C. Downstream of the HDS reactor, the gas needs to be cooled to allow H_2S capture. The heat removed can be used, for example, to raise the gas temperature between the OLGA exit and the HDS entrance.

Downstream of the HDS reactor the only sulfur compounds remaining are H_2S and COS. A ZnO based solid adsorbent can be applied to obtain the sub-ppmv sulfur concentration required. Operating costs may be reduced by bulk sulfur removal using a scrubbing technology or an Fe-based adsorbent. The bulk sulfur removal can be positioned upstream of the ZnO adsorbent or upstream of the HDS reactor. Downstream of the ZnO adsorbent a mixed-metal guard bed can be installed to remove the remaining trace compounds.

9.2.4   Reformer

MILENA producer gas contains about 5% unsaturated aliphatic compounds and 1% BTX. Nickel methanation catalysts exposed to these compounds deactivate by deposition of carbon or polymerization products. Hydrogenation of unsaturated compounds by the HDS catalyst reduces the risk. The reformer contains a dedicated steam reforming catalyst to take care of BTX. A large amount of steam and a temperature above 400 °C are needed to prevent or reduce catalyst coking.

The steam reforming catalyst also promotes the water gas shift, hydrogenation, and methanation reactions. Effectively, the reformer acts as first methanation reactor. Although reforming reactions are endothermic, overall the reactions lead to a temperature rise. The reformer exit temperature is determined by the methanation equilibrium. The mild pressure applied limits the temperature rise, obviating the need for gas recycling. The gas must be cooled for the next step in the process, CO_2 removal. Depending on process conditions and requirements, a methanation reactor can be inserted between the reformer and CO_2 removal. At this stage, the gas contains only CH_4, H_2O, CO_2, H_2, CO, and N_2. Some NH_3 and side products may be present as well.

9.2.5 CO_2 Removal

Conventional technology for CO_2 removal is scrubbing with aqueous solutions of a physical or chemical solvent. Physical solvents require high pressure and/or low temperature, which lead to considerable removal of hydrocarbons such as methane. Therefore, they are not optimal for bioSNG purposes.

Chemical solvents based on amines can be used at mild pressure. First, the gas must be cooled to their operating temperature of about $40\,°C$. Condensed water must be removed to prevent dilution of the amine solution. Here again some NH_3 will be removed. In order to tune the gas composition to the desired $(H_2 - CO_2)/(CO + CO_2) = 3$ ratio, the amine unit can be bypassed with part of the producer gas stream. The amine absorbent is regenerated by heating. The CO_2 stream can be vented, stored, or used. A guard bed may be needed to prevent deactivation of downstream catalysts by traces of the chemicals used in the CO_2 removal process.

Another option is CO_2 removal via regenerative solid adsorbents. The process can be operated at $350–450\,°C$. This novel technology, in development at ECN, is based on the adsorption of CO_2 on a solid material at pressures of typically $10–30$ bar. After saturation of the adsorbent, low pressure regeneration with steam is applied. As such, the technology constitutes a hot pressure-swing adsorption (PSA) system for CO_2 removal [7]. Compared to amine scrubbing, the energy requirement for CO_2 removal with solid sorbents is 25% lower [8,9]. Because of the inherent WGS activity of the solid sorbent, this hot-PSA system usually has a very high carbon removal rate. To obtain the desired $(H_2 - CO_2)/(CO + CO_2) = 3$ ratio for the downstream methanation, part of the producer gas can be bypassed. Alternatively, the PSA cycle can be tuned to decrease the carbon removal ratio.

9.2.6   Methanation and Upgrading

The final methanation is performed at high pressure, typically $30–50$ bar, using commercial nickel methanation catalysts. Several fixed beds in series are used, with intermediate cooling, until the residual H_2 concentration is below the level required. If upstream conditions are well chosen, gas recycling or steam addition is not needed to moderate the temperature or to prevent carbon deposition.

After the final methanation step, the gas has to be dried to a level that prevents condensate formation in the most severe usage conditions expected. The nearly finished product contains about 95% CH_4, some CO_2, trace amounts of H_2 and CO, and 2.5 times the N_2 concentration in the MILENA producer gas. Further treatment will be needed to bring it within the specifications required. Such treatment may involve N_2 or LPG addition to lower or raise the Wobbe index, odorant addition to allow quick leak detection, and further reduction of the H_2 concentration. If a separate H_2 removal step is required anyhow, the number of methanation steps may be reduced. Recovered H_2 can be recycled to the HDS reactor or reformer.

9.3   PROCESS EFFICIENCY AND ECONOMY

The economy of bioSNG production has been calculated using different methods and published by various authors. All results are projections and expectations, since there is no full-scale reference bioSNG plant existing at this moment. In any case however, the bioSNG production costs heavily depend on energy efficiency and feedstock costs.

At ECN, an estimate of the investment costs of future large-scale bioSNG plants has beenmade, based on costs of existing plants operating on coal and gas [10]. Differences in technology and scale have been quantified to arrive at a final average estimate: US$ 1.1 billion total capital investment costs for a 1 GW_{th} (input capacity) wood-based SNG plant on the long term, say 2030. In Figure 9.4, this investment has been translated to different scales using an 0.7 scaling factor. Two bioSNG references are included in the graph: the GoBiGas plant (started up early 2014) in Göteborg in Sweden and the E.On initiative for a larger bioSNG plant (FEED phase in 2014). The investment has been used to determine the production costs of bioSNG. From the figure it becomes clear that scale and efficiency from biomass to bioSNG are major factors.

The important question is how the cost of bioSNG compares to the alternative of bioSNG. This is not as easy as it sounds, since bioSNG can serve different markets, such as power production, transport, heat production, and chemical feedstock. One might argue that bioSNG replaces natural gas and therefore should be compared with natural gas, but that is only partly true. Obviously, bioSNG has a low CO_2 footprint, which can even be negative (see Section 9.5), and which makes it a different product than natural gas. But there is more. BioSNG serving as a biofuel instead of CNG or LNG (i.e., compressed or liquefied natural gas) replaces other biofuels such as ethanol rather than natural gas. This is true since the European Union created a biofuels market through the "biofuels directive" aiming at a certain fraction of fuels to come from renewable sources. BioSNG for power production also has no easy comparison. In regions where the fraction of intermittent renewable power

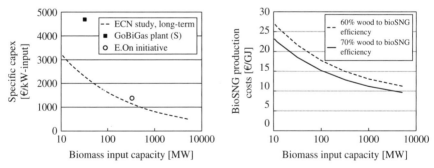

FIGURE 9.4   Specific investment costs (left) and bioSNG production costs (right) as a function of scale, assuming biomass costs of 5 €/GJ, capital costs of 10% per annum, and 8000 h/a.

production is significant (high share of solar and wind power), there is a need for both flexible and renewable power production. BioSNG perfectly fits that picture and therefore creates additional value.

In summary, the costs of producing bioSNG can be and has been estimated by different parties. Clearly the costs depend heavily on scale, efficiency, and biomass costs. The question whether it is attractive or not is more difficult to answer, since it concerns a new product that serves many markets, each with its typical requirements and alternatives. Looking at the various bioSNG initiatives of parties around the globe (but mainly in Europe), one might say that there clearly are attractive markets for bioSNG [11].

9.4   RESULTS AND STATUS

9.4.1   MILENA

The MILENA development is aimed at commercial application at a size between five and several hundred MW of biomass input. That is why first a design was made for a 10 MW$_{th}$ plant, from which designs for smaller systems are derived. Experimental work for the MILENA development is performed in a 30 kW$_{th}$ (~6 kg/h biomass) laboratory scale facility and an 800 kW$_{th}$ (~160 kg/h biomass) pilot plant.

The laboratory scale installation was taken in operation in 2004 and has been used for more than 6000 h. The reactor wall can be heated externally to compensate for the relatively high heat loss. The fuel particle size has to be in the range of 1–3 mm, because of the size of the feeding screw and riser reactor.

The MILENA pilot plant was to replace a 500 kW$_{th}$ CFB gasifier, which was used for 10 years. The goal was to realize an installation for experiments under realistic "commercial" conditions, that is, no external heat supply to the reactor wall and larger fuel particle size. The fuel particle size was limited to a 15 × 15 mm sieve fraction, based on experiments with the 500 kW$_{th}$ CFB gasifier. The pilot plant was taken into operation in 2008 and has been used for approximately 1500 h, in combination with the pilot scale OLGA gas cleaning.

Several fuels were tested in both installations. Most of the experimental work was done with wood and demolition wood, but several other fuels, such as RDF, soya residue, sunflower husk, lignite, sewage sludge, and high ash coals were tested as well.

After the first tests with the MILENA pilot plant, the HVC Group (a Dutch waste-processing company) joined ECN to develop the MILENA technology for bioSNG production. The first plan was to build a 10 MW$_{th}$ MILENA gasifier in combination with OLGA gas cleaning and a gas engine. The aim was to demonstrate the gasification technology and to supply a continuous slipstream of cleaned gas to a methanation test rig. Subsidy for the production of electricity from renewable sources was expected to help cover the cost of the demonstration plant. The plant would be located in Alkmaar, next to an HVC site and not far from the ECN site in Petten. The Dutch company Royal Dahlman was involved in the engineering of the demonstration plant for both the MILENA gasification and the OLGA tar removal technology.

The development involved several duration tests with demolition wood in the ECN pilot plant. These tests were performed in 2010 and 2012 and resulted in a number of modifications and an updated design for the commercial scale gasifier. When the subsidy scheme for renewable energy production was changed, plans for the demonstration plant had to be adapted. In 2013 Gasunie (a large Dutch gas company) joined the consortium to demonstrate the MILENA technology for SNG production. It was decided to go for a slightly smaller (4 MW) demonstration plant producing SNG. In 2014, a subsidy for bioSNG production and injection into the grid was granted. This provides a sound economic base for realization and operation of the demonstration plant.

In 2013 Royal Dahlman acquired a license for the MILENA gasification technology. In 2013 a large experimental program was carried out for Royal Dahlman within the framework of a project financed by the British company ETI. The data is used for basic engineering of a commercial scale demonstration plant using RDF fuel. In 2014 it will be decided if the project will be continued. Several other projects using different fuels are in preparation as well.

The Indian company Thermax is constructing a demonstration plant based on the MILENA and OLGA technology in India. The fuel is a residue from the soya crop. Fuel tests were performed in the ECN laboratory scale installation. The gas will be used in a gas engine. Commissioning of the demonstration plant is scheduled for 2014.

9.4.2 OLGA

Currently, four OLGA systems have been built and been in operation on different scales and downstream various gasifiers. From 2001, Dahlman has been involved in the technology development and construction of OLGA systems. In 2006, Dahlman and ECN signed a license agreement. An overview of systems realized is given Table 9.2.

The laboratory scale OLGA has the highest accumulated number of operating hours. It has also been tested with the largest variety in product gases. Ranging from pyrolysis gas with several hundred g/Nm^3 tar to high temperature gasification with

TABLE 9.2 Overview of OLGA Systems.

Location	Laboratory ECN (NL)	Pilot ECN (NL)	Moissannes (Fr)	Tondela (Pt)
Capacity (Nm^3/h)	2	200	2000	2000
Construction	2001	2004	2006	2010
Front end	BFB/MILENA	CFB/MILENA	PRMe gasifier	CFB
Application	Fuel cell SNG test rig	Boiler Gas engine Micro gas turbine	Gas engine	Gas engine
ORS included	No	No	No	Yes
ESP included	No	Yes	Yes	Yes

tar levels down to 10g/Nm^3. In all cases the overall removal efficiency of the lab scale OLGA is in the range of 95–99%. The laboratory scale system has been modified several times. It is still used, especially for research on difficult fuels and cost reduction.

In 2004 the pilot OLGA was taken into operation downstream of an air-blown CFB gasifier. The gas flow was 100 times larger than in the laboratory scale OLGA. Notable difference is the wet ESP in the line-up, which is not present in the laboratory scale OLGA. In a 700-h duration test with the CFB gasifier, tar removal efficiencies of up to 99% were achieved [12]. The installation was partly modified to accommodate gas from the MILENA indirect gasifier, which produces the same gas volume but about twice as much tar. In 2010 and 2012 duration tests of 250 and 500 h were performed. Despite tar concentrations of $60–70 \text{g/Nm}^3$, OLGA managed to effectively remove 97–99% of the total amount of tar.

In 2006, the first commercial OLGA demonstration system was commissioned by the French company ENERIA for a plant in Moissannes [13]. The system showed good performance when wood and wine residue were used as fuels. Upstream OLGA dust concentrations up to 1500mg/Nm^3 and tar concentrations up to $11\,000 \text{mg/Nm}^3$ were measured. Downstream OLGA a series of aerosol (oil, fine dust) measurements were carried out with a filter at 70 °C. The total amount of aerosols (dust, tar, oil) was far below the detection limit of 25mg/Nm^3. The key tar compounds phenol and naphthalene were sufficiently removed, that is, the remaining phenol was below the detection limit and naphthalene reduced by 99%. The cleaned gas was used in a Caterpillar gas engine to produce 1.1MW_e.

In 2010, the second commercial OLGA system was installed in Tondela for the Portuguese company Iberfer. Gas produced by a CFB gasifier from chicken manure and wood chips was cleaned by an OLGA system and applied in again a 1MW_e Caterpillar gas engine. The OLGA performance has been further improved. The gas tar content (excluding BTX) is reduced from 16g/Nm^3 dry gas to 63mg/Nm^3, that is, the system removes 99.6%. The key tar compounds phenol and naphthalene are reduced by more than 99.9%.

In 2014, another OLGA system will be realized in India. The other projects mentioned in Section 9.4.1, notably the RDF gasifier and the Alkmaar SNG demonstration plant, will also include an OLGA system.

9.4.3 HDS, Reformer, and Methanation

In 2006 ECN completed a test rig consisting of the laboratory scale MILENA, OLGA, sulfur and chloride adsorbents, hydrogenation, reformer, and methanation reactors [14]. The system operated at atmospheric pressure and was designed for a producer gas flow of $1 \text{Nm}^3/\text{h}$, that is, about 5 kW. Test runs of up to 200 h were performed. The first experiments showed quick catalyst deactivation and plugging by soot formation. The problems were reduced considerably by changes in operating conditions.

Another problem encountered was slip of organic sulfur compounds, that is, mainly thiophene. Adsorption of these compounds by active carbon was studied and proved feasible. However, the use of active carbon is not an attractive solution because of the large amounts of benzene and toluene which are adsorbed too. Regeneration of active carbon is an option if the recovered stream of BTX and thiophene can be used or sold [4]. In the present market conditions, ECN prefers to keep BTX in the producer gas and convert BTX into SNG.

The alternative for thiophene removal is conversion to H_2S, followed by H_2S removal. To that end, in 2007 an HDS reactor with $CoMoO(S)$ catalyst was added to the SNG test rig. The reactor is externally heated to obtain approximately adiabatic conditions. Tests were performed using $0.3–1.0 Nm^3/h$ gas containing 100–200 ppmv H_2S, 5–10 ppmv COS, about 10 ppm C_4H_4S and about 1 or 2 ppm each of CH_3SH and C_2H_5SH.

At atmospheric pressure, catalytic reactions start when the gas temperature reaches about 350°C. The temperature rises quickly, due to heat produced by exothermic reactions. Figure 9.5 shows the effect of temperature on the reactions. Between 450 and 500°C, the water gas shift (WGS) reaction reaches thermodynamic equilibrium. Conversion of thiols and COS takes place in the same temperature window. Hydrogenation of C_2H_4 needs a slightly higher temperature and clearly depends on gas hourly space velocity (GHSV). Thiophene conversion follows the same trend.

The tests showed that thiophene in producer gas could be reduced to below 0.5 ppmv, but the allowed GHSV was too low for practical purposes. In 2011 a test rig was built to study HDS performance at higher pressures. To simulate conditions for a pressurized gasifier, gas was first dried by cooling to 5°C, then compressed to the required pressure, and finally steam was added to restore the moisture content. Results for thiophene conversion differed from those shown in Figure 9.5 only by the GHSV values, which could be increased linearly with pressure.

In 2013 the test rig was extended to allow the testing of reforming and methanation at higher pressure too. The test rig is used to obtain and check design parameters for the SNG demonstration plant to be built in Alkmaar. The equipment also allows simulation of gas recycling and power to gas (P2G) operating conditions.

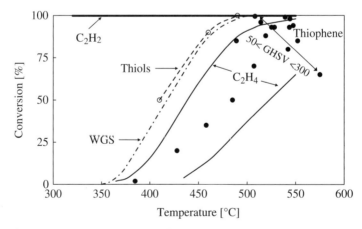

FIGURE 9.5 Temperature dependence of reactions in HDS reactor for MILENA producer gas at 1 bar and gas velocities of $50–300 h^{-1}$.

9.5 OUTLOOK

Although indirect gasification of biomass for the production of bioSNG is in its first phase of implementation, developments for the longer term are ongoing. These developments are targeted to make the process even more efficient in terms of energy, economy, and CO_2 reduction.

9.5.1 Pressure

The laboratory and pilot MILENA gasifiers operate near atmospheric pressure. Demonstration and commercial size gasifiers will operate at slightly elevated pressure. In general, pressurized gasification improves the energy efficiency, as pressurizing solid fuels consumes less energy than producer gas compression. Specific investment costs can be reduced, as reactor size decreases with pressure.

In case of a bioSNG plant, gasification at elevated pressure could result in a higher methane yield. That would benefit the overall efficiency to bioSNG, but tests showed no significant increase in methane yield at pressures up to 5 bar. The main benefit is the lower demand for steam to suppress coking in the reforming reactor (see Figure 9.1), as water formed in the gasifier remains in the gas phase. In case of atmospheric gasification, water vapor in the gas inevitably is condensed in the gas cooling/compression stage prior to reforming and thus has to be added again as steam.

The benefits of higher pressure come with a few penalties. A pressurized indirect gasifier needs air compression for the combustion reactor. The disadvantage can be turned into an advantage if an expansion turbine is fitted in the flue gas outlet, provided an appropriate turbine is available. Another penalty relates to the practical problems of pressurized biomass feeding, which translate into reduced availability and additional investments.

As explained in Section 9.2.1, there is an inherent limit to the operating pressure of an indirect gasifier. That limit is incorporated in the ECN bioSNG process design shown in Figure 9.1. A design requiring higher pressure downstream OLGA, would still require condensation of water vapor and thus lose an important advantage of indirect gasification at elevated pressure. At present, there is insufficient experience with the process to be able to tell at what size the theoretical cost and efficiency advantages of pressurized gasification will outweigh the increased technical complexity.

9.5.2 Co-production

Indirect gasification shows superior efficiency to bioSNG since a large part of the gas produced by the gasifier itself already is methane. The gas also contains significant amounts of other hydrocarbon molecules, such as BTX and ethylene, which are converted into methane in the system shown in Figure 9.1. BTX and ethylene add up to 25% of the heating value of the raw gas, but they require special attention, because of their tendency to form coke on catalysts. However, BTX and ethylene are valuable chemicals too [4]. That is why ECN is developing technology to separate BTX and ethylene rather than convert them to methane. Research is ongoing to increase the yield of valuable chemicals in gasification, with the aim to further improve the business case of co-production of bioSNG and chemicals.

9.5.3   Bio Carbon Capture and Storage

BioSNG production results in a large flow of essentially pure CO_2. The amount of CO_2 roughly equals the bioSNG production on a volume basis. The CO_2 stream may be vented to the atmosphere, from which it was recently taken by growing plants, but it may also be stored. When applied to reduce CO_2 emissions from processes using fossil fuel, Carbon Capture and Storage (CCS) involves considerable costs and efficiency loss. As CO_2 separation is an integral part of the bioSNG process, CCS combined with bioSNG production (BioCCS; or bioenergy with carbon capture and storage = BECCS) hardly involves an energy penalty. That makes BioCCS a relatively efficient and cheap method to reduce CO_2 concentrations in the atmosphere. BioCCS is a way to go far beyond CO_2-neutral [15].

9.5.4   Power to Gas

Balancing power production and demand is becoming more difficult when a growing share of power production comes from intermittent renewable sources such as wind and solar PV. One of the options for large-scale storage of temporary excess power is the production of hydrogen by water electrolysis, a concept called power to gas (P2G). Hydrogen can be used as such, or made to react with CO_2 to produce fuels, which fit more easily in the existing infrastructure.

As a bioSNG plant produces large quantities of essentially pure CO_2 as a co-product in the process (see above), it offers a smart way to accommodate hydrogen. In a bioSNG plant, hydrogen can be added to produce additional methane and consume CO_2 that otherwise would have to be separated and be lost as a carbon source. The P2G concept integrated with the bioSNG process involves limited additional costs. It only requires additional capacity and flexibility in the last part of the bioSNG process, and it saves on the expense of CO_2 separation [16].

ACKNOWLEDGEMENTS

The research and results described in this chapter have been made possible thanks to co-operation with and financial support from a number of parties: Royal Dahlman, HVC, the Dutch Ministry of Economic Affairs, Senter Novem, Agentschap NL, the European Union. Recently, research on tar and sulfur measurement and catalyst performance was performed within the BRISK project, funded by the European Commission Seventh Framework Programme (Capacities). SNG research is financed by a grant of the Energy Delta Gas Research (EDGaR) program. EDGaR is co-financed by the Northern Netherlands Provinces, the European Fund for Regional Development, the Ministry of Economic Affairs, and the Province of Groningen.

REFERENCES

[1] Rabou LPLM, Deurwaarder EP, Elbersen HW, Scott EL. 2006. *Biomass in the Dutch Energy Infrastructure 2030*. Report WUR, Wageningen, The Netherlands; 2006. Available at http://library.wur.nl/way/bestanden/clc/1871436.pdf (accessed 15 December 2015).

[2] Van der Meijden CM, Veringa HJ, Rabou LPLM. The production of synthetic natural gas (SNG): A comparison of three wood gasification systems for energy balance and overall efficiency. *Biomass and Bioenergy* **34**: 302–311; 2010.

[3] Van der Drift A, Zwart RWR, Vreugdenhil BJ, Bleijendaal LPJ. *Comparing the Options to Produce SNG from Biomass*. Proceedings of the 18th European Biomass Conference and Exhibition, Lyon, pp. 1677–1681; 2010.

[4] Rabou LPLM, Van der Drift A. *Benzene and Ethylene: Nuisance or Valuable Products*. Proceedings of the International Conference on Polygeneration Strategies, Vienna, pp. 157–162; 2011.

[5] Paisley MA, Slack W, Farris G, Irving J. *Commercial Development of the Battelle/FERCO Biomass Gasification Process – Initial Operation of the McNeil Gasifier*. Proceedings of the Third Biomass Conference of the Americas, Montreal, pp. 579–588; 1997.

[6] Hofbauer H, Stoiber H, Veronik G. *Gasification of Organic Material in a Novel Fluidization Bed System*. Proceedings of the First SCEJ Symposium on Fluidization, Tokyo, pp. 291–299; 1995.

[7] Van Selow ER, Cobden PD, Verbraeken PA, Hufton JR, Van den Brink RW. Carbon capture by sorption-enhanced water–gas shift reaction process using hydrotalcite-based material. *Industrial and Engineering Chemistry Research* **48**: 4184–4193; 2009.

[8] Gazzani M, Macchi E, Manzolini G. CO_2 capture in natural gas combined cycle with SEWGS. Part A: Thermodynamic performances. *International Journal of Greenhouse Gas Control* **12**: 493–501; 2013.

[9] Manzolini G, Macchi E, Gazzani M. CO_2 capture in natural gas combined cycle with SEWGS. Part B: Economic assessment. *International Journal of Greenhouse Gas Control* **12**: 502–509; 2013.

[10] Aranda Almansa G, Van der Drift B, Smit R. *The Economy of Large Scale Biomass to Substitute Natural Gas (bioSNG) Plants*. Report ECN-E–14-008. ECN, Petten, The Netherlands; 2014. Available from https://www.ecn.nl/publications/ (accessed 15 December 2015).

[11] Van der Drift A, Biollaz S, Waldheim L, Rauch R, Manson-Whitton C. *STATUS and FUTURE of bioSNG in EUROPE*. Report ECN-L–12-075. ECN, Petten, The Netherlands; 2012. Available from https://www.ecn.nl/publications/ (accessed 15 December 2015).

[12] Verhoeff F, Rabou LPLM, Van Paasen SVB, Emmen R, Buwalda RA, Klein Teeselink H. *700 Hours Duration Test with Integral 500 kW Biomass Gasification System*. Proceedings of the 15th European Biomass Conference and Exhibition, Berlin. pp. 895–900; 2007.

[13] Könemann HWJ, Van Paasen SVB. *OLGA Tar Removal Technology, 4 MW Commercial Demonstration*. Proceedings of the 15th European Biomass Conference and Exhibition, Berlin. pp. 873–878; 2007.

[14] Zwart RWR, Boerrigter H, Deurwaarder EP, Van der Meijden CM, Van Paasen SVB. *Production of Synthetic Natural Gas (SNG) from Biomass*. Report ECN-E–06-018. ECN, Petten, The Netherlands; 2006. Available from https://www.ecn.nl/publications/ (accessed 15 December 2015).

[15] Carbo MC, Smit R, Van der Drift A, Jansen D. Bio energy with CCS (BECCS): Large potential for BioSNG at low CO_2 avoidance cost. *Energy Procedia* **4**: 2950–2954; 2011.

[16] Saric M, Dijkstra JW, Rabou LPLM, Walspurger S. *Power-to-Gas Coupling to Biomethane Production*. Report ECN-L–13-061. ECN, Petten, The Netherlands; 2013. Available from https://www.ecn.nl/publications/(accessed 15 December 2015).

10

HYDROTHERMAL PRODUCTION OF SNG FROM WET BIOMASS

Frédéric Vogel

10.1 INTRODUCTION

Biomass is a renewable resource but its potential for sustainable use is limited. Land-based biomass can be roughly divided into (i) grown biomass, such as wood, straw, energy crops, and (ii) residual and waste biomass, such as sewage sludge, manure, crop and food processing residues, among others. These categories have a total technical potential of circa 104 EJ/a world-wide [1]. Other sources forecast an increase to 300 EJ/a by 2050 [2]. Compared to a total primary energy demand of 550 EJ/a, if assumed to remain constant by 2050, these biomass categories could increase their potential of today's 19% to contribute a maximum of 55% to the primary energy needs by 2050.

A substantial extension of the land based potential may be realized via algae. Both micro- and macro-algae have been shown to grow under a variety of environmental conditions, even in waste water. From a wide perspective, algae convert CO_2, water, nutrients, and sunlight into biomass at a much higher rate than land-based biomass (Figure 10.1). Sugarcane exhibits similarly high rates of around 75 t dry mass per hectare and year in a tropical climate. However, if only the actual sugar fraction is used to produce bioethanol, the hectare yield drops considerably. In a future bioenergy system, it is therefore imperative to achieve a full utilization of the biomass, and not just of parts of it. A sustainable bioenergy system will thus consist of different kinds of highly integrated biorefineries that allow for a very efficient use of the

Synthetic Natural Gas from Coal, Dry Biomass, and Power-to-Gas Applications, First Edition.
Edited by Tilman J. Schildhauer and Serge M.A. Biollaz.
© 2016 John Wiley & Sons, Inc. Published 2016 by John Wiley & Sons, Inc.

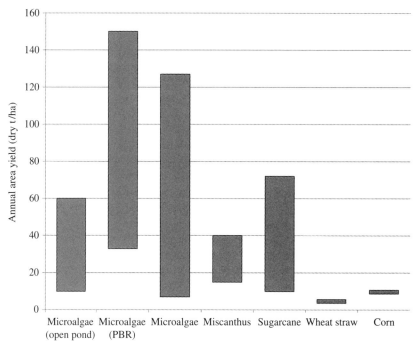

FIGURE 10.1 Productivity range of micro- and macro-algae cultivation and land crops used for biofuel production [3]. Adapted from Carlsson AS et al. 2007. PBR: Photobioreactor.

biomass resource to produce valuable chemicals and fuels, as well as co-generating heat and electricity from residues that cannot be used otherwise.

Such a highly integrated bioenergy concept is the SunCHem process developed at PSI [4, 5]. This allows for CO_2 capture by algae in a photobioreactor (closed reactor or open pond) and its transformation, together with water and nutrients, into algal biomass. After mechanical dewatering of the harvested algae slurry, it can be efficiently gasified to a methane-rich gas using hydrothermal methanation, while recovering and recycling most of the nutrients to the photobioreactor (Figure 10.2).

Brandenberger et al. [6] have performed a detailed techno-economic study of the SunCHem process. Their conclusion is that algae production for energy purposes is still too expensive by a factor of at least ten to compete with fossil alternatives. Coproduction of high value chemicals may reduce the economic pressure on algae growth. More importantly, they showed that only a well designed overall process chain will have a positive energy balance and thus a GHG mitigation effect.

A significant fraction of the total raw and of the processed biomass has a high water content. Thermochemical processing usually requires water contents to be below 20 wt% for a high process efficiency. Drying, however, requires a lot of thermal energy, and its efficiency is not very high, even if heat recovery is applied. Therefore, processes that do not require drying of the wet biomass prior to its processing may be a priori more efficient. Biochemical processes belong to this category. Their low processing

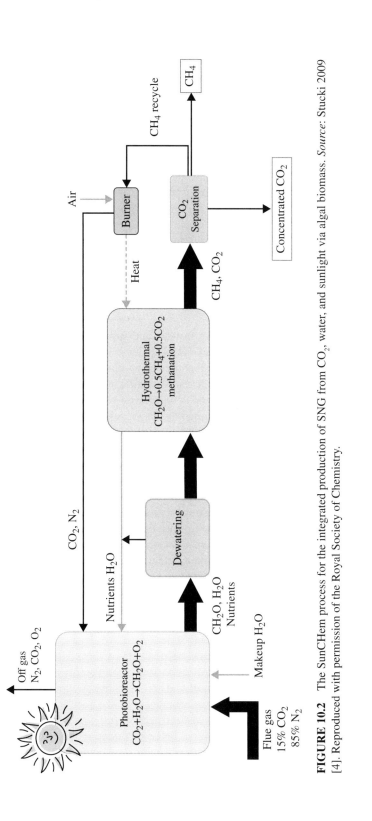

FIGURE 10.2 The SunCHem process for the integrated production of SNG from CO_2, water, and sunlight via algal biomass. *Source:* Stucki 2009 [4]. Reproduced with permission of the Royal Society of Chemistry.

temperature, however, leads to low space–time yields. Combining high processing temperatures of the wet biomass with high pressures is the rationale of hydrothermal processing. The pressure is kept such as to keep the water in a liquid or supercritical state but not in a vapor state. The water does not evaporate and thus no heat of vaporization needs to be supplied to the system to reach a high temperature [7].

The solvent properties of near- and supercritical water differ substantially from liquid water at ambient conditions and resemble the properties of common organic solvents at ambient conditions [8]. These properties play a very important role for the liquefaction of the biomass and the separation of salts and will be discussed in more detail in Subsections 10.3.1 to 10.3.3.

In this chapter, we present and discuss mainly new results from our own work at PSI obtained since the appearance of our review article in 2008 [9] as well as key findings from other researchers in the area of hydrothermal production of SNG.

10.2 HISTORICAL DEVELOPMENT

The gasification of woody biomass in near- and supercritical water was first proposed and studied by Modell at MIT [10, 11]. Their yield in gases and particularly methane was low, although they also studied the effect of adding catalysts. Their most important discovery was that the biomass did not form any tars or coke around the critical point. A few years later Elliott at Pacific Northwest National Laboratory (PNNL) developed a process for producing a methane-rich gas from wet biomass and liquid wastes [12]. They called their process "Thermochemical Environmental Energy System" or TEES. The preferred operating conditions for TEES are 350 °C and 20 MPa. A reduced metal catalyst was used to achieve a good conversion to gases. Elliott and coworkers screened many different catalysts able to both gasify and methanate the organic feed. Their initial choice fell on Raney nickel and a supported nickel catalyst from BASF. After testing them for several weeks in a continuous flow system, some catalyst deactivation by sintering was observed but it could be controlled by adding other metals as dopants. They then began using a catalyst based on ruthenium on an activated carbon support, Ru/C, and found it to be very active, selective, and stable. A number of studies with the Ru/C catalyst documented the ability of TEES to methanate a variety of feedstocks such as DDGS, dairy manure solids, corn ethanol stillage, destarched wheat millfeed, several types of algae, and highlighted at the same time the remaining challenges of such a technology, that is, mineral and salt precipitation and catalyst deactivation by sulfur compounds [13, 14].

The work on catalytic hydrothermal gasification is reviewed in [9, 13, 15]. Another use of catalysts in hydrothermal gasification aims at producing hydrogen as the main gaseous product. This can be accomplished at relatively low subcritical temperatures and pressures, as in aqueous phase reforming (APR [16]), or at relatively high temperatures (>500 °C) in supercritical water to achieve full conversion of difficult feedstocks. In this chapter, we focus on SNG production only. Since methane is thermodynamically stable up to circa 400 °C, this temperature sets the limit for SNG processes that want to avoid recycling gases (H_2, CO_2).

10.3 PHYSICAL AND CHEMICAL BASES

The typical approach chosen in industry for SNG production from biomass is the sequential one, that is, the process is subdivided into well known processing units and each of these is optimized individually. Then, all units are linked together and some process integration is realized with a limited number of degrees of freedom. In catalytic hydrothermal gasification, a "chemical integration" is realized by avoiding separate process steps for the gasification and the methanation. This allows for very high process efficiencies because the heat from the methanation step is used in situ on the catalyst for bond breaking and decomposition of the larger organic molecules, that is, gasification and steam reforming. In an idealized way, the conversion of biomass to SNG may be represented by the following equation[1], highlighting its nearly thermoneutral character:

$$CH_{1.49}O_{0.68}(s) + 0.29\,H_2O(g) \rightarrow 0.52\,CH_4(g) + 0.48\,CO_2(g) \quad \Delta_r H° = -26\,kJ/mol$$
$$(10.1)$$

Thus the catalytic reactor can be operated nearly isothermally without the need for efficient heat removal and heat integration, which is always accompanied by losses. A theoretical maximum efficiency of 95% is calculated from this equation. This represents the "chemical" efficiency alone. At the process level, two major sources of efficiency losses need to be introduced:

- The heat exchanger for preheating the feed with an assumed efficiency of circa 80%;
- The heating of the salt separator by hot flue gases, generated by combusting some of the produced SNG.

The feed heat exchanger may theoretically be designed to approach 100% efficiency but only at a prohibitive cost, that is, with a huge heat exchange surface area. The salt separator will always produce some losses, as the flue gases cannot be cooled to below the lowest temperature in the separator, that is, circa $380\,°C$. Further heat integration and/or polygeneration may, however, increase the overall process efficiency [18]. Therefore, typical overall process efficiencies for biomass to SNG range from circa 60 to 70% for wet feedstocks with circa 20 wt% dry solids.

To realize the "dream reaction," Equation (10.1), in an industrial process as efficiently as possible, several key aspects of chemical engineering and process development must be considered, of which catalysis is the most important one. In the following sections, we discuss the bases of these key aspects in some detail.

---

[1] This equation is written for standard conditions, that is, solid biomass and gaseous water, methane, and carbon dioxide. For hydrothermal conditions, the enthalpies would need to be calculated at the respective temperature and pressure (see [17]). We have included this simplified equation for illustration purposes only.

10.3.1 Catalysis

Biomass exposed to supercritical water will decompose to form gases only at temperatures well above 500 °C if no "reaction enhancer" is used. One proposal is to use relatively high concentrations of NaOH to achieve a substantial increase in gasification rate. A beneficial side effect of using an alkaline reaction medium is the absorption of CO_2 to form carbonates, which increases the concentration of combustible gases such as H_2 and CH_4 in the product gas. At 450 °C, 34 MPa, and 1.67 M NaOH, 1.0 g of rice straw yielded 0.12 g of combustible gases [19].

Complete gasification of biomass to combustible gases can be achieved by either working at temperatures around or above 700 °C [20] or by using a catalyst at lower temperatures [8, 21]. One benefit of using a catalyst is its ability to influence the composition of the gases. For the production of a methane-rich gas, that is, SNG, a methanation catalyst is needed. When working at temperatures below 700 °C, this catalyst must also catalyze the decomposition of the biomass to gases such as CO, CO_2, and H_2. We have found that the catalytic reactions in supercritical water on a ruthenium catalyst do not follow a sequential pathway, that is, first forming only CO, CO_2, and H_2 which subsequently react to CH_4 and H_2O by methanation. Peterson et al. [22] showed that ruthenium is able to decompose (or "scramble") the molecules on its surface into atomic adsorbates. These adsorbates then recombine statistically with others to form new stable compounds such as CH_4 and CO_2. Thus, CH_4 and CO_2 are formed simultaneously. Figure 10.3 shows the calculated free energy pathway for the

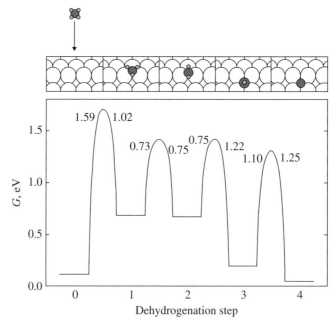

FIGURE 10.3 Calculated free energy levels for the dissociation of methane on a stepped ruthenium surface. *Source*: Peterson et al. 2012 [22]. Reproduced with permission of Wiley.

dehydrogenation of methane over a stepped ruthenium surface. The most stable adsorbates are atomic carbon plus four hydrogen atoms. Water adsorbs and dissociates similarly, providing further hydrogen and oxygen atoms. From this pool of atomic adsorbates, the stable molecules CH_4, CO_2, and H_2 are formed and desorb from the catalyst's surface. This hypothesis was proven by experiments with CH_4 and D_2O, where a substantial amount of CD_4 was formed, but only in the presence of the Ru/C catalyst.

For methane to be formed, Ru sites with a particular geometrical arrangement, the so-called "B5 sites", are required. Czekaj et al. [23] recently calculated the stability of ruthenium clusters on a carbon surface for different cluster sizes and geometries (Figure 10.4). Small, nanoscopic clusters have the highest density of B5 sites, which explains why a catalyst with such small clusters exhibits a high activity for the methanation. In the same work, EXAFS data from a commercial Ru/C catalyst matched well the theoretical calculations using a $Ru_{11}C_{54}$ assembly.

Rabe et al. [24] were the first to show that, under hydrothermal gasification conditions, the ruthenium on a commercial Ru/C catalyst is present in its reduced, metallic form. The ruthenium oxide (RuO_2), present on the commercial catalyst in its delivered state, is actually quickly reduced around 125 °C by a liquid stream of ethanol in water. The presence of nanoscopic Ru clusters was ascertained by HAADF-STEM analyses as well as EXAFS data [25, 26]. The very small initial Ru particles (<1 nm) sintered to a stable size of circa 2 nm after exposure to supercritical water. A stable activity and selectivity of this catalyst over 220 h was demonstrated by Waldner et al. [25]. Elliott et al. [27] have used similar Ru/C catalysts at subcritical conditions for even longer times on stream without any noticeable deactivation. Ru/C is thus an excellent choice for the production of SNG from various types of biomass

(a)   (b)

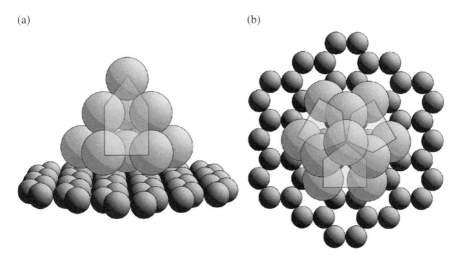

FIGURE 10.4 Stable configuration of a cluster of 11 ruthenium atoms on a carbon surface, $Ru_{11}C_{54}$, calculated by DFT; (a) side view, (b) view from above. The pentagons denote the B5 sites required for methanation. *Source*: Czekaj 2013 [23]. Reproduced with permission of ACS.

in a hydrothermal environment. Peng et al. [28] studied the effect of the preparation method on the activity of Ru/C and found lower Ru loadings to result in a higher dispersion of the metal and in consequence also to a higher activity. The choice of the solvent for the impregnation is also an important synthesis parameter.

Processing biomass in hydrothermal medium using a catalyst poses several challenges to the stability of the catalyst. Hot compressed water per se is a harsh reaction environment, dissolving all kinds of materials, even quartz [29]. Salts and acids present in the biomass feedstock enhance the chemical attack of the feed stream on materials of construction and catalysts. Almost any known mode of deactivation may also be observed during processing in hydrothermal medium, for example:

- Sintering of nickel-based catalysts [25];
- Leaching of the active metal [30];
- Dissolution of the support [31];
- Phase change of the support [31];
- Coking [32, 33];
- Fouling by precipitated mineral matter (salts) [33, 34];
- Poisoning by sulfur [25, 26, 35].

When using Ru/C catalysts, only the last three modes of deactivation are relevant. In PSI's hydrothermal SNG process, a dedicated salt separator is foreseen to capture most of the salts and other mineral particles present in the feed stream or formed during its heating up (see Subsection 10.3.2). Therefore, fouling by mineral matter is not a major challenge in this process. While coking is very difficult to unequivocally detect due to the carbon support, much work has been performed on poisoning by sulfur. Osada et al. [36] showed that different forms of sulfur lead more or less to the same deactivation of a Ru/C catalyst. This supports a hypothesis in which any form of sulfur is converted quickly to the same species eventually poisoning the ruthenium catalyst. An EXAFS analysis of Ru/C poisoned by DMSO by Dreher et al. [26] revealed this species to be sulfidic S(-II). The sulfur atoms would preferentially block the B5 sites, reducing the rate of the methanation reaction to very low values. The excess carbon on the Ru surface would then also recombine with oxygen atoms, yielding CO in the product gas of a deactivating catalyst. The excess hydrogen would recombine to molecular H_2. This was indeed observed by Waldner et al. [25] and Dreher et al. [26]. The latter found a significant shift in the gas composition when gasifying a solution of 7.5 wt% ethanol in water at 400 °C and 24.5 MPa. The fresh catalyst yielded 22% CO_2, 18% H_2, 60% CH_4, and 0.2% CO. The same catalyst, poisoned by adding 200 ppm of DMSO, yielded 4% CO_2, 60% H_2, 20% CH_4, and 16% CO. The overall conversion of ethanol to gases dropped from 99.9% (fresh) to 30% (poisoned). Assuming that CO stems only from the C–O bond in ethanol, the theoretical gas composition, arbitrarily fixed to the measured value for CH_4, would yield: 6% CO_2, 49% H_2, 20% CH_4, and 26% CO. The higher concentration of CO and the lower concentration of H_2 suggest that some watergas shift activity of the poisoned catalyst remained. However, a direct comparison is complicated by the fact that the experimental data were obtained at 30% conversion.

An in situ XAS study on catalyst poisoning by Dreher et al. [26] revealed a surface coverage with sulfur of circa 40%, corresponding to a stoichiometry of $RuS_{0.33}$ for the fully poisoned catalyst. By the same analysis a bulk sulfidation of the ruthenium was ruled out. This surface coverage is in remarkable agreement with the cluster geometries suggested by Czekaj et al. [23]. The cluster $Ru_{11}C_{54}$ shown in Figure 10.4 would be poisoned by three sulfur atoms blocking the three B5 sites. This would involve all 10 surface atoms. The surface stoichiometry would thus be $Ru_{10}S_3$ or $RuS_{0.3}$. For the largest cluster studied, $Ru_{50}C_{96}$ with 37 Ru atoms at the surface, a complete blocking of the twelve B5 sites would affect all the 36 B5 site atoms. The surface stoichiometry would thus be $Ru_{36}S_{12}$ or $RuS_{0.33}$. This analysis can be reversed to determine the number of active B5 sites by fully saturating the Ru/C catalyst with sulfur and performing an EXAFS analysis to determine the surface coverage of sulfur for a known cluster size and assumed shape.

A DFT study conducted for a fresh and poisoned stepped Ru surface [26] revealed that, already at low sulfur coverage, the energy barrier for the first dehydrogenation step of methane is increased, which explains the drop in overall conversion to gases. To understand the strong drop in methane selectivity from ethanol one must assume that sulfur poisoning blocks the cleavage of C–C bonds. Support for this assumption comes from the observation that, during catalyst deactivation by sulfur poisoning, the concentration of C_{2+} hydrocarbons increases [37, 38]. Dreher [39] devised a reaction mechanism for the methanation of ethanol in supercritical water and suggested "site isolation" to be the main effect caused by sulfur (S) poisoning that explains all experimental observations.

In fact, an S-poisoned Ru catalyst exhibits a strongly decreased ability to break C–C, C–H, and C–O bonds, whereas the dissociation of water is facilitated [39]. The early appearance of higher hydrocarbons, C_{2+}, and the later appearance of CO for a deactivating catalyst suggests that S-poisoning first affects C–C bond cleavage but not C–O bonds. As the S-coverage reaches saturation, also C–O bond cleavage and methane formation are inhibited, resulting in an increase of CO and a strong decrease of CH_4 in the gas. Theoretically, the amount of CO formed would then be proportional to the amount of C–O bonds in the feedstock. Because S-poisoned catalysts never reach full conversion of the feedstock to gases, the chemical composition of the intermediates in the liquid phase must also be accounted for in a discussion of the fate of the C–O bonds.

In most cases, catalyst deactivation is inevitable and the reactor is designed to allow for sufficient time on stream at full conversion before the catalyst must be replaced. In industrial processes this "life time" may be several years, or it may be in the order of seconds, as in the FCC process. The latter requires fast on-stream regeneration of the coked catalyst by circulating it from the riser reactor to a combustor to burn off the coke. In hydrothermal methanation, catalyst deactivation will also be ubiquitous because biomass always contains a number of detrimental compounds that cannot be removed completely from the feed stream. To achieve commercially reasonable life times, a short periodic regeneration of the (partially) deactivated catalyst is a viable option.

We have focused on regenerating S-poisoned ruthenium catalysts. Catalysts deactivated by coke have not been studied much so far. Waldner [40] reported a full and partial recovery of the catalyst's activity and stability, respectively, after treating it with a dilute solution of H_2O_2 at mild conditions. Building on these first positive results, Dreher et al. [41] performed a detailed study on the oxidative regeneration of S-poisoned Ru/C with H_2O_2. Besides regaining the full activity of the catalyst, the carbon support must not be attacked by the oxidative treatment to avoid a loss of stability. Figure 10.5 shows the effect of treatment with 3% H_2O_2 at 75 °C for twice 20 min. The initial activity was fully recovered, while the methane selectivity was increased further. However, the high activity dropped after several hours on stream. The cause of this behavior has not been identified yet.

To facilitate regeneration of heavily poisoned and/or fouled catalysts, we studied a number of ruthenium catalysts based on refractory oxide supports. These supports would allow for harsher regeneration conditions, for example, higher temperatures, higher H_2O_2 concentrations, and/or longer exposure times. Such harsher conditions may also be required to regenerate catalysts deactivated by coke.

Monoclinic zirconia, a stabilized variant of tetragonal zirconia, and TiO_2 in the rutile form were found to exhibit both good physical and crystallographic stability during a static treatment at 430 °C and 30–35 MPa for 20 h. The BET surface area did

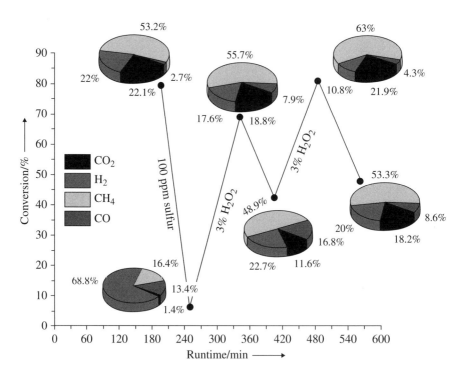

FIGURE 10.5 Effect of poisoning and regeneration procedures on the conversion of a Ru/C catalyst. The pie charts represent the gas composition measured during each step. *Source*: Dreher 2014 [41]. Reproduced with permission of Wiley.

not change significantly for these samples, even after a second aging test at the same conditions [31]. These supports were then loaded with 2 wt% ruthenium and tested in a continuous flow reactor for their activity, selectivity, and stability by gasifying a glycerol solution. A catalyst made from tetragonal zirconia, stabilized by hafnia and lanthana, showed the best overall performance. However, at higher feed concentrations its performance degraded, but it could be restored after switching back to a lower feed concentration. This class of ruthenium catalysts shows a promising performance, yet still requiring some optimization, and may be an alternative to carbon-supported catalysts if harsher conditions are needed.

10.3.2   Phase Behavior and Salt Separation

A key challenge in processing wet biomass streams is dealing with the mineral fraction. This inorganic fraction consists mostly of oxides (e.g., SiO_2, Al_2O_3, TiO_2, Fe_2O_3) as well as dissolved and precipitated salts (carbonates, chlorides, phosphates, nitrates, sulfates, hydroxides of potassium, sodium, calcium, magnesium, ammonium, and others). These inorganics may cause plugging, fouling, corrosion, or even catalyst deactivation. The best option would be to remove them from the biomass feed stream before processing. However, in many cases this is not feasible or impracticable. Furthermore, some of the inorganic compounds are released during hydrothermal processing from hetero atoms such as N and S, for example, chemically bound in proteins. This "mineralization" of the organic fraction is an important prerequisite to the complete removal of hetero atoms (other than oxygen) from the reacting biomass stream. This aspect is discussed in more detail in Subsection 10.3.3.

Because supercritical water exhibits a low solubility for polar (or charged) compounds, the cations and anions present in a hydrated form at subcritical conditions will form ion pairs as the critical point of the mixture is approached. Upon further lowering the fluid density a salt-rich phase will precipitate from the supercritical fluid. This salt-rich phase may be a solid salt or one or more salt-saturated liquid phase(s). We have studied the formation of such salt-rich phases from K_2HPO_4 solutions using high pressure differential scanning calorimetry (HP-DSC). Figure 10.6 shows the three-phase equilibrium line (G–L1–L2) for an average fluid density of $300 \, kg/m^3$ [42]. The knowledge when these salt-rich phases precipitate can be used to tune the process conditions, such as to recover the salts as a liquid brine without plugging the equipment, whereas the formation of a solid salt phase can be avoided.

Often, the behavior of the aqueous salt mixtures are described as either of "Type 1" or of "Type 2", which is a simplified typology of the complex phase behavior these mixtures exhibit [43]. Figure 10.7 shows solid deposits of the Type 2 salt K_2SO_4 in a salt separator vessel after an experiment. The picture was taken with an endoscopic camera after cooling and depressurizing the vessel.

While both Na_2SO_4 and K_2SO_4 alone form Type 2 mixtures with water, Schubert et al. [44] have shown that ternary mixtures of Na_2SO_4-K_2SO_4-H_2O exhibit Type 1 behavior for certain compositions below 470 °C at 30 MPa. Such a mixture can be continuously withdrawn from a separator device, as opposed to the "sticky" solid deposits formed by Type 2 salts (cf. Figure 10.7). The formation of two immiscible *liquid* phases of the system Na_2SO_4-K_2SO_4-H_2O was confirmed by HP-DSC [42].

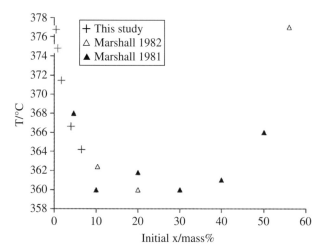

FIGURE 10.6 Temperature versus initial composition plot for the liquid at the three-phase equilibrium (G–L1–L2) of the system $K_2HPO_4 + H_2O$, determined by HP-DSC. Reproduced from [42] with permission.

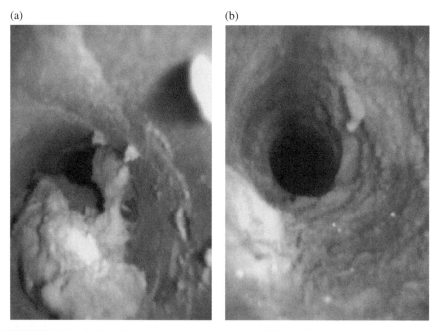

FIGURE 10.7 K_2SO_4 deposits in a salt separator at different heights, operated at super-critical conditions. Pictures (a) and (b) were taken with an endoscopic camera inserted from the bottom of the vessel. The inner diameter of the vessel is 12 mm. Source: Müller JB, 2012 [38].

The apparatus used for these continuous salt separation experiments was a reverse-flow vessel similar to the design developed for the MODAR supercritical water oxidation process [45]. The preheated feed stream enters the vessel from the top via a dip tube, mixing with the supercritical contents of the vessel. The salts precipitate and settle to the cooler bottom of the vessel, while the main fluid flow reverses to flow upward, exiting the vessel from a side port at the top. The salt brine can be removed continuously from the bottom of the separator. Schubert et al. [46] showed that, by adjusting the brine flow rate, an increase in salt concentration in the brine by a factor of more than 10 can be achieved.

To understand how the subcritical salt mixture behaves inside the separator vessel before and during mixing with the supercritical fluid inside, a special separator vessel was built allowing for fluid visualization by neutron radiography [47]. Figure 10.8 shows a time progression of the cooler tracer fluid (H_2O) entering at the top, exiting the dip tube and mixing with the hot fluid (D_2O) inside the vessel. At supercritical conditions inside the vessel the fluid jet entering the vessel spreads quickly radially

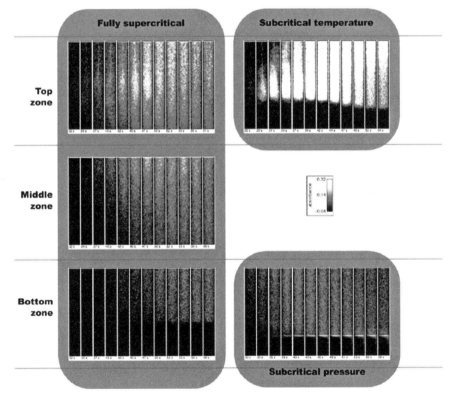

FIGURE 10.8 Neutron radiographs visualizing the dispersion of a tracer (H_2O) in a reverse-flow salt separator vessel continuously flushed with D_2O. The interface between the hotter top section and the cooler bottom section is visible in the bottom zone. *Source*: Peterson et al. 2010 [47]. Reproduced with permission of Elsevier.

while reaching the cooler pool of salt brine at the bottom of the vessel. At subcritical temperatures (and supercritical pressure), the fluid jet reverses and mixes only within the top section of the vessel.

Using the same technique, Peterson et al. [48] studied the precipitation of Type 2 salts inside the vessel and blockages caused by salt plugs. Figure 10.9 visualizes salt deposits just below and around the dip tube at the top of the vessel. The knowledge gained by these studies enabled us to optimize the salt separation in the reverse-flow vessel [38]. The length of the dip tube had only a minor influence, but it is important to have one to avoid short-circuiting of flow. The pressure has a larger influence because at lower pressure, that is, at lower fluid densities, the solubility of salts decreases, which increases the salt separation efficiency.

A very important task of the salt separator is to reduce the load of sulfur flowing into the catalytic reactor. Ideally, all chemically bound sulfur would be mineralized to inorganic sulfur species, that is, sulfide, sulfite, sulfate, which can be precipitated as salts in the salt separator. Schubert et al. [49] extrapolated their salt separation efficiencies to estimate the temperature needed in the salt separator to achieve a residual sulfate concentration of 1 ppm at the entrance to the reactor (Figure 10.10). For sodium sulfate at a pressure of 30 MPa, this temperature must be at least 468 °C, which is within reach of the design of the salt separator vessel.

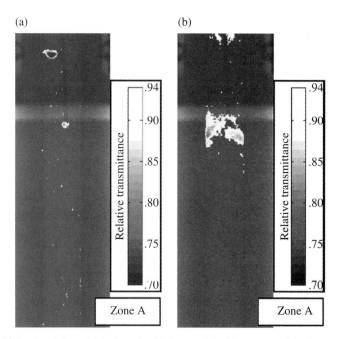

FIGURE 10.9 Precipitated salt deposits (light spots) inside a supercritical water salt separator vessel. Images (a) and (b) are neutron radiographs of a $Na_2SO_4/Na_2B_4O_7$ solution in D_2O at 450 °C and 30 MPa. Shown is the top entrance section, with the dip tube slightly visible as a darker, vertical strip. *Source*: Peterson et al. 2010 [47]. Reproduced with permission of Elsevier.

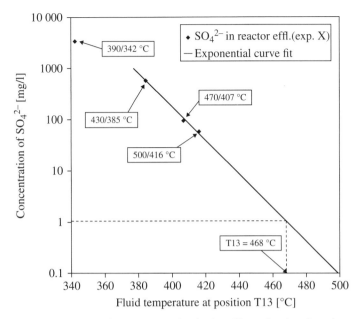

FIGURE 10.10 Residual sulfate concentration in the effluent leaving the salt separator to the catalytic reactor at 30 MPa. "Position T13" is the temperature measured at the salt separator exit. *Source*: Schubert 2010 [49]. Reproduced with permission of Elsevier.

The precipitation of ammonium salts would be an ideal means to recover the nitrogen as nutrient. Schubert et al. [46], however, have shown that ammonium salts decompose at hydrothermal conditions. NH_4Cl, $(NH_4)_2SO_4$, and $(NH_4)_2CO_3$ form NH_3 and HCl, H_2SO_4, and CO_2, respectively. Because NH_3 is completely miscible with supercritical water, it will flow out of the separator vessel into the catalytic reactor. During cooling down it will form NH_4HCO_3 and $(NH_4)_2CO_3$ with the CO_2 from the product gas, which might lead to a plugging of the lines in the downstream section. In a well designed cooling section, however, complete recovery of nitrogen as ammonium salt should be feasible. A partial recovery of a solid ammonium salt from the brine leaving the salt separator was found by Zöhrer et al. [50] in the form of magnesium ammonium phosphate (MAP or struvite).

10.3.3 Liquefaction of the Solid Biomass, Tar, and Coke Formation

Before the solid biomass particles and the large macromolecules can be catalytically converted to methane, they have to be transformed into smaller fragments that can access the active sites of the catalyst. Thus a "liquefaction" step must precede any catalytic conversion of solid biomass particles. Liquefaction is achieved by pyrolysis and hydrolysis of the solid structures. Mosteiro-Romero et al. [51, 52] studied the dissolution of wood particles in hot compressed water experimentally and theoretically. They modeled the main chemical and physical processes of liquefaction and mass transfer using a shrinking spherical particle approach. The main idea was to model the diffusion of water through the oily film of liquefaction products sticking around the particles. Hydrolysis was shown to be the dominant reaction, yielding more water- and methanol-soluble (or oily) products than pyrolysis at the same temperature and residence time. The transformation of the solid biomass into solid char by dehydration reactions proceeds in parallel to the liquefaction reactions, although at a slower rate. The ability of near- and supercritical water to dissolve the oily liquefaction products sticking around the particle is key to a fast and complete hydrolysis of the biomass particle. Figure 10.11 shows the yields of water-soluble products (WSP), methanol-soluble products (MSP), gases (G), and the remaining solid residue (SR) for the liquefaction of spruce particles at 350 °C and 25 MPa in a batch reactor. The heating time was 10 min. After a fast initial formation of liquid and gaseous products, the liquefaction stops and the product yields remain constant (within experimental reproducibility). Only the charring of the solid residue progresses further. This may be due to the static conditions in the batch reactor, leading to an accumulation of liquid products over time, which inhibits the mass transfer of the liquid products from the surface of the particles into the solution. Bobleter and Binder [53] were able to solubilize 83% of spruce in a flow system, where hot water (340 °C, 23 MPa) was flushed through a fixed bed of wood particles, continuously renewing the particles' surface. In our static batch experiments, we reached 70–75% on a carbon basis at 350 °C and 25 MPa (Figure 10.11, upper left panel).

The more charred the solid residue becomes, the less reactive it is for hydrothermal liquefaction and consequently also for gasification. Therefore, solids from hydrothermal carbonization (HTC) and coal are non-reactive feedstocks for low-temperature (catalytic) hydrothermal gasification.

In an ideal hydrothermal liquefaction (HTL) process, the liquefied biomass products would be stable and could be collected from the process as bio-oil. In reality, some of these products are very reactive and either decompose to smaller fragments or condense with other products to larger molecules, that is, tars, eventually forming coke. We showed that the water-soluble products (WSP) collected from batch liquefaction experiments of spruce, subjected to the same reaction conditions once more, were no longer reactive, as the yield of gases and methanol-soluble products from these WSP were very low. The formation of condensed products must thus happen during the early stages of liquefaction from reactive intermediates with half-lives of less than a few minutes at temperatures of 250 °C and higher. These intermediates may only be studied in situ with fast spectroscopic techniques such as Raman spectroscopy [54].

Müller and Vogel [55] studied the formation of tars and secondary coke from potential precursor molecules, glycerol and glucose, with feed concentrations in the range of 10–30 wt%. The highest coke yields were found in a small temperature range of 350–370 °C at circa 30 MPa. At supercritical conditions, no coke was formed at all (Figure 10.12), supporting the earlier report of Modell [11]. They speculated that probably the dehydration of the parent molecule to an aldehyde was the key step for initiating further condensation reactions. But their experiments did not allow the study of fast reactions and short-lived intermediates. In all experiments with glycerol and glucose, the pH of the aqueous solution decreased to around 3. When the initial pH was set in a range of 10–12 by adding small amounts of NaOH, neither tars nor coke were formed, even though the pH was acidic after the experiment.

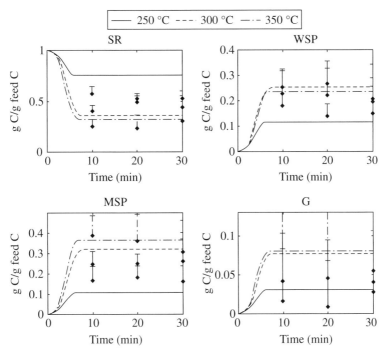

FIGURE 10.11 Experimental data and model results for the carbon yields of water-soluble products (WSP), methanol-soluble products (MSP), gases (G), and the conversion of solid residue (SR) from spruce under hydrolysis conditions at 25 MPa and three different temperatures. The heating time was 10 min. *Source*: Mosteiro 2014 [52]. Reproduced with permission of Elsevier.

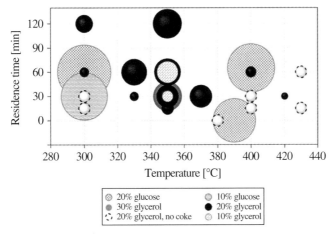

FIGURE 10.12 Amount of coke formed from glycerol and glucose solutions heated to the indicated temperature in a batch reactor reaching a pressure of circa 30 MPa. The area of the "bubbles" is proportional to the amount of solid carbon formed. The circles with dashed contours denote experiments *without* coke formation. *Source*: Müller 2012 [55]. Reproduced with permission of Elsevier.

This supports the hypothesis that fast, early reactions, requiring acidic conditions, determine the formation of tars and coke. Phenol and hydroquinone, proxies for lignin moieties, were unreactive towards tar and coke formation.

Besides forming small molecules that can access catalytic sites, liquefaction has a second important goal: the splitting off of the heteroatoms N, S, P bound in organic molecules such as proteins, chlorophyll, or phospholipids. These atoms are undesired both in the final products (bio-oil or SNG) and in the process itself due to corrosion, precipitation and plugging, and catalyst deactivation by fouling or poisoning. Ideally, they can be converted to inorganic salts, removed from the process and reused as fertilizers. Without any additive or catalyst, Brandenberger [56] showed that deamination of the protein BSA (bovine serum albumin) occurred in two rate regimes in subcritical water: a fast initial regime, followed by a much slower regime. After 10 min at 360 °C, about 60% of the nitrogen in the BSA was deaminated, while after 360 min the conversion had reached only 70–75%. A chemical analysis of the liquid phases revealed that pyridines, alkylamines, acetamide, aromatic amines, and alkyl-pyrrolidones had been formed, among others, which are persistent to deamination at these conditions.

For the catalytic production of SNG the incomplete deamination of the liquefied biomass is not a major problem. The remaining organic N-compounds should be gasified over Ru/C catalysts, releasing NH_3 which is not known to be a catalyst poison. Since ammonium salts are not stable at the conditions in the salt separator, they decompose to NH_3 which is fed into the catalytic reactor.

The mineralization of organic sulfur compounds to inorganic species such as sulfides, sulfites, or sulfates is a major challenge of any hydrothermal process, because organic S is highly undesired both in bio-oils and in SNG. Not much work has been done on hydrothermal desulfurization of biomass. Zöhrer and Vogel [35] have shown that circa 75% of the sulfur bound in the solid fraction of fermentation residues is released to form liquid S-containing products during liquefaction in a batch reactor at 406 °C, 31 MPa, 70 min. However, most of this sulfur is present in the methanol- and hexane-soluble products, that is, it is still organically bound. Circa 40% of the sulfur in the initial biomass was not recovered, most likely because it formed volatile compounds such as H_2S.

As opposed to deamination, removal of S from the feed to the catalytic reactor is of paramount importance for a commercial SNG process. Any remaining organic S-compound would decompose on the Ru/C catalyst and immediately poison the active sites. Research on sulfur removal from liquefied biomass feeds should thus be strongly fostered.

10.4 PSI'S CATALYTIC SNG PROCESS

10.4.1 Process Description and Layout

The biomass feedstock is prepared as a pumpable slurry, if needed by macerating and wet grinding. An organic dry matter content of at least 10 wt% should be targeted for an economical process. The slurry is pumped from the feed tank to the preheater by

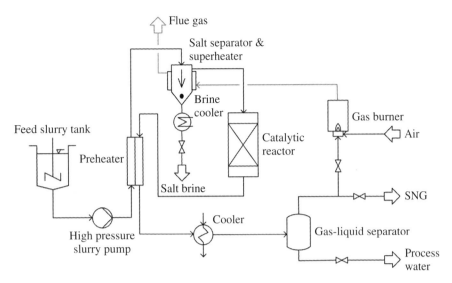

FIGURE 10.13 Simplified process flowsheet of PSI's catalytic hydrothermal SNG process.

a high pressure slurry pump (Figure 10.13). The discharge pressure is typically in the range of 25–35 MPa. A tubular heat exchanger preheats the slurry to a temperature close to the critical one, circa 350–360 °C. During heating up, the solid biomass fraction is liquefied, while the salts present in the feedstock are kept dissolved. The heating rate in the preheater is chosen such as to minimize the formation of secondary coke. The preheated and liquefied biomass stream enters then a separator vessel to precipitate and remove solid minerals and salts. This separator vessel is typically designed analogous to the reverse-flow vessel described in Subsection 10.3.2. This vessel is heated to process temperature indirectly by hot fumes from a gas burner. The salt brine, along with some liquefied biomass, is removed continuously from the bottom of the separator. The remaining feed stream, cleaned from solid matter, enters the catalytic reactor, typically designed as a fixed-bed reactor, filled with the Ru/C catalyst and operated in an adiabatic mode. The reactor does not need to be heated nor cooled because the overall heat of reaction is close to zero (see Section 10.3). The catalytic reactions take place at 400–450 °C and lead to a mixture of CH_4, CO_2, H_2, and some higher hydrocarbons such as ethane and propane. It is assumed that this gas mixture forms a homogeneous phase with the supercritical water inside the reactor.

After leaving the reactor, the hot effluent is cooled in the preheater heat exchanger. The water-gas mixture becomes now a two-phase mixture. It is further cooled to below 100 °C by an additional cooler. Here, care must be taken not to cool the mixture below circa 20 °C because at the high partial pressures of methane gas hydrates (or "clathrates") may form and cause a blockage of the cooler.

The two-phase mixture is then separated in a vessel under pressure. The gas and liquid phases are depressurized independently. One stream is a process water, containing most of the feedstock's nitrogen as ammonia, or ammonium carbonate. It will also contain some small amounts of residual salts not separated in the salt separator.

Depending on the level of feedstock conversion the catalytic reactor is designed for, some small amounts of residual organics are present in the process water. These organics do not normally cause any strong coloration or odor of the water.

A fraction of the gas phase from the G-L separator is fed to a gas burner to provide the hot fumes needed to heat the salt separator to process temperature. These fumes leave the salt separator still at a high temperature (ca. 550 °C) and, in a larger plant, may be used to co-produce some electricity. The raw SNG contains about 45–55 vol% CH_4, 40–50 vol% CO_2, and 3–4 vol% H_2, as well as 1–2 vol% of higher hydrocarbons. If needed, the content of CO_2 can be reduced by scrubbing it with pressurized water, or with other available processes such as pressure swing adsorption.

The experimental proof of the design concept with the simultaneous gasification and salt separation, as well as autothermal operation of the catalytic reactor, was reported by Schubert et al. [57].

10.4.2 Mass Balance

The mass balance of a plant processing 4.4 t/h of wet sewage sludge with 22% dry solids was calculated using Aspen plus [58]. The results are shown in Table 10.1.

The composition of the SNG after G-L phase separation was calculated to be: 46.2 vol% CH_4, 48.2 vol% CO_2, 4.4 vol% H_2, 1.0 vol% H_2O, and 0.2 vol% NH_3. The latter is an undesired component and its concentration in the SNG can be controlled by adjusting the pH in the G-L separator.

These data are based on idealized calculations assuming complete chemical equilibrium at the exit of the catalytic reactor and perfect salt separation. It should be used with caution, especially regarding the composition of the brine. One can see that the dry solids are mostly converted to SNG, together with a small fraction of the water. Considerable amounts of air are needed for the gas burner, resulting in a large

TABLE 10.1   Calculated Input and Output Mass flows for a Hypothetical Hydrothermal Methanation Plant Processing 4.4 t/h of Wet Sewage Sludge to SNG. Source: Vogel F, 2012 [58].

	Input (t/h)	Output (t/h)
Water (from sludge)	3.4	
Dry solids (from sludge)	1.0	
Air	2.2	
SNG		0.7
Process water		3.0
Flue gas		2.3
Brine – water		0.3
Brine – organics		0.07
Brine – inorganics		0.23
Total	6.6	6.6

amount of hot flue gas. In this simulation, the flue gas was used to preheat the combustion air. It may alternatively be used for generating electricity via a steam cycle, but this option would make sense only for larger plants.

The brine would typically contain most of the dissolved and suspended minerals from the feed. Although nitrogen will be mostly recovered in the process water as NH_4HCO_3, we found that some nitrogen may also be removed from the salt separator as solid magnesium ammonium phosphate (MAP or struvite [50]). An important aspect of the hydrothermal methanation is that nitrogen in the feed is not lost to a flue gas but recovered as a nutrient salt. This applies of course also to most of the other nutrient elements such as K, P, Mg, Ca, and so on. An element that has not been well studied at hydrothermal conditions is silicon. Its hydrothermal chemistry is complicated by the formation of oligomeric and polymeric species. Sewage sludge and other waste biomass are xenobiotic sources of organosilicon compounds, besides the natural sources from plants and aquatic biomass.

10.4.3 Energy Balance

The energy flow diagram for the same process simulation is shown in Figure 10.14. A considerable amount of energy is recirculated within the process itself in the feed heat exchanger (preheater). This piece of equipment is therefore key to a high thermal efficiency of the process. The high pressure slurry pump has only an insignificant power uptake because the feed is considered incompressible. Some losses occur with the hot brine and the heating value of the organics therein. For the well insulated, adiabatic catalytic reactor, some heat loss to the surrounding air has been assumed. The process water contains most of the feed nitrogen in the form of NH_3 (or NH_4HCO_3). Since nitrogen is included in the calculation of the feed's heating value, it also has to be accounted for in the balance, contributing circa 10% of the feed's heating value.

The thermal efficiency of this particular case is 66%, based on the lower heating value of the sewage sludge's dry solids and that of the SNG produced. This is considered an upper limit for this process when using a feedstock with a high minerals content such as sewage sludge, animal manure, or algae. It may be increased by thermally integrating the hot brine stream and by minimizing reactor heat losses. Also, with feedstocks low in minerals, such as wood or spent coffee grounds, thermal process efficiencies higher than 70% may be achieved (cf. Figure 10.16).

10.4.4 Status of Process Development at PSI

PSI's hydrothermal SNG process has been demonstrated at the laboratory scale with a throughput of 1 kg/h, using a number of model and real biomass feeds. As an important milestone, the simultaneous salt separation and methanation at autothermal conditions was successfully demonstrated by Schubert [37, 57]. Operational difficulties for processing solids-containing streams owing to the small orifices in such a small-scale plant make it necessary to scale up the technology within reasonable limits. In 2010, the company Hydromethan AG has been founded to realize

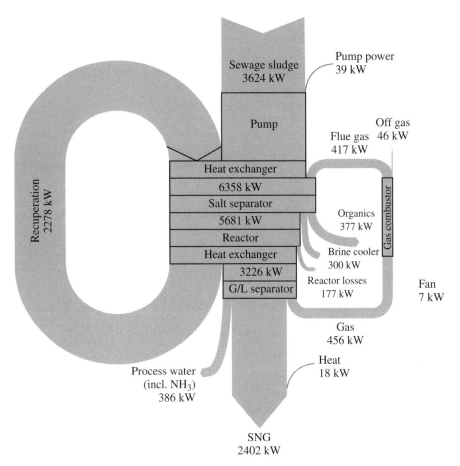

FIGURE 10.14 Energy flow diagram of PSI's catalytic hydrothermal SNG process for sewage sludge as feedstock. All enthalpies were calculated at 0.1 MPa and 25 °C [58]. *Source*: Vogel F. Hydrothermale Vergasung von Klärschlamm zu Methan. Final technical report. Villigen (Switzerland): Paul Scherrer Institut; 2012 Apr 2. Project no. 100831-06_HTV-KS.

the scaling-up and commercialization of the process. To this aim, a prototype salt separator vessel with an industrial design has been built and tested at a scale of 50 kg/h in KIT's pilot plant VERENA [59]. Furthermore, an engineering study for a demonstration plant with a capacity of 1 t/h has been performed. The main conclusion was that such a technology can be realized at an industrial scale in a safe and economically viable way.

In parallel, the mobile hydrothermal methanation plant KONTI-C with a capacity of 1 kg/h has been built at PSI to allow for longer campaigns with a variety of real feeds, in particular algae (Figure 10.15). It was operated successfully on an algae feedstock for about 100 h of continuous operation in August 2014.

A similar concept was realized at PNNL with a trailer-mounted reactor system for their subcritical catalytic gasification process with a capacity of 10 kg/h [34].

FIGURE 10.15 PSI's mobile hydrothermal SNG plant KONTI-C.

10.4.5 Comparison to other SNG Processes

PSI's catalytic hydrothermal SNG process has a number of features such as:

- Efficient production of SNG from wet biomass streams (sludges) with a content in organic matter as low as circa 10 wt%.
- No need for drying the feedstock; mechanical dewatering is sufficient.
- High thermal process efficiencies for biomass to SNG of 60–70% (typical) and beyond 70% for polygeneration of SNG, electricity, and heat.
- Feedstock flexibility: The process can be designed to accommodate a relatively wide range of feedstock compositions regarding water content, mineral content, chemical composition, viscosity, pH. The most severe limits concern the pumpability (feeding), the sulfur content (catalyst lifetime), and the chloride content and pH (corrosion).
- Reduced effort for SNG purification due to the absence of dust and tars, and only very low levels of HCl, NH_3, alkali, SO_2, H_2S, and so on, because these compounds mostly remain dissolved in the process water or end up in the salt brine.
- The SNG produced can be made available at high pressure, avoiding additional compression losses in downstream processing.
- Recovery of the nutrients, including N, P, K, with high efficiency, to be used as fertilizers.
- No formation of a solid byproduct (e.g. coke or char).
- A very favorable environmental performance.

- Small plant footprint.
- Scalable from a few MW to circa 20 MW of thermal input.

As with any novel bioenergy process, the hydrothermal production of SNG presents some challenges:

- Increased risk of corrosion due to the aqueous electrolyte environment.
- Higher capital and maintenance cost due to the high pressure equipment and materials of construction[2].
- Potentially higher risk of societal acceptance due to high pressures and temperatures.
- Long-term performance at industrial scale not yet proven.
- Only very limited industrial experience available from only a few companies for the design, engineering, construction, and operation of such plants.
- Organic fraction in the salt brine must be separated from the salts before their use as fertilizer. If it cannot be recycled within the process, this organic fraction will have to be disposed of, creating additional costs.
- Heavy metals will end up in the salt brine, making its work-up to a fertilizer more costly.

10.4.5.1   Process Economics   Gassner et al. [18] estimated the specific investment costs for a catalytic hydrothermal methanation plant of the 20 MW class for a number of different feedstocks. The general rule is that the lower the organic content of the feedstock, that is, the higher the moisture and mineral contents, the lower the SNG efficiency (Figure 10.16). The specific investment costs depend on the desired SNG efficiency: up to a certain value the specific investment costs remain quite insensitive to the efficiency. When a certain efficiency level is reached, a further increase of this level is very costly. Thus there is an optimum for every feedstock, and no general value can be given. For the seven feedstocks studied, the lowest specific investment costs (per kW) were around US$ 600. This would correspond to US$ 12 M for a 20 MW plant (based on the thermal biomass input) and can be regarded as a lower bound for an easy feedstock such as spent coffee grounds. A study conducted for a plant with a capacity of 3.6 MW of sewage sludge, which is a more challenging feed than spent coffee grounds, estimated the investment costs to be circa CHF 22 M [60].

10.4.5.2   Ecological Performance   An LCA study performed by Luterbacher et al. [61] showed the hydrothermal methanation process to have a very low environmental impact. In fact, if manure is used as feedstock, a net reduction of GHG emissions is achieved because it avoids the emission of N_2O and CH_4 from the fermentation residues spread on the field as fertilizer. Another study, based on the same LCI data,

---

[2]This is alleviated somewhat by the relatively simple process design with only few key process units. For example, there is no need for an intermediate gas cleaning step.

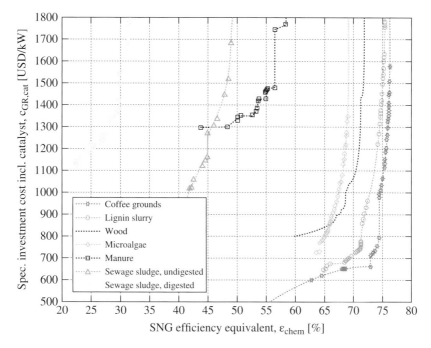

FIGURE 10.16 Specific investment cost and overall process efficiency without power recovery from the high pressure vapor phase for PSI's catalytic hydrothermal methanation process. Reproduced from [18] with permission.

calculated the total environmental impact (UBP, 2006) and compared it to a number of other biofuel pathways for Switzerland. Biomethane from manure or wood chips via the hydrothermal methanation route ranked among the best options [62]. For the SunCHem process based on algal biomass, Brandenberger et al. [6] found the combination of raceway ponds and hydrothermal methanation to exhibit an energy return on energy invested (EROEI) of 1.8–5.8, depending on the assumptions regarding future developments in algae productivity.

10.5   OPEN QUESTIONS AND OUTLOOK

As with any catalytic biomass process, it is a long way to a full-scale demonstration under "real life" conditions. Catalysis at hydrothermal conditions is a largely underresearched topic, and thus, most concepts known from gas phase catalytic processes, such as steam reforming, had to be adapted and restudied first under hydrothermal conditions [8]. This process is not yet completed, and we keep discovering new aspects of this interesting topic in our daily work. The aspect of the behavior of minerals at hydrothermal conditions bears many parallels to geochemical and geothermal situations. On the other hand, our findings related to SNG production may be of value to the geoprocess field as well. The management of the hot brine from a geothermal production well in a surface power plant is a key topic in geothermal energy generation. Our findings and results on continuous salt separation from supercritical water may hopefully inspire our colleagues in the geothermal field. A closer cooperation could lead to fruitful discussions and novel ideas in both areas, hydrothermal processing of biomass and geothermal energy.

Several aspects of catalytic hydrothermal SNG production from biomass have not yet been studied in enough detail. Besides catalysis and minerals behavior, hydrodynamic flow simulations at supercritical water conditions relevant to SNG production have only been tackled by a few groups. The non-ideal phase behavior of supercritical fluids and mixtures and the steep gradients in physical properties near the critical point make these computations time-consuming. The limited accuracy of current equations of state, especially for mixtures, are a big obstacle in using such simulations as a predictive tool. Further phenomena that are not well described by physical models for practical application at hydrothermal conditions include salt nucleation, precipitation kinetics, and reaction kinetics for mixed feeds. A promising approach may be the use of a distributed activation energy model (DAEM), based on carbon conversion. Although it was applied to supercritical water oxidation [63], it has not yet been applied within the context of heterogeneous catalysis at these conditions. Other topics that need further investigation include the recovery of nutrients from salt brine as a marketable fertilizer, the detailed analysis of trace compounds in process water and SNG, and the deep desulfurization of the feed to the catalytic reactor. The latter is a topic of current research in our group at PSI, and we have obtained first promising results with a commercial sulfur trapping material.

A feedstock meriting a deeper study for its suitability to hydrothermal methanation is black liquor, produced in large quantities from the Kraft pulping of wood. This is a very challenging feedstock due to its high sulfur and minerals content, but its conversion to SNG may potentially replace the currently used black liquor boilers if the pulping chemicals can be recovered in a form suitable for recycling within the Kraft process.

From a process point of view, the coupling of hydrothermal gasification to a high temperature fuel cell (MCFC or SOFC) may result in a very efficient, small- to medium-scale power generation process with estimated electrical efficiencies of up to 43% [8].

REFERENCES

[1] Kaltschmitt M, Hartmann H, Hofbauer H (eds). *Energie aus Biomasse, 2nd edn*. Springer, Berlin; 2009.

[2] Tester JW, Drake EM, Driscoll MJ, Golay MW, Peters WA. *Sustainable Energy – Choosing Among Options*. The MIT Press, Cambridge, MA; 2005.

[3] Carlsson AS, van Beilen JB, Möller R, Clayton D. *Outputs from the EPOBIO Project*. CPL Press, London; 2007.

[4] Stucki S, Vogel F, Ludwig C, Haiduc AG, Brandenberger M. Catalytic gasification of algae in supercritical water for biofuel production and carbon capture. *Energy and Environmental Science* 2: 535–541; 2009.

[5] Haiduc AG, Brandenberger M, Suquet S, Vogel F, Bernier-Latmani R, Ludwig C. SunCHem: an integrated process for hydrothermal production of methane from microalgae and CO_2 mitigation. *Journal of Applied Phycology* **21**: 529–541; 2009.

[6] Brandenberger M, Matzenberger J, Vogel F, Ludwig C. Producing synthetic natural gas from microalgae via supercritical water gasification: A techno-economic sensitivity analysis, *Biomass and Bioenergy* **51**: 26–34; 2013.

[7] Vogel F, Waldner MH, Rouff AA, Rabe S. Synthetic natural gas from biomass by catalytic conversion in supercritical water. *Green Chemistry* **9**(6): 616–619; 2007.

[8] Vogel F. *Catalytic Conversion of High-Moisture Biomass to Synthetic Natural Gas in Supercritical Water*. In: Anastas P, Crabtree R (eds) Handbook of Green Chemistry, Vol. 2, Heterogeneous Catalysis. Wiley-VCH, Weinheim, pp. 281–324; 2009.

[9] Peterson AA, Vogel F, Lachance RP, Fröling M, Antal MJ, Tester JW. Thermochemical biofuel production in hydrothermal media: A review of sub- and supercritical water technologies. *Energy and Environmental Science* **1**(1): 32–65; 2008b.

[10] Modell M, Reid C, Amin SI. *Gasification Process*. US Patent 4 113 446; 1978.

[11] Modell M. *Gasification and Liquefaction of Forest Products in Supercritical Water*. In: Overend RP, Milne TA, Mudge LK (eds) Fundamentals of Thermochemical Biomass Conversion. Elsevier Applied Science, London, pp. 95–120; 1985.

[12] Elliott DC, Sealock LJ Jr. *Low Temperature Gasification of Biomass under Pressure*. In: Overend RP, Milne TA, Mudge LK (eds) Fundamentals of thermochemical biomass conversion. Elsevier Applied Science, London, pp. 937–950; 1985.

[13] Elliott DC. Catalytic hydrothermal gasification of biomass. *Biofuels Bioproduction Biorefining* **2**: 254–265; 2008.

[14] Elliott DC, Hart TR, Neuenschwander GG, Rotness LJ, Olarte MV, Zacher AH. Chemical Processing in High-Pressure Aqueous Environments. 9. Process Development for Catalytic Gasification of Algae Feedstocks. *Industrial Engineering and Chemical Research* **51**: 10768–10777; 2012.

[15] Kruse A, Vogel F, van Bennekom J, Venderbosch R. *Biomass Gasification in Supercritical Water*. In: Knoef HAM (ed.) Handbook Biomass Gasification, 2nd edn. Biomass Technology Group, Berlin, pp. 251–280; 2012.

[16] Davda RR, Shabaker JW, Huber GW, Cortright RD, Dumesic JA. A review of catalytic issues and process conditions for renewable hydrogen and alkanes by aqueous-phase reforming of oxygenated hydrocarbons over supported metal catalysts. *Applied Catalysis B: Environmental* **56**: 171–186; 2005.

[17] Gassner M, Vogel F, Heyen G, Maréchal F. Process design of SNG production by hydrothermal gasification of waste biomass: Thermoeconomic process modelling and integration. *Energy and Environmental Science* **4**: 1726–1741; 2011a.

[18] Gassner M, Vogel F, Heyen G, Maréchal F. Optimal process design for the polygeneration of SNG, power and heat by hydrothermal gasification of waste biomass: Process optimisation for selected substrates. *Energy and Environmental Science* **4**: 1742–1758; 2011b.

[19] Onwudili JA, Williams PT. Role of sodium hydroxide in the production of hydrogen gas from the hydrothermal gasification of biomass. *International Journal of Hydrogen Energy* **34**: 5645–5656; 2009.

[20] D'Jesus P, Artiel C, Boukis N, Kraushaar-Czarnetzki B, Dinjus E. Influence of educt preparation on gasification of corn silage in supercritical water. *Industrial Engineering and Chemical Research* **44**: 9071–9077; 2005.

[21] Waldner MH, Vogel F. Renewable production of methane from woody biomass by catalytic hydrothermal gasification. *Industrial Engineering and Chemical Research* **44**(13): 4543–4551; 2005.

[22] Peterson AA, Dreher M, Wambach J, Nachtegaal M, Dahl S, Nørskov JK, Vogel F. Evidence of scrambling over ruthenium-based catalysts in supercritical-water gasification. *ChemCatChem* **4**(8): 1185–1189; 2012.

[23] Czekaj I, Pin S, Wambach J. Ru/active carbon catalyst: improved spectroscopic data analysis by density functional theory. *Journal of Physical Chemistry C* **117**: 26588–26597; 2013.

[24] Rabe S, Nachtegaal M, Ulrich T, Vogel F. Towards understanding the catalytic reforming of biomass in supercritical water. *Angewandte Chemie International Edition* **49**: 6434–6437; 2010.

[25] Waldner M, Krumeich F, Vogel F. Synthetic natural gas by hydrothermal gasification of biomass Selection procedure towards a stable catalyst and its sodium sulfate tolerance. *Journal of Supercritical Fluids* **43**: 91–105; 2007.

[26] Dreher M, Johnson B, Peterson AA, Nachtegaal M, Wambach J, Vogel F. Catalysis in supercritical water: Pathway of the methanation reaction and sulfur poisoning over a Ru/C catalyst during the reforming of biomolecules. *Journal of Catalysis* **301**: 38–45; 2013.

[27] Elliott DC, Hart TR, Neuenschwander GG. Chemical processing in high-pressure aqueous environments. 8. Improved catalysts for hydrothermal gasification. *Industrial Engineering and Chemical Research* **45**: 3776–3781; 2006.

[28] Peng G, Steib M, Gramm F, Ludwig C, Vogel F. Synthesis factors affecting the catalytic performance and stability of Ru/C catalysts for supercritical water gasification. *Catalysis Science and Technology* **4**(9): 3329–3339; 2014.

[29] Crerar D, Hellmann R, Dove P. Dissolution kinetics of albite and quartz in hydrothermal solutions. *Chemical Geology* **70**(1/2): 77; 1988.

[30] Yu J, Savage PE. Catalyst activity, stability, and transformations during oxidation in supercritical water. *Applied Catalysis B: Environment* **31**: 123–132; 2001.

[31] Zöhrer H, Mayr F, Vogel F. Stability and performance of ruthenium catalysts based on refractory oxide supports in supercritical water conditions. *Energy and Fuels* **27**(8):4739–4747; 2013.

[32] Wambach J, Schubert M, Döbeli M, Vogel F. Characterization of a spent Ru/C catalyst after gasification of biomass in supercritical water. *Chimia* **66**(9): 706–711; 2012.

[33] Bagnoud-Velásquez M, Brandenberger M, Vogel F, Ludwig C. Continuous catalytic hydrothermal gasification of algal biomass and case study on toxicity of aluminum as a step toward effluents recycling. *Catalysis Today* **223**: 35–43; 2014.

[34] Elliott DC, Neuenschwander GG, Hart TR, Butner RS, Zacher AH, Engelhard MH, Young JS, McCready DE. Chemical processing in high-pressure aqueous environments. 7. Process development for catalytic gasification of wet biomass feedstocks. *Industrial Engineering and Chemical Research* **43**:1999–2004; 2004.

[35] Zöhrer H, Vogel F. Hydrothermal catalytic gasification of fermentation residues from a biogas plant. *Biomass and Bioenergy* **53**: 138–148; 2013.

[36] Osada M, Hiyoshi N, Sato O, Arai K, Shirai M. Effect of sulfur on catalytic gasification of lignin in supercritical water. *Energy and Fuels* **21**: 1400–1405; 2007.

[37] Schubert M. *Catalytic Hydrothermal Gasification of Biomass – Salt Recovery and Continuous Gasification of Glycerol Solutions*. Dissertation, Thesis no. 19039, ETH Zürich, Switzerland; 2010.

[38] Müller JB. *Hydrothermal Gasification of Biomass – Investigation on Coke Formation and Continuous Salt Separation with Pure Substrates and Real Biomass.* Dissertation, Thesis no. 20458, ETH Zürich, Switzerland; 2012.

[39] Dreher M. *Catalysis under Extreme Conditions: In Situ Studies of the Reforming of Organic Key Compounds in Supercritical Water.* Dissertation, Thesis no. 21531, ETH Zürich, Switzerland; 2013.

[40] Waldner M. *Catalytic Hydrothermal Gasification of Biomass for the Production of Synthetic Natural Gas.* Dissertation, Thesis no. 17100 ETH Zürich, Switzerland; 2007.

[41] Dreher M, Steib M, Nachtegaal M, Wambach J, Vogel F. On-stream regeneration of a sulfur-poisoned Ru/C catalyst under hydrothermal gasification conditions. *ChemCatChem* **6**: 626–633; 2014.

[42] Reimer J, Vogel F. High pressure differential scanning calorimetry of the hydrothermal salt solutions K_2SO_4-Na_2SO_4-H_2O and K_2HPO_4-H_2O. *RSC Advances* **3**:24503–24508; 2013.

[43] Valyashko VM. *Phase Equilibria in Binary and Ternary Hydrothermal Systems.* In: Valyashko VM (ed.) Hydrothermal Experimental Data. John Wiley & Sons Ltd, Chichester, pp. 1–133; 2008.

[44] Schubert M, Aubert J, Müller JB, Vogel F. Continuous salt precipitation and separation from supercritical water. Part 3: Interesting effects in processing type 2 salt mixtures. *Journal of Supercritical Fluids* **61**(1): 44–54; 2012.

[45] Killiliea WR, Hong GT, Swallow KC, Thomason TB. *Supercritical Water Oxidation: Microgravity Solids Separation*, SAE Paper 881038, SAE, London; 1988.

[46] Schubert M, Regler JW, Vogel F. Continuous salt precipitation and separation from supercritical water. Part 1: Type 1 salts. *Journal of Supercritical Fluids* **52**(1): 99–112; 2010a.

[47] Peterson AA, Tester JW, Vogel F. Water-in-water tracer studies of supercritical-water reversing jets using neutron radiography. *Journal of Supercritical Fluids* **54**(2): 250–257; 2010.

[48] Peterson AA, Vontobel P, Vogel F, Tester JW. In situ visualization of the performance of a supercritical-water salt separator using neutron radiography. *Journal of Supercritical Fluids* **43**: 490–499; 2008a.

[49] Schubert M, Regler JW, Vogel F. Continuous salt precipitation and separation from supercritical water. Part 2: Type 2 salts and mixtures of two salts. *Journal of Supercritical Fluids* **52**(1): 113–124; 2010b.

[50] Zöhrer H, De Boni E, Vogel F. Hydrothermal processing of fermentation residues in a continuous multistage rig – operational challenges for liquefaction, salt separation, and catalytic gasification. *Biomass and Bioenergy* **65**: 51–63; 2014.

[51] Mosteiro Romero M, Vogel F, Wokaun A. Liquefaction of wood in hot compressed water. Part 1: Experimental results. *Chemical Engineering Science* **109**: 111–122; 2014a.

[52] Mosteiro Romero M, Vogel F, Wokaun A. Liquefaction of wood in hot compressed water. Part 2: Modeling of particle dissolution. *Chemical Engineering Science* **109**: 220–235; 2014b.

[53] Bobleter O, Binder H. Dynamischer hydrothermaler Abbau von Holz. *Holzforschung* **34**: 48–51; 1980.

[54] Masten DA, Foy BR, Harradine DM, Dyer RB. In-situ Raman spectroscopy of reactions in supercritical water. *Journal of Physical Chemistry* **97**(33): 8557–8559; 1993.

[55] Müller JB, Vogel F. Tar and coke formation during hydrothermal processing of glycerol and glucose. Influence of temperature, residence time and feed concentration. *Journal of Supercritical Fluids* **70**(10): 126–136; 2012.

[56] Brandenberger M. *Process Development for Catalytic Supercritical Water Gasification of Algae Feedstocks*. Dissertation, EPFL, Lausanne, Switzerland; 2014.

[57] Schubert M, Müller JB, Vogel F. Continuous hydrothermal gasification of glycerol mixtures: Autothermal operation, simultaneous salt recovery, and the effect of K_3PO_4 on the catalytic gasification. *Industrial and Engineering Chemistry Research* **53**: 8404–8415; 2014.

[58] Vogel F. *Hydrothermale Vergasung von Klärschlamm zu Methan. Final Technical Report.* Project no. 100831-06_HTV-KS. Paul Scherrer Institut, Villigen, Switzerland; 2012.

[59] Boukis N, Galla U, D'Jesus P, Müller H, Dinjus E. *Gasification of Wet Biomass in Supercritical Water. Results of Pilot Plant Experiments*. Proceedings 14th European Biomass Conference, Paris, France; 2005.

[60] Vogel F, Heusser P, Lemann M, Kröcher O. Mit Hochdruck Biomasse zu Methan umsetzen. *Aqua and Gas* **4**: 30–35; 2013.

[61] Luterbacher JS, Fröling M, Vogel F, Maréchal F, Tester JW. Hydrothermal gasification of waste biomass: Process design and life cycle assessment. *Environmental Science and Technology* **43**(5): 1578–1583; 2009.

[62] Faist Emmenegger M, Gmünder S, Reinhard J, Zah R, Nemecek T, Schnetzer J, Bauer C, Simons A, Doka G. *Harmonisation and Extension of the Bioenergy Inventories and Assessment. Final Report*. Bundesamt für Energie, Bern, Switzerland; 2012.

[63] Vogel F, Smith KA, Tester JW, Peters WA. Engineering kinetics for hydrothermal oxidation of hazardous organic substances. *AIChE Journal* **48**(8):1827–1839; 2002.

11

AGNION'S SMALL SCALE SNG CONCEPT

Thomas Kienberger and Christian Zuber

The SNG concept offered by the German bioenergy supplier agnion energy GmbH [1] is based on agnion's allothermal biomass gasification technology, the heatpipe reformer. This technology, well described in the literature [2–4], is particularly designed for small and medium scale syngas generation in decentralized applications. In relation to other biomass gasification technologies applicable for SNG production, the thermal firing capacity of the heatpipe reformer is for instance around 10 times smaller than the Güssing FICFB technology [5] or around 100 times smaller than the capacity of entrained flow gasification plants [6]. Due to the plant size, the paradigms for decentralized power plant engineering must be fulfilled. In comparison to large scale plants (>100 MW_{SNG}) in the power range of agnion's plants (1–5 MW_{SNG}), no economy of scale effects can be taken into account. In order to reach capital expenses (CAPEX) which allow being competitive in decentralized operations, a simplified plant design becomes necessary. At this juncture agnion operates commercial CHP units at sites in Grassau, Germany, and Auer, Italy. A pilot plant installation, used as a test bed among others also used for SNG research, is located in Pfaffenhofen/Ilm, Germany. Basic research is done by the subsidiary agnion Highterm research based at the Technical University Graz, Austria. All measurements shown here reflect results obtained in the Pfaffenhofen pilot plant.

In order to produce SNG from biomass, four process steps need to be carried out (Figure 11.1a). In the first step, the solid feedstock is converted into a hydrogen-rich synthesis gas via thermochemical gasification. Besides the permanent gases H_2, CO,

Synthetic Natural Gas from Coal, Dry Biomass, and Power-to-Gas Applications, First Edition.
Edited by Tilman J. Schildhauer and Serge M.A. Biollaz.
© 2016 John Wiley & Sons, Inc. Published 2016 by John Wiley & Sons, Inc.

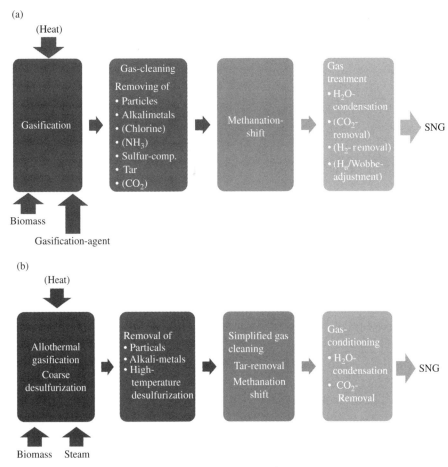

FIGURE 11.1   (a) Fundamental process chain for SNG production. (b) Agnion's simplified biomass to SNG process.

CO_2, CH_4, and H_2O, sulfur components like H_2S, COS, and thiophene as well as higher hydrocarbons, referred to as biomass tar, are formed according to Equation (11.1). The fuel alkali components like Na and K vaporize. Depending on the gasification process and water content of the fuel, the H_2O vapor content of the producer gas can roughly be estimated between 10 and 50 vol%. [5, 7, 8].

$$CH_xO_y + v_1 \times H_2O \rightarrow v_2 \times CO + v_3 \times H_2 + v_4 \times CO_2 + v_5 \times H_2O + v_6 \times CH_4$$
$$+\, tar + sulfur\ components + \dots \tag{11.1}$$

In the gas-cleaning step subsequent to the gasification step, the major gas impurities such as alkali components, sulfur, and tar must be removed. The third step,

Equation (11.2), represents the catalytic conversion of the synthesis gas to a methane-rich raw SNG [9]:

$$v_1 \cdot CO + v_2 \cdot H_2 + v_3 \cdot CO_2 + v_4 \cdot H_2O + v_5 \cdot CH_4 \rightleftarrows v_6 \cdot CH_4 + v_7 \cdot H_2O + v_8 \cdot CO_2 \tag{11.2}$$

In order to meet the regulations, for example in Austria [10] or Germany [11], for feeding the gas into the natural gas grid, the raw SNG has to be purified. Basically therefore the gas has to be dried and CO_2 and H_2 traces have to be removed. Figure 11.1a shows the fundamental process chain for producing SNG from biomass.

As depicted in Figure 11.1b, agnion's simplified biomass to SNG fulfills all necessary steps for SNG production. In order to reduce the number of process units and therefore the investment costs of the SNG plant, the idea behind the concept is to combine as many process functions as possible.

For the first process step – allothermal gasification – agnion uses heatpipe reformer technology [4]. The gasification reactor, designed as a bubbling fluidized bed, gives the opportunity to operate in such a way that a discrete bed of biomass char is formed above the fluidized sand bed. After devolatilization during the gasification process, the inserted fuel becomes biomass char. The particles in this char bed shrink gradually due to erosion till they get ejected from the gasification reactor. Via a filtration system they are separated from the syngas stream and transferred to the combustion reactor in order to be burned. The char partially contributes to the heating of the allothermal gasification.

Appropriate operation leads to a maximization of the charcoal-specific BET surface area. Surfaces in a range from 320 to 680 m²/g were measured [12, 13]. Biomass char particles with specific BET surface areas in that dimension are referred to in the literature as activated charcoal [14]. Using activated charcoal for the removal of H_2S at room temperature is a state of the art process, for instance used for the desulfurization of gases from anaerobic digestion. High temperature desulfurization on activated charcoal is not applied that intensively, but is also well described in the literature [15, 16]. Puri et al. [15] specify that there are two mechanisms of high temperature sulfur adsorption on activated charcoal:

- Physisorption on empty spaces on the activated charcoal;
- Chemisorptive exchange of sulfur and oxygen atoms at the activated charcoal's surface.

Cal et al. [16] and Garcia et al. [17] investigated the temperature influence on high temperature H_2S removal. Garcia et al. see no significant influence between 600 and 800 °C.

The described mechanisms, combined with the heatpipe reformer's discrete char bed, allow a coarse desulfurization inside the gasification reactor, ending up with a low sulfur content in the syngas, compared to other fluidized bed systems. Table 11.1 shows the mean composition of heatpipe reformer's syngas.

The measurement values shown in Table 11.1 represent mean values. Figure 11.2a shows the dry syngas composition of the heatpipe reformer over an interval of 190 h.

TABLE 11.1 Mean Syngas Composition (Operation at Agnion's Pfaffenhofen Pilot Plant with DIN Wood Pellets, Water Content 6%, S Content 0.03 wt%).

Component	Gas Composition [vol%]	
	Dry	Wet
y_{N2}	7.5	4.2
y_{CO2}	21.9	12.4
y_{CO}	18.8	10.6
y_{CH4}	8.9	5.0
y_{H2}	42.9	24.3
y_{H2O}	—	43.4
H_2/CO	~2.3	
Tar content	4–8 g/Nm³	
H_2S	24 ppm	
Organic sulfur	~1 ppm	

Figure 11.2b depicts the same syngas in a ternary plot of the system C–H–O. In the CHO ternary plot the elementary balance of the C, H, and O atoms of the biomass fuel and feed water is normalized to 100%. The plot contains temperature dependent boundary lines. In areas above these lines solid carbon can occur thermodynamically due to the Boudouard reaction. By the addition of water, the syngas' CHO composition can be shifted below the boundary line. Operation outside of the thermodynamic carbon deposition boundary is fundamental to limit coking on the methanation catalyst. Even if this is guaranteed, coke depositions on the catalyst can occur due to kinetic reasons. According to the definition by Seemann et al. [18] and Czekaj et al. [19], so-called carbon depositions are formed due to syngas compositions within the carbon boundaries, whereas coke depositions result from the decomposition of hydrocarbons on the catalyst surface.

For decentralized biomass SNG production processes, a reduced number of process units is crucial. Therefore gas cleaning cannot be performed at room temperature or at low temperature levels as is done in medium to large scale units. Cooling the gas down for instance to room temperature for removing particulate matter, sulfur, and tar and heating it again to the temperature needed for methanation would imply tremendous effort and therefore leads to a much too complicated process design. Also high temperature gas cleaning, which means gas cleaning at temperatures of 800 °C and higher, does not seem ideal for the small scale approach. At those temperatures basically two issues lead to the fact that the paradigms for decentralization are not fulfilled:

- High temperatures lead to high volume flows. High volume flows cause high pressure drops and/or complex apparatus designed specially for high temperature applications, very expensive.

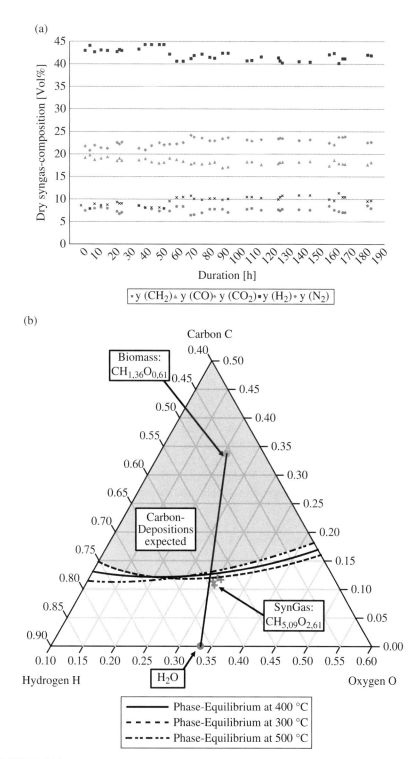

FIGURE 11.2 (a) Dry syngas composition (operation at agnion's Pfaffenhofen pilot plant with DIN wood pellets, water content 6%, S content 0.03 wt%). (b) Carbon boundaries in the CHO ternary plot (operation at agnion's Pfaffenhofen pilot plant with DIN wood pellets, water content 6%, S content 0.03 wt%).

- At temperatures higher than 800 °C, the biomass alkali components are partially vaporized; and the remaining solid particles begin to change their adhesive surface properties. The elutriated matter from the gasification reactor can therefore lead to distinct fouling in subsequent high temperature reactors.

For small scale application, agnion's approach is to apply a SNG generation system at medium temperatures. Temperatures between 300 and 600 °C allow a simple apparatus design and avoid the above-mentioned problems with sticky alkali components.

In the proposed system, downstream from the heatpipe reformer's gasification reactor, the syngas is cooled in a steam generator to about 300 °C. The generated steam is used as a fluidization as well as a gasification medium in the gasification reactor. A mid temperature particle filter, subsequently installed, removes ash, bed material, and as mentioned before, biomass char from the gas.

Considering sulfur removal for medium temperature applications, sulfur adsorption on various metal oxides is state of the art [20, 21], see Equations 11.3 and 11.4.

$$Me_{v_1}O_{v_2} + v_1 \cdot H_2S + (v_2 - v_1) \cdot H_2 \rightleftarrows v_1 \cdot MeS + v_2 \cdot H_2O \tag{11.3}$$

$$Me_{v_1}O_{v_2} + v_1 \cdot COS + (v_2 - v_1) \cdot CO \rightleftarrows v_1 \cdot MeS + v_2 \cdot CO_2 \tag{11.4}$$

Medium temperature desulfurization is commonly used in fixed bed reactors at temperatures between 300 and 500 °C. Meng et al. [20] consider zinc and copper oxide based adsorbents as most suitable in this temperature range. In the observed temperature range, especially with high H_2O contents in the gas, one must consider restrictions in the sulfur adsorption equilibrium [22]. The higher the water content is, the less H_2S can be adsorbed in the chemical equilibrium. Besides the main reactions, unwanted side reactions such as the Boudouard reaction, the shift reaction (10), the methanation reaction (9), as well as Fischer–Tropsch reactions may also occur [23].

Considering ZnO as the adsorbent, hydrogen sulfide (H_2S) can be removed under the conditions described in Table 11.1 down to the thermodynamic equilibrium concentration of around 1.3 ppm at 300 °C. Thereby mainly the sulfidation reaction according to Equation (11.5) takes place.

$$ZnO + H_2S \rightleftarrows ZnS + H_2O \tag{11.5}$$

Removing significant amounts of organic sulfur components is not possible using ZnO. In the designated temperature range (temperature downstream from the particle filter is at 300 °C) a certain carbonization of ZnO [Equation (11.6)] can occur, if CO_2 is present in the gas.

$$ZnO + CO_2 \rightleftarrows ZnCO_3 \tag{11.6}$$

$ZnCO_3$ formation leads to a reduced sulfur adsorption capacity. The adsorption capacity depends also on the synthesis gas water content. For the application in mind, capacities between 0.18 and 0.32 g S/g ZnO seem to be possible.

The other feasible adsorbents are copper based. They exist either in form of copper(I)oxide, Cu_2O, or in form of copper(II)oxide, CuO. H_2S adsorption follows Equations (11.7) and (11.8); the mechanism of adsorption is again chemisorption.

$$Cu_2O + H_2S \rightleftarrows Cu_2S + H_2O \qquad (11.7)$$

$$2 \cdot CuO + H_2S + H_2 \rightleftarrows Cu_2S + 2 \cdot H_2O \qquad (11.8)$$

With copper oxide based absorbent materials, one can reach H_2S concentrations far below 1 ppm [20]. Furthermore it should be possible to remove organic sulfur components. If hydrogen and/or carbon monoxide are present in the feed gas, a reduction of copper oxide to elementary copper becomes possible. Elementary copper does not provide good adsorption behavior. In order to enhance stability and/or reactivity, most copper based adsorbents are doped with other metals, such as titanium, iron, zinc, or chromium. For agnion's applications, a theoretic adsorption capacity of 0.22 g S/g CuO can be reached.

In agnion's process development, both adsorbent materials were tested. As expected, with a combined ZnO/CuO adsorbent used in a fixed bed reactor, we measured desulfurization deep into the ppb range. With increasing duration of the experiment, the H_2S content in the synthesis gas rose. After around 95 h, the H_2S content reached a value which corresponded to the value measured with ZnO only. This is explained by the fact that the high H_2 content in the synthesis gas acts as a reducing agent for CuO. Therefore CuO desulfurization with the tested adsorbent materials, despite the initially low sulfur content downstream from the adsorber, cannot be used in the process.

By the use of pure ZnO as H_2S adsorber in a fixed bed reactor, one can reach thermodynamic equilibrium concentrations for H_2S as long as the operation instructions of the adsorbent supplier are fulfilled. In the case of the agnion process, one ends up with a H_2S concentration of around 1.3 ppm in the off gas of the adsorber. The restrictions due to the high water and CO_2 content of the syngas seem not to lead to a significant reduction of the adsorption capacity. With the conditions of the agnion process, an adsorption capacity of approximately 0.06 g S/g ZnO can be measured. Concerning organic sulfur, the gas contains the main components thiophene and benzothiophene. According to Table 11.2, the influence of ZnO on those components is within measurementtolerance.

In order to reduce the amount of organic sulfur, hydrodesulfurization (HDS) should be considered. HDS is a catalytic process which converts organic sulfur components by means of hydrogen to H_2S. As previously mentioned, H_2S removal is rather simple. HDS is industrially applied, mainly for the desulfurization of liquid fuel. In those applications reaction pressures of 80 bar (hydrogen partial pressure 30 bar) are usual. The conditions for HDS in biomass applications are much milder. Rabou et al. [7] for instance did HDS tests on biomass based syngas at atmospheric pressure and 350 °C. They achieved a significant hydrogenation of thiols and olefins. Agnion's system also includes a HDS step using a commercial cobalt molybdenum oxide catalyst. At optimized process conditions, the HDS catalyst reduces the concentration of organic sulfur components significantly (cf. Table 11.2).

TABLE 11.2   Sulfur Reduction on ZnO (Operation at Agnion's Pfaffenhofen Pilot Plant with DIN Wood Pellets, Water Content 6%, S Content 0.03 wt%). Sampling by means of SPA, Measurement with GC-PFPD. LoD = Level of Detection.

	Concentration [$ppm_{V,dry}$]		
Component	Before ZnO	Downstream ZnO without HDS	Downstream ZnO with HDS
H_2S	22–24	0.9	0.9
Thiophene	0.7–1.0	0.7–1.0	0.3–0.8
Benzothiophene	<0.2	<0.2	<0.2
2-Methylthiophene	Below LoD to 0.05	Below LoD	Below LoD

At the temperature level of desulfurization (300 °C), the total tar load of the agnion heatpipe reformer is in gaseous phase, no tar condensation takes place. This temperature is also the inlet temperature of agnion's methanation reactor. This fixed bed reactor, filled with a commercial nickel based catalyst, basically contains two stages. The purpose of the adiabatic first stage is to convert the biomass tar into combustible products (mainly CO and H_2). The second stage is cooled and converts the gas into a thermodynamic condition with a maximum CH_4 concentration. Basically it is possible to divide those two stages in two separate reactor vessels.

The nearly sulfur-free synthesis gas enters the first stage at about 300 °C. Due to the highly exothermic methanation reaction in Equation (11.9) and the slightly exothermic water gas shift reaction in Equation (11.10) – both catalyzed by the nickel catalyst – the temperature rises to an adiabatic reaction temperature of about 560 °C. Till the end of stage 1, the temperature is constant on this level. This is necessary to provide enough residence time for the requested rate of tar conversion.

$$CO + 3 \cdot H_2 \rightleftarrows CH_4 + H_2O \quad \Delta H_R = -206 \frac{kJ}{mol} \tag{11.9}$$

$$CO + H_2O \rightleftarrows CO_2 + H_2 \quad \Delta H_R = -42 \frac{kJ}{mol} \tag{11.10}$$

Various works [22, 24–26] show that temperatures as they occur in the first stage of agnion's methanation reactor allow a conversion of biomass tar by the mechanism of catalytic steam reforming. Thereby reactions according to Equation (11.11) take place.

$$C_{v_1} H_{v_2} + v_1 \cdot H_2O \rightarrow \left(\frac{v_3}{2} + v_2 \right) \cdot H_2 + v_3 \cdot CO \tag{11.11}$$

Vosecky et al. [27] for instance demonstrate that biomass tar can also be converted at temperatures in the range of 500 °C.

Table 11.3 shows the tar components before and after stage 1 of the methanation reactor. The concentrations were obtained by means of the SPA sampling and GC-FID analysis. In the raw synthesis gas naphthalene is the major tar component.

TABLE 11.3   Tar Reforming in the First Stage of the Agnion Methanation Reactor (Operation at Agnion's Pfaffenhofen Pilot Plant with DIN wood Pellets, Water Content 6%, S Content 0.03 wt%). Sampling by means of SPA, Measurement with GC-FID.

Component	Concentration [g/Nm³]	
	Before First Stage Methanation Reactor	Downstream First Stage Methanation Reactor
Naphthalene	3.16	Below LoD
1-Methylnaphthalene	0.14	Below LoD
2-Methylnaphthalene	0.06	Below LoD
Biphenyl	0.09	Below LoD
Acenaphthylene	0.79	Below LoD
Acenaphthene	0.10	Below LoD
Fluorene	0.27	Below LoD
Phenanthrene	0.63	Below LoD
Anthracene	0.16	Below LoD
Fluoranthene	0.38	Below LoD
Pyrene	0.37	Below LoD
Indene	0.94	Below LoD
Phenol	0.43	0.07
Total	7.52	0.07

Beside naphthalene and naphthalene derivates like acenaphthene and acenaphthylene, also the three- and four-ring aromatics phenanthrene and fluoranthene are measured in relevant concentrations. Due to allothermal gasification also some phenolic components can be found. However, most of the measured components can be dedicated to ECN class 4 [28]. In the sample shown here, the sum of tar concentration in the raw synthesis gas amounts to 7.52 g/Nm³, whereas no BTX components are included.

Downstream from the first stage of the reactor, within this sample a total tar concentration of 70 mg/Nm³ was determined after a test duration of 310 h. The only partly unconverted component is phenol. The BTX components are also below the detection limit after the Ni catalyst. According to Equation (11.12), the measurement values shown in Table 11.3 lead to rate of total tar conversion of higher than 99%. The almost complete tar conversion, as shown by the data in Table 11.3, has been proved by more than 80 comparisons between the tar contents of the raw syngas and the gas downstream from stage 1.

$$X_{\text{Tar}} = \frac{W_{\text{Tar,Syngas}}\left[\dfrac{g}{\text{Nm}^3}\right] - W_{\text{Tar,Raw-SNG}}\left[\dfrac{g}{\text{Nm}^3}\right]}{W_{\text{Tar,Syngas}}\left[\dfrac{g}{\text{Nm}^3}\right]} \tag{11.12}$$

Disadvantageous in the medium temperature tar reforming process is the fact that nickel based catalysts are very sensitive to sulfur deactivation in the designated temperature range.

In order to determine the sulfur and coke based catalyst deactivation, a model for determining a deactivation rate of the catalyst was introduced, see Equation (11.13). This model calculates a specific catalyst consumption from the loss of catalytic activity, derived from changes in the axial temperature profile of stage 1 within the test duration.

The specific catalyst consumption describes the amount of catalyst (in g), which is deactivated by the conversion of 1 kWh of raw syngas into tar-free syngas.

$$\Sigma \left[\frac{g}{kWh^3} \right] = \frac{m_{Cat}\,[g]}{H_{SynGas}\left[\dfrac{kWh}{Nm^3} \right] \cdot \dot{V}_{Syngas}\left[\dfrac{Nm^3}{h} \right] \cdot \tau_{Deac}\,[h]} \cdot \Delta A_I\,[-] \quad (11.13)$$

$$\Delta A_I = \frac{A_{Start} - A_{End}}{A_{Start} - A_{Inert}} \quad (11.14)$$

The factor ΔA_I [Equation (11.14)], needed for the calculation of Σ, considers the area loss in the axial temperature profile of the adiabatic stage 1 (cf. Figure 11.3). It describes the area loss from the start of an experiment till its end (A_{Start} to A_{End}), in relation to the area loss measured with a nonreactive gas (A_{Start} to A_{Inert}). Hence, this value corresponds to the loss in catalytic activity during the experiment, and together with the test duration τ_{Deac}, the total mass of inserted catalyst m_{Cat}, the synthesis gas volume flow $\dot{V}_{Syngas}$, and the heating value H_{Syngas}, the specific catalyst consumption Σ can be calculated.

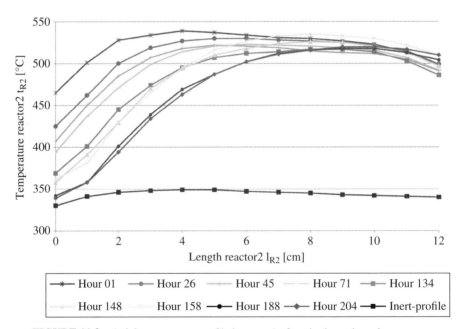

FIGURE 11.3   Axial temperature profile in stage 1 of agnion's methanation reactor.

Czekaj et al. [19] did intensive research on carbon and coke formation processes on Ni catalysts. A major result is that hydrocarbons with double and triple bonding like ethene (C_2H_4) or ethyne (C_2H_2) are highly involved in forming coke deposits. This coke-forming process is enhanced at higher temperatures [29]. With a catalyst entrance temperature of 300 °C in agnion's process, significant carbon depositions are avoided.

In standard operating conditions, a specific catalyst consumption Σ of around 0.2 g/kWh$_{Syngas}$ is obtained. This leads to economically competitive gas cleaning costs compared to state of the art FAME scrubbing systems. In order to furthermore reduce the catalyst consumption, work is being done on regenerating the ZnO adsorbent and Ni catalyst. Fine desulfurization also can help reduce the sulfur based Ni catalyst deactivation.

The permanent gas composition downstream from stage 1 is, as shown in Table 11.4, in a thermodynamic equilibrium corresponding to the stage's exit temperature.

As shown in Table 11.4, the H_2/CO ratio is far from the desired value of 3 [cf. Equation (11.9)]. In order to reach a maximum H_2, respectively a full CO conversion, the H_2/CO ratio must be adjusted to the stoichiometric value by means of the reverse water gas shift (WGS) reaction, see Equation (11.10). If a synthesis gas, nonstoichiometric concerning the methanation reaction, see Equation (11.9), is fed to a nickel-containing catalytic bed, the water gas shift reaction in Equation (11.10) takes place as a parallel reaction to the methanation reaction in Equation (11.9). This implies that, if sufficient water and CO_2 is in the gas, for synthesis gas compositions over and under stoichiometric regarding the methanation reaction, a complete H_2 and CO conversion can be reached without the need of a separate reactor for the WGS reaction.

TABLE 11.4 Mean Syngas Composition Downstream from Stage 1 of the Methanation Reactor (Operation at Agnion's Pfaffenhofen Pilot Plant with DIN Wood Pellets, Water Content 6%, S Content 0.03 wt%).

	Gas Composition [vol%]	
	Dry	Wet
Component		
y_{N2}	7.8	4.3
y_{CO2}	23.2	12.7
y_{CO}	13.1	7.2
y_{CH4}	10.3	5.6
y_{H2}	45.4	24.9
y_{H2O}	—	45.3
H_2/CO	3.46	
Tar content	~70 mg/Nm^3	
H_2S	Below LoD	
Organic sulfur	Below LoD	

According to Le Chatelier's law, the methane yield increases with lower temperatures. The reactivity of the used catalyst depends on the reaction temperature as well as on the gas water content. With higher H_2 conversion, the H_2O content in the gas increases. The higher the H_2O content is, the lower is the catalyst activity. Due to that fact, a maximum CH_4 yield, respectively a maximum H_2 conversion rate, is achievable under the discussed reaction conditions at a temperature of 250 °C at the outlet of reactor stage 2. If higher temperatures are applied, the CH_4 yield decreases due to thermodynamic equilibrium reasons. Setting lower temperatures leads also to a decreased CH_4 yield as a result of insufficient catalytic activity.

Due to the heat release of the highly exothermic methanation reaction, see Equation (11.9), the gas hourly space velocity (GHSV) value necessary for an optimal CH_4 yield strongly depends on the heat transport from the catalyst to the cooling media. Agnion's air cooled test system allows reaching the optimum gas composition with a maximum H_2 conversion of 89.6% at a GHSV value of around $1000\,h^{-1}$. In commercial applications, other cooling agents are preferentially used. Changes in the necessary GHSV value and therefore in the design of stage 2 are possible. Table 11.5 shows the permanent gas composition downstream from the second stage of agnion's methanation reactor. The values were obtained in agnion's test facility. In order to avoid N_2 being in the gas, all rinsing and locking streams were in this test done by means of CO_2. The raw SNG must be nearly free from nitrogen – a very small amount of fuel based nitrogen cannot be avoided – due to the fact that N_2 removal is only possible with uneconomic effort.

Discussing agnion's methanation system, one could argue that cooling down the gas to 300 °C followed by a heat-inducing process leading to a temperature of 560 °C at the end of stage 1, is disadvantageous due to an exergetic loss. Following the process chain for the production of SNG, in the second stage of agnion's methanation reactor, the gas composition is as mentioned before, adjusted towards a maximum CH_4 yield. This methanation leads also to an exergetic loss. The total loss of exergy and therefore the efficiency of agnion's SNG process only depends on the difference of the chemical potential between educts and products of the reactor. Thereby it does not matter if the gas heats up in stage 1, or not; see Equation (11.15).

TABLE 11.5 Mean Syngas Composition Downstream from Stage 2 of the methanation reactor (Operation at Agnion's Pfaffenhofen Pilot Plant with DIN Wood Pellets, Water Content 6%, S Content 0.03 wt%).

	Gas Composition [vol%]	
Component	Dry	Wet
y_{CO2}	43.8	15.8
y_{CO}	Below LoD	Below LoD
y_{CH4}	51.2	18.4
y_{H2}	2.5	0.9
y_{H2O}	—	64

$$\eta_{SNG} = \frac{\dot{n}_{RawSNG} \cdot H_{RawSNG}}{\dot{n}_{DeSulf} \cdot H_{DeSulf}} = \frac{\dot{n}_{Tar-free} \cdot H_{tar-free}}{\dot{n}_{DeSulf} \cdot H_{DeSulf}} \cdot \frac{\dot{n}_{RawSNG} \cdot H_{RawSNG}}{\dot{n}_{Tar-free} \cdot H_{tar-free}} \qquad (11.15)$$

As shown previously, all permanent gas reactions in the methanation reactor reach their thermodynamic equilibrium. The total efficiency of the methanation process mainly depends on that equilibrium and reaches a value η_{SNG} of around 86%. This value represents exergetic efficiency – no heat losses are taken into account. As shown in Equation (11.15), η_{SNG} also does not include possible losses in the CO_2 separation unit.

In the last process step the raw SNG must be dried, CO_2 has to be removed, and the gas quality has to be adjusted to the grid operator's regulations. Downstream from a condensation unit, the gas contains mainly CO_2 and CH_4 (cf. Table 11.5). Compared to the biogas from anerobic digestion, the differences are minor. The plant size of such biogas plants is also comparable. Using CO_2 separation units developed for that application within agnion's process seems to be the better approach instead of developing one's own system. The most suitable for the heatpipe reformer process seem to be the CO_2 removal systems based on membrane technology. This system would benefit from a pressurized heatpipe reformer.

REFERENCES

[1] Agnion. *Homepage*. http://www.agnion.de/(accessed 16 September 2014).

[2] Karl J. *Vorrichtung zur Vergasung biogener Einsatzstoffe*. DE Patent 19926202 C1; 1999.

[3] Karl J. *Fluidized Bed Reactor*. DE Patent 19926201 C2; 1999.

[4] Gallmetzer G, Ackermann P, Schweiger A, Kienberger T, Gröbl T, Walter H, Zankl M, Kröner M. The agnion heatpipe reformer – operating experiences and evaluation of fuel conversion and syngas composition. *Biomass Conversion Biorefinery* **2**: 207–215; 2012.

[5] Rauch R. *Biomass CHP Güssing Biomass Steam Reforming*. IEA Bioenergy Task 33, Thermal Gasification of Biomass, IEA, Berlin; 2009.

[6] Watanabe H, Otaka M. Numerical simulation of coal gasification in entrained flow coal gasifier. *Fuel* **85**: 1935–1943; 2006.

[7] Rabou LPLM, Bos L. High efficiency production of substitute natural gas from biomass. *Applied Catalysis B Environment* **111/112**: 456–460; 2012.

[8] Kienberger T, Zuber C, Novosel K, Baumhakl C, Karl J. Desulfurization and in situ tar reduction within catalytic methanation of biogenous synthesis gas. *Fuel* **107**: 102–112; 2013.

[9] Hayes RE, Thomas WJ, Hayes KE. A study of the nickel catalyzed methanation reaction. *Journal of Catalysis* **92**: 312–326; 1985.

[10] ÖVGW. *ÖVGW G 31, Erdgas in Österreich – Gasbeschaffenheit*. Österreichische Vereinigung für das Gas und Wasserfach, Vienna; 2001.

[11] DVGW. *DVGW G 260, Gasbeschaffenheit*. Deutscher Verein des Gas und Wasserfaches e.V., Bonn; 2000.

[12] Zuber C. *Untersuchung von Schwefelverbindungen und deren Entfernung beim Prozess der Biomassevergasung.* Dissertation, Graz University of Technology, Graz; 2012.

[13] Kienberger T. *Methanierung biogener Synthesegase mit Hinblick auf die Umsetzung von höheren Kohlenwasserstoffen.* Dissertation, Graz University of Technology, Graz; 2010.

[14] Bansal RC, Goyal M. *Activated Carbon Adsorption.* CRC Press, London; 2010.

[15] Puri BR, Hazra RS. Carbon sulfur surface complexes on charcoal. *Carbon* **9**: 123–134; 1971.

[16] Cal MP, Strickler BW, Lizzio AA, Gangwal SK. High temperature hydrogen sulfide adsorption on activated carbon: II. Effects of gas temperature, gas pressure and sorbent regeneration. *Carbon* **38**: 1767–1774; 2000.

[17] García G, Cascarosa E, Ábrego J, Gonzalo A, Sánchez JL. Use of different residues for high temperature desulfurisation of gasification gas. *Chemical Engineering Journal* **174**: 644–651; 2011.

[18] Seemann M. *Methanation of Biosyngas in a Fluidized Bed Reactor.* Dissertation, Eidgenössische Technische Hochschule Zürich, Zürich; 2007.

[19] Czekaj I, Loviat F, Raimondi F, Wambach J, Biollaz S, Wokaun A. Characterization of surface processes at the Ni based catalyst during the methanation of biomass derived synthesis gas: X-ray photoelectron spectroscopy (XPS). *Applied Catalyis General* **329**: 68–78; 2007.

[20] Meng X, De Jong W, Pal R, Verkooijen AHM. In bed and downstream hot gas desulfurization during solid fuel gasification: A review. *Fuel Processing Technology* **91**: 964–981; 2010.

[21] Elseviers WF, Verelst H. Transition metal oxides for hot gas desulfurisation. *Fuel* **78**: 601–612; 1999.

[22] Schweiger A. *Reinigung von heißen Produktgasen aus Biomassevergasern für den Einsatz in Oxidkeramischen Brennstoffzellen.* Dissertation, Graz University of Technology, Graz; 2008.

[23] Irschara F. *Entschwefelung von biogenen Produktgasen.* Dissertation, Graz University of Technology, Graz; 2009.

[24] Rostrup Nielsen JR. Activity of nickel catalysts for steam reforming of hydrocarbons. *Journal of Catalysis* **31**: 173–199; 1973.

[25] Pfeifer C, Hofbauer H. Development of catalytic tar decomposition downstream from a dual fluidized bed biomass steam gasifier. *Powder Technology* **180**: 9–16; 2008.

[26] Korre SC, Klein MT, Quann RJ. Polynuclear aromatic hydrocarbons hydrogenation. 1. Experimental reaction pathways and kinetics. *Industrial Engineering and Chemical Research* **34**: 101–117; 1995.

[27] Vosecky M, Kameníková P, Pohořelý M, Skoblja S, Punčochář M. *Efficient Tar Removal from Biomass Producer Gas at Moderate Temperatures via Steam Reforming on Nickel Based Catalyst.* In: Proceedings of the 17th European Biomass Conference, From Research to Industry and Markets, ETA – Renewable Energies. pp. 862–866; 2009.

[28] Kiel JHA, Van Paasen SVB, Neeft JPA, Devi L, Ptasinski KJ, Janssen F, Meijer R, Berend RH, Temmin HM, Brem G. *ECN: Primary Measures to Reduce Tar Formation in Fluidised Bed Biomass Gasifiers.* Available at ftp://130.112.2.101/pub/www/library/report/2004/c04014.pdf; 2004 (accessed 18 September 2014).

[29] Bartholomew CH. Mechanisms of catalyst deactivation. *Applied Catalysis General* **212**: 17–60; 2001.

12

INTEGRATED DESULFURIZATION AND METHANATION CONCEPTS FOR SNG PRODUCTION

CHRISTIAN F.J. KÖNIG, MAARTEN NACHTEGAAL, AND TILMAN J. SCHILDHAUER

12.1 INTRODUCTION

All biomass contains sulfur, for instance in the form of sulfurous proteins (cystein and methionine) or sulfur-containing metalloproteins [1]. The sulfur concentration in lignocellulosic biomass is approximately 0.1 wt%, which is significantly less than in various types of coal (for instance, lignite contains 0.8 wt% S [2], bituminous coal contains 1.7 wt% S [3]). The first step in the thermochemical conversion of biomass to synthetic natural gas (SNG) is gasification, where sulfur-containing proteins are broken down and are converted to small gaseous molecules. In low temperature gasification (~800 °C) the majority of the biomass is converted to H_2, CO, CO_2, CH_4, H_2O, and tars [4]; and sulfur is mostly converted to H_2S, but also to organic sulfur species such as COS, C_4H_4S, or others [5]. In contrast, in coal gasification, where the temperature is typically higher than in biomass gasification, larger molecules (such as sulfur-containing heterocyclic compounds) are broken down into H_2, CO, CO_2, and H_2S [4].

The synthesis step in the thermochemical conversion of biomass to SNG is methanation, where H_2 and CO in the cleaned producer gas from gasification are converted to CH_4 over a catalyst. The catalysts typically used for methanation are Ni or Ru based [6], while also W [7], Mo [8], and Fe [8] are reported to have

Synthetic Natural Gas from Coal, Dry Biomass, and Power-to-Gas Applications, First Edition.
Edited by Tilman J. Schildhauer and Serge M.A. Biollaz.
© 2016 John Wiley & Sons, Inc. Published 2016 by John Wiley & Sons, Inc.

methanation activity. These metal catalysts are sensitive to sulfur [9, 10]. There are several strategies to remove sulfur compounds from the gas before the methanation reactor. One such option is fixed bed adsorption by metal oxides [11]. One of the most widely reported oxides, ZnO removes H_2S by the reaction $ZnO + H_2S = ZnS + H_2O$ [12]. When the ZnO is fully converted to ZnS, it can be regenerated in an O_2 atmosphere at temperatures between 590 and 680 °C. Other sulfur species, such as COS or organic sulfur-containing molecules, are however not absorbed by ZnO [2]. In addition, the sulfur uptake capacity of ZnO is reduced by steam [13, 14], which is present in large amounts during the steam gasification of biomass. This is why low temperature scrubbers are currently used in large-scale coal to SNG plants and in biomass to SNG pilot plants to remove sulfur species [15]. Scrubbing is a physical and/or chemical process to remove certain molecules from the gas, depending on the scrubbing liquid that is used. For example, scrubbing with methanol (rectisol process) results in less than 0.1 ppmv H_2S in the effluent gas, while scrubbing with MDEA (methyl diethanolamine) leaves 10–20 ppmv H_2S in the effluent [2]. In addition, scrubbing removes CO_2, NH_3, tars, and olefins [16]. The OLGA process, which is an oil-based gas washing developed at the Energy Research Centre of the Netherlands (ECN), was reported to also remove thiophene [17]. Rectisol scrubbing operates at temperatures between –60 and –35 °C, and MDEA scrubbing operates at ambient temperature [2]. This means that the producer gas, coming from the gasifier at ~ 800 °C needs to be cooled to ambient temperature or less for the scrubbing, where all steam is condensed. After scrubbing, the gas needs to be heated and water needs to be evaporated and added to the cleaned producer gas, to adjust the H_2:CO ratio by the water–gas shift reaction ($CO + H_2O \rightarrow CO_2 + H_2$) to the desired value of three for methanation.

It seems advantageous to omit scrubbing from the biomass to SNG process chain, since it causes additional costs for equipment, such as heat exchangers and water evaporators, and for the handling and regeneration of the scrubbing liquid. In the context of biomass to ethanol conversion, technoeconomic analysis showed that state of the art gas cleaning represents 31% of the minimum selling price of the produced ethanol [18]. Reducing the cost of gas cleaning by avoiding low temperature scrubbing therefore would improve the economic competitiveness of the process. While no such study focussing on the economic impact of different gas cleaning options is currently available for the biomass to SNG process, it can be assumed that the potential savings are on a similar range.

Much effort has been made towards developing dedicated processes and materials for high temperature desulfurization, which are summarized in Chapter 3. To further reduce plant complexity and number of unit operations, it is also possible to integrate desulfurization with methanation, in order to reduce overall costs, and consequently make SNG more economically viable. While such processes are still in an early stage of development, this chapter presents two such integrated process concepts and discusses their advantages and disadvantages, and highlights some research questions which need to be answered for such novel integrated desulfurization and methanation process to become applicable.

12.2 CONCEPTS FOR INTEGRATED DESULFURIZATION AND METHANATION

In order to omit low temperature gas cleaning by scrubbing, the scrubbing could be replaced by a combination of high temperature H_2S removal (for bulk desulfurization) and a methanation reactor, where the methanation catalyst must be either resistant to sulfur poisoning (by organic sulfur species, which are not removed by H_2S sorbents), or must be periodically regenerated after activity has fallen below a certain threshold. Such a process is schematically depicted in Figure 12.1. The overall process would start with gasification of the biomass or coal (here shown as a steam-blown gasifier), followed by high temperature filtration to remove particulate matter [19]. A subsequent high temperature reforming unit converts tars at temperatures above 700 °C [20] (see also Chapter 3), while a high temperature sorbent (such as ZnO) removes most inorganic sulfur in the gas [11]. ZnO is however not reactive towards thiophene (C_4H_4S), which is shown to chemisorb only weakly [21]. The following methanation unit then faces high steam loading and organic sulfur species, which is in contrast to typical coal gasification conditions (low steam content, no organic sulfur species). This methanation unit then should include a sulfur-resistant methanation catalyst, which is tolerant towards sulfur species that were not removed by the sulfur sorbent, or a methanation catalyst which is prone to sulfur poisoning, but can be regenerated multiple times to enable long catalyst lifetime. In the case of sulfur-resistant methanation, the H_2S absorption bed could be placed after the methanation reactor, if the catalyst activity is not too much limited by competitive adsorption of sulfur. In both cases, sulfur must be removed from the gas before it is injected into the gas grid. As many of the process steps shown in Figure 12.1 are still under active development, the temperature levels are meant to indicate only the high temperature character of the whole process chain, where ideally no intermediate heating is required.

In the case of a nonsulfur-resistant methanation catalyst (such as Ni or Ru), adjusting the H_2:CO ratio to the desired value of three by the water gas shift (WGS: $H_2O + CO = H_2 + CO_2$) reaction might be necessary. The high steam content in gasified biomass provides already a source of H_2O for the WGS reaction. Since most methanation catalysts also show WGS activity [22], it is assumed in the simplified picture in Figure 12.1 that WGS and methanation occur in parallel over the same catalyst, and therefore a dedicated WGS reactor is omitted. Finally, H_2O is condensed out and CO_2 is removed from the gas, so that only SNG is left. While high temperature filtration and gas cleaning steps are discussed in Chapter 3 of this book, the two different options for integrated desulfurization and methanation (sulfur-resistant methanation or periodic regeneration of the methanation catalyst) will be discussed in the following.

12.2.1 Sulfur-Resistant Methanation

Direct methanation from synthesis gas at low H_2:CO ratios around one in the presence of sulfur is referred to as "sulfur-resistant methanation" (SRM). Typical catalysts used for SRM are molybdenum based and were first tested in the 1970s and 1980s

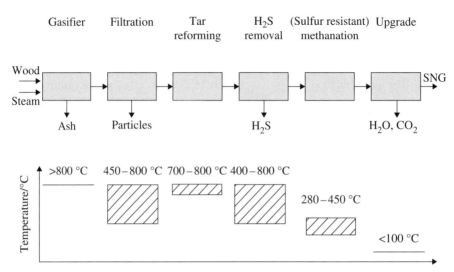

FIGURE 12.1 Schematic of an integrated high temperature process at ambient pressure for the conversion of biomass to SNG without the need for scrubbing.

with coal as feedstock for the synthesis gas. While the conversion of coal-derived synthesis gas is in many ways comparable to that of biomass-derived synthesis gas, some important differences should be kept in mind. Coal typically has higher sulfur content than biomass, and the sulfur compounds are different, owing to the different gasification conditions of biomass and sulfur. The higher moisture of most biomass (compared to coal) leads to significant concentrations of H_2O in the producer gas, whereas the lower gasification temperature (compared to coal) leads to the additional presence of organic sulfur compounds.

MoS$_2$ catalysts were studied for the synthesis of CH$_4$ from syngas at low H$_2$:CO ratios, and the reaction for SRM was found to be: $2\,CO + 2\,H_2 = CH_4 + CO_2$ [23]. According to this reaction, all hydrogen is converted to CH$_4$, and using a H$_2$:CO ratio of unity at the inlet is possible. Since the catalysts are typically metal sulfides, or become sulfided in the process, they are active in the presence of sulfur species, or are even more active when the H$_2$S concentration is increased [23,24]. A process for producing methane-rich gas from sulfur-containing hydrocarbonaceous feedstock (in this example, oil) by gasification and subsequent SRM over a Co-Mo/Al$_2$O$_3$ catalyst was patented in 1975 [25]. In a subsequent patent, the modification of the Mo catalyst with lanthanides or actinides was revealed, which resulted in higher activities compared to unmodified Mo catalysts [26]. Using a Mo or V catalyst on a TiO$_2$ support and the associated process for SRM of sulfur-containing synthesis gas was patented in 1985 [27]. In this patent, it was reported that using a TiO$_2$ support increased the activity for methanation, compared to an Al$_2$O$_3$ support, while possible reasons for this improvement were not discussed. A long-term test (1080 h) was performed with a process development unit (PDU), using the producer gas of an entrained flow gasifier, and an industrial catalyst of undisclosed composition. The feedstocks were different coals. The results showed almost 100% conversion and good stability in the presence of H_2S, typically in a concentration of 1000 ppm [28]. A multicomponent catalyst, including Mo, Ni, V, Al, Cr, Co, and Zr, was used for the methanation of a raw gas directly from a coal gasifier and showed good stability over 1000 h [29].

In recent years, research was resumed to optimize SRM catalysts, especially in terms of activity, and the catalyst systems, conditions, and parameters that were investigated are summarized in Table 12.1. Novel catalyst formulations for SRM were developed, such as Mo supported on Ce-modified Al_2O_3, which was reported to enhance methanation activity [24, 30]. The increased activity of the Ce-modified system was attributed to suppression of SO_4^{2-} species by Ce, which are supposed to have a negative impact on methanation activity [24]. A systematic study of the effect of Mo loading and calcination temperature of Mo/γ Al_2O_3 on the catalyst activity was performed [31]. The influence of the catalyst support on the activity of Co-Mo SRM catalysts was studied, comparing pure Al_2O_3 and Al_2O_3 modified with Ce, Zr, Ti, or Mg [32]. The Ce-modified support resulted in the highest catalyst activity, which was attributed to the improved dispersion of the Mo on the Ce-Al support. Different methods for the preparation of the Ce-Al support and their influence on the methanation activity were compared [33]. A comparison of Mo impregnated on different supports [γ-Al_2O_3, SiO_2, SiO_2-Al_2O_3, ZrO_2, CeO_2, TiO_2, and yttria-stabilized zirconia (YSZ)] showed highest activity for the ZrO_2 supported Mo catalyst [34]. Different methods for sulfidation of the Mo catalyst were also compared and their effect on catalyst stability were discussed [35,36]. To summarize the studies presented in this paragraph, it can be stated that a highly dispersed Co-Mo catalyst on a Ce-Al support that was sulfided in a stepwise manner is a promising system for SRM.

A lot of work has been done on the material science side of sulfur-resistant methanation, including preparation methods, sulfidation methods, choice of promotors, or the effect of supports. All these parameters can affect catalyst activity, selectivity, and stability, and it can be expected that there is still room for improvement. In the context of biomass to SNG conversion however, utilization of SRM catalysts under real conditions remains to be proven, as no scale-up of SRM reactors to convert real gasified biomass has been reported to date. Therefore, process development in terms of modeling and up-scaling is required to test the applicability of SRM for biomass conversion.

12.2.2 Regeneration of Methanation Catalysts

Instead of having a sulfur-resistant methanation catalyst, one could alternatively allow slow poisoning of the catalyst, as long as regeneration of the catalyst is possible. This would allow simplifying the process even further by omitting high temperature gas cleaning that removes inorganic and organic sulfur species [37], while a highly active Ni or Ru based catalyst for methanation can be used. In such a configuration, a ZnO bed that removes H_2S but is ineffective towards organic sulfur species could still

TABLE 12.1 Overview of Various SRM Catalyst Systems from Literature.

Catalyst Systems	Investigated Parameters	Activity/CO Conversion of Best System	Conditions	Ref.
Mo/Al_2O_3	$c(H_2S)$, $c(CO)$, $c(H_2)$, $c(CO_2)$, T	0.45 vol% min^{-1} g$_{cat}^{-1}$	$p=100$ kPa, $T=527$°C, CO:H$_2$=1, 0.2 vol% H$_2$S	[23]
Mo/CeO_2-Al_2O_3	Sulfidation temperature profile	60% X (CO) (stepwise sulfidation)	GHSV=5000h^{-1}, $p=3$MPa, $T=550$°C, CO:H$_2$=1, 1.2vol% H$_2$S, 3ml catalyst	[35]
Co-Mo/Al_2O_2	Sulfidation temperature	56% X (CO) (sulfided at 400°C)	GHSV=5000h^{-1}, $p=3$MPa, $T=550$°C, CO:H$_2$=1, 1.2vol% H$_2$S, 3ml catalyst	[36]
Mo/CeO_2-Al_2O_3	Preparation of composite support	60% X (CO) (co-precipitation of Ce-Al support)	Same as in [31]	[30]
Mo/Al_2O_3 Mo/SiO_2 Mo/SiO_2-Al_2O_3 Mo/ZrO_2 Mo/YSZ Mo/CeO_2 Mo/TiO_2	Effect of support	1.77×10^3 mmol g^{-1} h^{-1} after 10 h (5 wt% Mo/ZrO$_2$)	$p=2$MPa, $T=500$°C, CO:H$_2$=1, 0.5 vol% H$_2$S, 100 mg catalyst	[34]
Co-Mo/Al_2O_3 Co-Mo/CeO_2-Al_2O_3	$c(H_2S)$, $c(CO_2)$, $c(CH_4)$, $c(H_2O)$, $c(CO)$:$c(H_2)$, GHSV, T, p	51% X (CO) at 563°C (Co-Mo/CeO$_2$-Al$_2$O$_3$)	GHSV=5000h^{-1}, $p=3$MPa, $T=560$°C, CO:H$_2$=1, 0.2vol% H$_2$S, 3ml catalyst	[24]
Mo/Al_2O_3	Mo loading and calcination temperature	47% X (CO) (25 wt% Mo)	GHSV=5000h^{-1}, $p=3$MPa, $T=560$°C, CO:H$_2$=1, 0.24vol% H$_2$S, 3ml catalyst	[31]
Co-Mo/CeO_2-Al_2O_3 Co-Mo/MgO-Al_2O_3 Co-Mo/TiO_2-Al_2O_3 Co-Mo/ZrO_2-Al_2O_3	Support effects	56% X (CO) at 560°C (Co-Mo/Ce-Al)	GHSV=5000h^{-1}, $p=3$MPa, CO:H$_2$=1, 0.2vol% H$_2$S	[32]
Co-Mo/Ce-Al	Preparation method	56% (CO) at 610°C (co-precipitation of Co-Mo)	GHSV=5000h-1, $p=3$MPa, CO:H$_2$=1, 0.24vol% H$_2$S	[33]

p Pressure, T temperature, X conversion, c concentration. GHSV: gas hourly space velocity.

be included to increase catalyst lifetime by removing the bulk of the sulfur load. However, such a step is in principle not necessary.

Since the sulfur content in biomass is relatively low, especially after a ZnO bed, it is likely that the catalyst lifetime is on the order of hundreds of hours. While this would not be sufficient for permanent operation, periodic regeneration of parts of the catalyst (e.g., in a swing reactor design) would be viable, given that the catalyst can be regenerated from sulfur poisoning in the first place.

12.2.2.1 *Regeneration of Ni Catalysts*

It is well known that sulfur poisons Ni methanation catalysts [10, 38]. Sulfur poisoning of Ni leads to the formation of Ni sulfides (NiS or Ni_3S_2) [39]. Different Ni catalysts were poisoned with 10 ppm sulfur, and regeneration was tested with sulfur-free reactant gases or in pure H_2 at 523 K and 773 K [40]. However, all regeneration attempts in this study failed. Different methods to regenerate Ni catalysts after sulfur poisoning, by treatments at elevated temperatures in pure H_2, subsequent exposure to H_2, O_2 and again H_2, or sulfur-free reactant mixture followed by pure H_2, only partially restored activity [38]. This permanent deactivation is probably caused by the formation of Ni sulfates, which are stable under the tested conditions.

Regeneration of a sulfur-poisoned Ni catalyst with H_2O and/or O_2-containing atmosphere forms $NiSO_4$, which is partially reduced to Ni sulfide, recovering activity for methane reforming [41]. $NiSO_4$ is stable to high temperatures [42], decomposing at temperatures between 700 and 800 °C, when exposed to air at ambient pressure [43]. Regeneration of $NiSO_4$ to Ni at such high temperatures typically leads to sintering of the Ni particles, due to agglomeration of the Ni nanoparticles, thus reducing surface area, and apparent activity [44, 45].

Using a sequence of reduction in H_2 and oxidation at low O_2 partial pressures (500 ppm O_2), Ni catalysts that had previously been poisoned with thiophene (C_4H_4S) were successfully regenerated [46]. Another process for the regeneration of a Ni catalyst by using very low O_2 partial pressures is also disclosed in [47]. In this patent, regeneration of a sulfur-poisoned Ni catalyst with O_2 is described (concentration of O_2 in inert gas was 1–10 ppm) at temperatures between 300–500 °C over several tens of hours. The catalyst activity was reported to be recovered up to 80% of the pre-poisoned level. The successful regeneration at relatively mild temperatures (<500 °C) is probably due to the fact that formation of the $NiSO_4$ phase is avoided at low O_2 pressures. From a practical point of view, this approach is challenging, since reaching and maintaining such low O_2 concentration requires very careful handling of the catalyst and good tightness of all equipment.

Another strategy for the regeneration of sulfur poisoned Ni catalysts that operates at higher O_2 partial pressures would need to prevent Ni sintering, or enable redispersion of the Ni particles. Modification of a Ni/Al_2O_3 catalyst with alkali metal oxides and/or rare earth metal oxides improved thermal stability and reduced carbon deposition during partial oxidation of methane [48]. Sintering of Ni catalysts was prevented by preparation of a $Ni/MgAlO_x$ catalyst from hydrotalcite precursors [49]. These catalysts showed high temperature stability up to 900 °C. A particle size of ~ 10 nm was maintained under dry reforming conditions ($CO_2 + CH_4 \rightarrow 2\,CO + 2\,H_2$) [49]. Stability at high temperatures was attributed to the embedding nature of the amorphous oxide matrix that successfully separated Ni particles. Such a catalyst could be used for high temperature regeneration for the removal of S by decomposing $NiSO_4$ to NiO and SO_2, opening up a route for the regeneration of a non-noble metal catalyst without the necessity for achieving very low O_2 partial pressures, but with the necessity to reach higher temperatures.

12.2.2.2 Regeneration of Ru Catalysts   Ru catalysts, like Ni catalysts, are very sensitive to sulfur poisoning of their methanation activity [9]. Unlike Ni however, Ru is not known to form sulfate species under typical methanation and regeneration conditions. Consequently, the regeneration of Ru is expected to be easier than that of Ni. A Ru/Al_2O_3 catalyst that was active for methanation was poisoned with H_2S, COS, and C_4H_4S, representing the various sulfur species in biomass-derived producer gas [50]. Regeneration with O_2 in rather high concentration (5%) was successful over multiple cycles, proving in principle that the process of integrated methanation and desulfurization is possible. However, the activity was not restored completely.

Using X-ray absorption spectroscopy (XAS) it was shown that sulfur poisoning under methanation conditions leads to the formation of RuS_x [50]. Catalyst activity was successfully regenerated by oxidation at moderate temperature (430–600 °C) where XAS showed no evidence of permanently poisoned Ru species or of Ru particle sintering. This opens up the way for a process in which sulfur poisoning is accepted and the catalyst is periodically regenerated by air, or air diluted with flue gas. In a follow-up study, it was further shown using sulfur K-edge XAS that the sulfur is removed from the Ru catalyst by oxidation, but is stored on the Al_2O_3 support as sulfate species. Upon subsequent re-activation, sulfur is transported back to the Ru catalyst, which is again poisoned, even with sulfur-free feed gas [51].

12.2.3 Discussion of the Concepts

Periodic regeneration of methanation catalysts has the advantage that high activity Ni or Ru catalysts can be used, as opposed to SRM catalysts with rather low methanation activity. A study from 1980 systematically investigated different catalysts' activity for CO hydrogenation [8]. It was found that Ni has a 123 times higher hydrogenation rate at 350 °C (970 μmol min^{-1} g^{-1}), compared to MoS_2 (7.9 μmol min^{-1} g^{-1}) on a weight basis, and 1235 times higher rate than MoS_2 on a surface-area normalized basis (420 μmol min^{-1} m^{-2} versus 0.34 μmol min^{-1} m^{-2}). This indicates that Ni, which is typically the methanation catalyst of choice, is much more active than Mo based catalysts. This is a significant advantage for Ni or Ru catalysts and the associated concepts (Ru has a CO methanation activity that is comparable to Ni [6]) over sulfur-resistant methanation. It has to be noted though, that the experiments in Reference [8] were performed in sulfur-free atmosphere, while the activity of Mo-based catalysts was shown to increase with sulfur concentration in the feed. On the other hand, oxidative regeneration of poisoned catalysts (Ni or Ru based) generates metal oxides (NiO or RuO$_2$) and requires subsequent reduction of the catalyst, as metallic Ni or Ru is the active phase for methanation. For this, H$_2$ is used, which is then no longer available for methanation, thus lowering the process efficiency. Generating and maintaining very low concentrations of O$_2$, which would be necessary for the regeneration of Ni catalysts at low temperatures [46, 47], could be challenging. Ru catalysts, which can be regenerated by using air, are very expensive, which might prohibit their use in production of a relatively low-value product like methane. Thus, the modification of Ni catalysts to withstand sintering at high temperatures that are necessary for decomposition of Ni sulfates, currently seems like a promising route for periodically regenerated methanation.

In terms of process design, sulfur-resistant methanation or periodic regeneration of methanation catalysts would be beneficial, since low temperature scrubbing could be omitted, simplifying the process and reducing the costs for equipment and operation. In addition, it could be possible to integrate CO$_2$ removal into the cyclic methanation/regeneration reactor by adding CO$_2$ sorbents (e.g., Ca based) to the catalyst. Such a highly integrated approach would allow higher reaction temperatures by shifting the thermodynamic limit [52]. Due to CO$_2$ removal by the sorbent, selectivity towards CH$_4$ is increased, which allows operating the process at higher temperatures (500–600 °C). Furthermore, integration of CO$_2$ sorbents would facilitate downstream upgrading of the gas, because no further CO$_2$ scrubbing would be required [53]. However, integrating such novel approaches of CO$_2$ removal with methanation and sulfur removal (by regeneration of the catalyst) requires careful consideration of kinetic (for methanation, CO$_2$ uptake, sulfur poisoning, and regeneration) and economic (reactor cost, H$_2$ cost due to reduction/oxidation cycling) aspects as well as the complexity of operation.

Sulfur-resistant methanation does not require catalyst regeneration, and thus does not use H$_2$ from the process. This simplifies the process and potentially increases efficiency. However, Mo-based SRM catalysts have much lower methanation activity than Ni or Ru, which would result in much larger catalyst use and reactor size. Integration into cyclic processes, for example, for CO$_2$ removal where a fraction of the catalyst/sorbent is continuously oxidized to remove CO$_2$ [52, 53], seems prohibited for SRM catalysts, since oxidation would quickly burn off the sulfur from the sulfide catalysts, which could destroy their catalytic activity by changing the surface chemistry and due to heat-induced sintering.

12.3   REQUIRED FUTURE RESEARCH

12.3.1   Sulfur Resistant Methanation

To make SRM catalysts competitive to Ni-based methanation catalysts, increasing the activity of Mo-based SRM catalysts is one main goal. This could for instance be achieved by systematic variation of single synthesis parameters (such as preparation method, promotors, or sulfidation procedure), as shown in the studies presented above, or by high-throughput experimentation [54].

Most of the SRM systems in the literature that are described above were tested under conditions for coal-derived synthesis gas. While it can be expected that catalysts that are active for SRM of coal-derived synthesis gas are in principle also active to convert biomass-derived synthesis gas, there are however some differences to be considered. As biomass generally contains much more moisture than coal, the water content in biomass-derived synthesis gas is much higher than for coal. Furthermore, the sulfur content of biomass is generally lower than that of coal, which can influence the activity of SRM catalysts, as a positive relationship between sulfur content in the gas and catalytic activity was reported [24]. This means that dedicated efforts to use SRM catalysts for biomass conversion are needed.

Furthermore, it should be noted that sulfur-resistant methanation has not yet received much attention from process modeling. While several options for the biomass to SNG process based on sulfur removal before Ni-based methanation have been compared to identify the optimum configuration [55–58], no such study has been performed so far with a sulfur-resistant methanation concept. Such a modeling study could point out if a process to produce SNG by sulfur-resistant methanation is economically viable, indicate further research needs by analyzing the sensitivities of different process parameters (catalyst activity, stability, plant size, etc.), and thus trigger more activity in this field.

12.3.2   Periodic Regeneration

Since Ru is a very expensive metal, optimization of the catalyst to achieve maximum usage of the Ru is required. The particle size of the Ru/Al_2O_3 catalyst that was used in the reported studies on periodic regeneration [50, 51] was rather large (>20 nm), and thus utilization of Ru was low. If small Ru particles are successfully synthesized, particle growth due to temperature increase upon oxidation is a possible effect. Therefore, dedicated catalyst development to create small and stable Ru catalysts is required. Furthermore, the effect of the catalyst support (such as Al_2O_3, ZrO_2, or active C) on the extent to which the catalyst can be regenerated was shown to be strong, as it stores and releases sulfur, which subsequently poisons the catalyst [51]. Therefore, development of catalyst supports that store less sulfur (ideally none), release it faster, or store it without releasing it to the catalyst over time would be a highly relevant research task. This development could include promotion of the catalyst support with other elements, such as Li or K. These elements were shown to decrease the temperature at which sulfur is desorbed from the sulfur-poisoned support, thereby facilitating regeneration of the catalyst [59]. On the other hand, Na-doped Al_2O_3 was shown to form more stable sulfites when contacted with SO_2, compared to pure Al_2O_3 [60]. This stability could prevent the release of sulfur from the support under reducing conditions, prolonging the catalyst lifetime. Such support development to minimize sulfur sensitivity not only applies to Ru catalysts, but would most likely be relevant for different catalysts, such as the more commonly used Ni catalysts. For regeneration of Ni catalysts, it remains to be shown that regeneration of stable Ni catalysts at high temperatures [49, 61] is possible after sulfur poisoning without loss of dispersion or activity.

REFERENCES

[1] De Kok LJ, Tausz M, Hawkesford MJ, Hoefgen R, McManus MT, Norton RM, Rennenberg H, Saito K, Schnug E, Tabe L (eds). *Sulfur Metabolism in Plants.* Springer, Heidelberg; 2012.

[2] Mondal P, Dang GS, Garg MO. Syngas production through gasification and cleanup for downstream applications – recent developments. *Fuel Processing Technology* **92**: 1395–1410; 2011.

[3] McKendry P. Energy production from biomass (part 1): overview of biomass. *Bioresource Technology* **83**: 37–46; 2002.

[4] Rabou LPLM, Zwart RWR, Vreugdenhil BJ, Bos L. Tar in biomass producer gas, the energy research centre of the netherlands (ECN) experience: an enduring challenge. *Energy and Fuels* **23**: 6189–6198; 2009.

[5] Cui H, Turn SQ, Keffer V, Evans D, Tran T, Foley M. Contaminant estimates and removal in product gas from biomass steam gasification. *Energy and Fuels* **24**: 1222–1233; 2010.

[6] Dalla Betta RA, Piken AG, Shelef M. Heterogeneous methanation: steady-state rate of CO hydrogenation on supported ruthenium, nickel and rhenium. *Journal of Catalysis* **40**: 173–183; 1975.

[7] Kelley RD, Madey TE, Yates JT. Activity of tungsten as a methanation catalyst. *Journal of Catalysis* **50**: 301–305; 1977.

[8] Saito M, Anderson RB. The activity of several molybdenum compounds for the methanation of CO. *Journal of Catalysis* **63**: 438–446; 1980.

[9] Agrawal PK, Katzer JR, Manogue, WH. Methanation over transition-metal catalysts, V. Ru/Al$_2$O$_3$ – kinetic behaviour and poisoning by H$_2$S. *Journal of Catalysis* **74**: 332–342; 1982.

[10] Fitzharris WD, Katzer JR, Manogue WH, Sulfur deactivation of nickel methanation catalysts. *Journal of Catalysis* **76**: 369–384; 1982.

[11] Cheah S, Carpenter DL, Magrini-Bair KA. Review of mid- to high-temperature sulfur sorbents for desulfurization of biomass- and coal-derived syngas. *Energy and Fuels* **23**: 5191–5307; 2009.

[12] Bu X, Ying Y, Zhang C, Peng W. Research improvement in Zn-based sorbent for hot gas desulfurization. *Powder Technology* **180**: 253–258; 2008.

[13] Novochinskii II, Song C, Ma X, Liu X, Shore L, Lampert J, Farrauto RJ. Low-temperature H$_2$S removal from steam-containing gas mixtures with ZnO for fuel cell application. 1. ZnO particles and extrudates. *Energy and Fuels* **18**: 576–583; 2004.

[14] Kim K, Jeon SK, Vo C, Park CS, Norbeck JM. Removal of hydrogen sulfide from a steam-hydrogasifier product gas by zinc oxide sorbent. *Industrial and Engineering Chemistry Research* **46**: 5848–5854; 2007.

[15] Kopyscinski J, Schildhauer TJ, Biollaz SMA. Production of synthetic natural gas (SNG) from coal and dry biomass – A technology review from 1950 to 2009. *Fuel* **89**: 1763–1783; 2010.

[16] Pröll T, Siefert IG, Friedl A, Hofbauer H. Removal of NH$_3$ from biomass gasification producer gas by water condensing in an organic solvent scrubber. *Industrial and Engineering Chemical Research* **44**: 1576–1584; 2005.

[17] Zwart RWR, van der Drift A, Bos A, Visser HJM, Cieplik MK, Könemann HWJ. Oil-based gas washing – flexible tar removal for high-efficient production of clean heat and power as well as sustainable fuels and chemicals. *Environmental Progress and Sustainable Energy* **28**: 324–335; 2009.

[18] Phillips SD. Technoeconomic analysis of a lignocellulosic biomass indirect gasification process to make ethanol via mixed alcohol synthesis. *Industrial and Engineering Chemistry Research* **46**: 8887–8897; 2007.

[19] Nagel FP, Ghosh S, Pitta C, Schildhauer TJ, Biollaz S. Biomass integrated gasification fuel cell systems – Concept development and experimental results. *Biomass and Bioenergy* **35**: 354–362; 2011.

[20] Berguerand N, Lind F, Israelsson M, Seemann M, Biollaz S, Thunan H. Use of nickel oxide as a catalyst for tar elimination in a chemial-looping reforming reactor operated with biomass producer gas. *Industrial and Engineering Chemistry Research* **51**: 16610–16616; 2012.

[21] Jirsak T, Dvorak J, Rodriguez JA. Chemistry of thiophene on ZnO, S/ZnO, and Cs/ZnO surfaces: effects of cesium on desulfurization processes. *Journal of Physical Chemistry B* **103**: 5550–5559; 1999.

[22] Grenoble DC, Estadt MM, Ollis DF. The chemistry and catalysis of the water gas shift reaction. *Journal of Catalysis* **67**: 90–102; 1981.

[23] Hou PY, Wise H, Kinetic studies with a sulfur-tolerant methanation catalyst. *Journal of Catalysis* **93**: 409–416; 1985.

[24] Li Z, Wang H, Wang E, Lv J, Shang Y, Ding G, Wang B, Ma X, Qin S, Su Q. The main factors controlling generation of synthetic natural gas by methanation of synthesis gas in the presence of sulfur-resistant Mo-based catalysts. *Kinetics and Catalysis* **54**: 338–343; 2013.

[25] Child ET, Robin AM, Slater WL Richter GN. *Production of a Clean Methane-Rich Fuel Gas from High-Sulfur Containing Hydrocarbonaceous Materials*. US Patent 3 928 000; 1975.

[26] Happel J, Hnatow MA. *Sulfur Resistant Molybdenum Catalyst for Methanation*. US Patent 4 151 191; 1979.

[27] Pedersen K, Andersen KJ, Rostrup Nielsen JR, Jorgensen IGH. *Process and Catalyst for the Preparation of a Gas Mixture having a High Content in Methane*. US Patent 4 540 714; 1985.

[28] Skov A, Pedersen K, Chen C-L, Coates RL. *Testing of a Sulfur Tolerant Direct Methanation Process*. American Chemical Society Division Fuel Chemistry, New York; 1986.

[29] Shufen L, Diyong W, Guizhi F, Quan Y. Reactivity and stability of a sulphur-resistant methanation catalyst. *Fuel* **70**: 835–837; 1991.

[30] Jiang M, Wang B, Yao Y, Wang H, Li Z, Ma X, Qin S, Sun Q. The role of the distribution of Ce species on MoO_3/CeO_2-Al_2O_3 catalysts in sulfur-resistant methanation. *Catalysis Communications* **35**: 32–35; 2013.

[31] Wang B, Ding G, Shang Y, Lv J, Wang H, Wang E, Li Z, Ma X, Qin S, Sun Q. Effects of MoO_3 loading and calcination temperature on the activity of the sulphur-resistant methanation catalyst MoO_3/γ-Al_2O_3. *Applied Catalysis A: General* **431/432**: 144–150; 2012.

[32] Wang H, Li Z, Wang E, Lin C, Shang Y, Ding G, Ma X, Qin S, Sun Q. Effect of composite supports on the methanation activity of Co-Mo-based sulphur-resistant catalysts. *Journal of Natural Gas Chemistry* **21**: 767–773; 2012.

[33] Wang B, Shang Y, Ding G, Lv J, Wang H, Wang E, Li Z, Ma X, Qin S, Sun Q. Effect of the ceria-alumina composite support on the Mo-based catalyst's sulfur-resistant activity for the synthetic natural gas process. *Reaction Kinetics, Mechanisms and Catalysis* **106**: 495–506; 2012.

[34] Kim MY, Ha SB, Koh DJ, Byun C, Park ED. CO methanation over supported Mo catalysts in the presence of H_2S. *Catalysis Communications* **35**: 68–71; 2013.

[35] Jiang M, Wang B, Yao Y, Wang H, Li Z, Ma X, Qin S, Sun Q. Effect of stepwise sulfidation on a $MoO_3/CeO_2-Al_2O_3$ catalyst for sulfur-resistant methanation. *Applied Catalysis A: General* **469**: 89–97; 2014.

[36] Jiang M, Wang B, Yao Y, Wang H, Li Z, Ma X, Qin S, Sun Q. Effect of sulfidation temperature on CoO-MoO_3/γ-Al_2O_3 catalyst for sulfur-resistant methanation. *Catalysis Science and Technology* **3**: 2793–2800; 2013.

[37] König CFJ, Schuh P, Schildhauer TJ, Nachtegaal M. High-temperature sulfur removal from biomass-derived synthesis gas over bifunctional molybdenum catalysts. *ChemCatChem* **5**: 3700–3711; 2013.

[38] Bartholomew CH, Weatherbee GD, Jarvi GA. Sulfur poisoning of nickel methanation catalysts I. In situ deactivation by H_2S of nickel and nickel bimetallics. *Journal of Catalysis* **60**: 257–269; 1979.

[39] Yung MM, Kuhn JN. Deactivation mechanisms of Ni-based tar reforming catalysts as monitored by X-ray absorption spectroscopy. *Langmuir* **26**: 16589–16594; 2010.

[40] Fowler RW, Bartholomew CH. Activity, adsorption, and sulfur tolerance studies of fluidized bed methanation catalysts. *Industrial and Engineering Chemistry Product Research and Development* **18**: 339–347; 1979.

[41] Yung MM, Cheah S, Magrini-Bair K, Kuhn JN. Transformation of sulfur species during steam/air regeneration on a Ni biomass conditioning catalyst. *ACS Catalysis* **2**: 1363–1367; 2012.

[42] Chughtai AR, Riter JR. Thermodynamic model for the regeneration of sulfur-poisoned nickel catalyst. 1. Using thermodynamic properties of bulk nickel compounds only. *Journal of Physical Chemistry* **83**: 2771–2773; 1979.

[43] Siriwardane RV, Poston JA, Fisher EP, Shen M-S, Miltz AL. Decomposition of the sulfates of copper, iron(II), iron(III), nickel, and zinc: XPS, SEM, DRIFTS, XRD, and TGA study. *Applied Surface Science* **152**: 219–236; 1999.

[44] Bartholomew CH. Sintering kinetics of supported metals: new perspectives from a unifying GPLE treatment. *Applied Catalysis A: General* **107**: 1–57; 1993.

[45] Sehested J, Gelten JAP, Remediakis IN, Bengaard H, Nørskov JK. Sintering of nickel steam-reforming catalysts: effects of temperature and steam and hydrogen pressure. *Journal of Catalysis* **223**: 432–443; 2004.

[46] Aguinaga A, Montes M, Regeneration of a nickel/silica catalyst poisoned by thiophene. *Applied Catalysis A: General* **90**: 131–144; 1992.

[47] Katzer JR, Windawi H. *Process for the Regeneration of Metallic Catalysts*. US Patent 4 260 518; 1981.

[48] Miao Q, Xiong G, Sheng S, Cui W, Xu L, Guo X. Partial oxidation of methane to syngas over nickel-based catalysts modified by alkali metal oxide and rare earth metal oxide. *Applied Catalysis A: General* **154**: 17–27; 1997.

[49] Mette K, Kühl S, Düdder H, Kähler K, Tarasov A, Muhler M, Behrens M. Stable performance of Ni catalysts in the dry reforming of methane at high temperature for the efficient conversion of CO_2 into syngas. *ChemCatChem* **6**: 100–104; 2013.

[50] König CFJ, Schildhauer TJ, Nachtegaal M. Methane synthesis and sulfur removal over a Ru catalyst probed in situ with high sensitivity X-ray absorption spectroscopy. *Journal of Catalysis* **305**: 92–100; 2013.

[51] König CFJ, Schuh P, Huthwelker T, Smolentsev G, Schildhauer TJ, Nachtegaal M. Influence of the support on sulfur poisoning and regeneration of Ru catalysts probed by sulfur K-edge X-ray absorption spectroscopy. *Catalysis Today* **44**: 23–32; 2013.

[52] Lebarbier VM, Dagle RA, Kovarik L, Albrecht KO, Li X, Li L, Taylor CE, Bao X, Wang Y. Sorption-enhanced synthetic natural gas (SNG) production from syngas: a novel process combining CO methanation, water-gas shift, and CO_2 capture. *Applied Catalysis B: Environmental* **144**: 223–232; 2014.

[53] Liu, K. Qin Q, inventors. National institute of clean-and-low-carbon energy, assignee. System for producing methane-rich gas and process for producing methane-rich gas using the same. International patent application WO 2012/051924 A1. 2012 Apr. 26.

[54] Potyrailo RA, Maier WF (eds) *Combinatorial and High-Throughput Discovery and Optimization of Catalysts and Materials*, CRC Press, Boca Raton; 2007.

[55] Gassner M, Maréchal F. Thermo-economic process model for thermochemical production of synthetic natural gas (SNG) from lignocellulosic biomass. *Biomass and Bioenergy* **33**: 1587–1604; 2009.

[56] Steubing B, Zah R, Ludwig C. Life cycle assessment of SNG from wood for heating, electricity, and transportation. *Biomass and Bioenergy* **35**: 2950–2960; 2011.

[57] Gassner M, Maréchal F. Thermo-economic optimization of the polygeneration of synthetic natural gas (SNG), power and heat from lignocellulosic biomass by gasification and methanation. *Energy and Environmental Science* **5**: 5768–5789; 2012.

[58] Rönsch S, Kaltenschmitt M. Bio-SNG production – concepts and their assessment. *Biomass Conversion and Biorefinery* **2**: 285–296; 2012.

[59] Matsumoto S, Ikeda Y, Suziki H, Ogai M, Miyoshi N. NO_x storage-reduction catalyst for automotive exhaust with improved tolerance against sulfur poisoning. *Applied Catalysis B: Environmental* **25**: 115–124; 2000.

[60] Mohammed Saad AB, Saur O, Wang Y, Tripp CP, Morrow BA, Lavalley JC. Effect of sodium on the adsorption of SO_2 on Al_2O_3 and its reaction with H_2S. *Journal of Physical Chemistry* **99**: 4620–4625; 1995.

[61] Guo J, Lou H, Zhao H, Chai D, Zheng X. Dry reforming of methane over nickel catalysts supported on magnesium aluminate spinels. *Applied Catalysis A: General* **273**: 75–82; 2004.

INDEX

Synthetic Natural Gas from Coal, Dry Biomass, and Power-to-Gas Applications, First Edition.
Edited by Tilman J. Schildhauer and Serge M.A. Biollaz.
© 2016 John Wiley & Sons, Inc. Published 2016 by John Wiley & Sons, Inc.